METHODS IN MOLECULAR BIOLOGY™

Series Editor
John M. Walker
School of Life Sciences
University of Hertfordshire
Hatfield, Hertfordshire, AL10 9AB, UK

For other titles published in this series, go to
www.springer.com/series/7651

Immunoelectron Microscopy

Methods and Protocols

Edited by

Steven D. Schwartzbach

Department of Biology, University of Memphis, Memphis, TN, USA

Tetsuaki Osafune

Department of Life Science, Nippon Sport Science University, Yokohama, Japan

Humana Press

Editors
Steven D. Schwartzbach
Department of Biology
University of Memphis
3774 Walker Avenue
Memphis, TN 38152
USA
sdschwrt@memphis.edu

Tetsuaki Osafune
Department of Life Science
Nippon Sport Science University
Kamosida 1221-1
227-0033 Yokohama
Japan
osafunet@aol.com

ISSN 1064-3745 e-ISSN 1940-6029
ISBN 978-1-4939-5758-3 ISBN 978-1-60761-783-9 (eBook)
DOI 10.1007/978-1-60761-783-9
Springer New York Dordrecht Heidelberg London

Cover illustration: A 3D solid surface rendering showing that the immunogold-labeled small subunit of ribulose-bis-phosphate carboxylase, yellow dots, is concentrated in the propyrenoid, blue, rather than being uniformly distributed throughout the prolamellar body, red, and proplastid. The 3D reconstruction of the distribution of ribulose-bis-phosphate carboxylase is shown in the left, while the 3D reconstruction of the propyrenoid and prolamellar body superimposed upon the 3D distribution of ribulose-bis-phosphate carboxylase is shown in the right.

Printed on acid-free paper

Humana Press is part of Springer Science+Business Media (www.springer.com)

Preface

Cell biology is the science of correlating cell structure and function. The electron microscopist obtains high-resolution pictures of the intricate structures found in cells. Electron tomography and serial sections can be used to reconstruct the three-dimensional structure of the cell and its organelles. Pictures are worth a thousand words, but they are unable to provide information regarding the function of the intricate structures found in cells and their macromolecular composition.

The biochemist and molecular biologist determine the functions of the molecules, macromolecular complexes, and organelles found within cells. They isolate individual cellular constituents and reconstruct vital cellular processes. These in vitro experiments provide a detailed understanding of cellular function. Organelle isolation provides a method to place macromolecular functions within a structural context. Understanding structure–function relationships through organelle isolation has restricted utility because organelles cannot be isolated from every organism, not every organelle can be isolated free of contamination by other organelles, suborganellular compartments often cannot be purified for biochemical characterization, and, when a protein is recovered in multiple organelles, it is often difficult to distinguish true localization from contamination artifacts.

Immunoelectron microscopy is the technique that bridges the information gap between biochemistry, molecular biology, and ultrastructural studies placing macromolecular functions within a cellular context. Immunoelectron microscopy can be used on virtually every unicellular and multicellular organism. The only requirements are suitable fixation protocols and the availability of an antibody to the molecule whose structural location is to be determined. Structure–function relationships can be determined even when it is impossible to purify the organelle or suborganellular compartment containing the macromolecules being studied. Most importantly, immunoelectron microscopy is a totally objective procedure that is not dependent on conjectures as to where the protein is localized and thus which organelles to isolate for biochemical studies.

Two examples from our own work demonstrate how immunoelectron microscopy provides unexpected insights into structure–function relationships. Immunoelectron microscopy first demonstrated that the *Euglena* light harvesting chlorophyll a/b binding protein of photosystem II (LHCPII) was found in the Golgi apparatus prior to its presence in the chloroplast. This finding was the impetus for detailed biochemical studies that elucidated a new mechanism for chloroplast protein import: transport from the ER to the Golgi apparatus to the chloroplast. Immunoelectron microscopy identified the pyrenoid as the site of the enzyme ribulose 1–5 bisphosphate carboxylase/oxygenase (RUBISCO) showing that the enzyme moves from the stroma to the pyrenoid at cell cycle phases when enzyme activity is high while the pyrenoid disappears and RUBISCO redistributes back to the stroma at cell cycle phases when enzyme activity is low. Biochemical studies identified the activity changes but organelle fractionation experiments never identified this change in suborganellular localization because pyrenoids could not be isolated. At all cell cycle stages, the enzyme was recovered in the soluble chloroplast fraction.

The successful application of immunoelectron microscopy requires combining the tools of the molecular biologist with those of the microscopist. From the molecular biology toolbox, this volume will present methods for antigen production by protein expression in bacterial cells and by expression of epitope tagged proteins in plant and animal cells. Methods for production of anti-peptide, monoclonal, and polyclonal antibodies will be presented. From the microscopy toolbox, this volume will present methods for cryoultramicrotomy and rapid freeze-replacement fixation which have the advantage of retaining protein antigenicity at the expense of ultrastructural integrity as well as chemical fixation methods that maintain structural integrity while sacrificing protein antigenicity. Plants and algae contain cell walls, vacuoles, and other structures which present barriers to antibody penetration and complicate fixation. Due to these problems, separate chapters will discuss fixation and immunolabeling protocols for animals, plants, and algae. Pre- and post-embedding immunogold labeling protocols will be presented. Pre-embedding methods perform immunogold labeling before ultrathin sections are prepared from resin-embedded samples resulting in greater sensitivity and better microstructure preservation. Post-embedding methods perform immunolabeling after ultrathin sections are prepared from resin-embedded samples resulting in decreased antigenicity. The detailed methods and notes will facilitate choosing the best method for the antibody and biological material to be studied. Finally, methods will be presented for immunogold labeling of two antigens for protein colocalization studies, for three-dimensional reconstruction of intracellular antigen distribution, for immunogold labeling of DNA, and for immunogold scanning electron microscopy. It is our hope that the toolbox created by this volume will facilitate an increased understanding of structure–function relationships.

Steven D. Schwartzbach
Tetsuaki Osafune

Contents

Preface . . . . *v*

Contributors . . . . *ix*

PART I MOLECULAR TOOLBOX

1. Protein Antigen Expression in *Escherichia coli* for Antibody Production 3
David M. Rancour, Steven K. Backues, and Sebastian Y. Bednarek

2. Expression of Epitope-Tagged Proteins in Plants 21
Takuya Furuichi

3. Expression of Epitope-Tagged Proteins in *Arabidopsis* Leaf Mesophyll Protoplasts 33
Young-Hee Cho and Sang-Dong Yoo

4. Transient Expression of Epitope-Tagged Proteins in Mammalian Cells 43
Melanie L. Styers, Jason Lowery, and Elizabeth Sztul

5. Production and Purification of Polyclonal Antibodies 63
Masami Nakazawa, Mari Mukumoto, and Kazutaka Miyatake

6. Production and Purification of Monoclonal Antibodies 75
Masami Nakazawa, Mari Mukumoto, and Kazutaka Miyatake

7. Production of Antipeptide Antibodies 93
Bao-Shiang Lee, Jin-Sheng Huang, G.D. Lasanthi P. Jayathilaka, Syed S. Lateef, and Shalini Gupta

8. Preparation of Colloidal Gold Particles and Conjugation to Protein A, IgG, $F(ab')_2$, and Streptavidin 109
Sadaki Yokota

PART II MICROSCOPY TOOLBOX

9. Immunoelectron Microscopy of Chemically Fixed Developing Plant Embryos . . 123
Tetsuaki Osafune and Steven D. Schwartzbach

10. Pre-embedding Immunogold Localization of Antigens in Mammalian Brain Slices 133
Thomas Schikorski

11. Pre-embedding Immunoelectron Microscopy of Chemically Fixed Mammalian Tissue Culture Cells 145
Haruo Hagiwara, Takeo Aoki, Takeshi Suzuki, and Kuniaki Takata

12. Immunoelectron Microscopy of Cryofixed and Freeze-Substituted Plant Tissues 155
Miyuki Takeuchi, Keiji Takabe, and Yoshinobu Mineyuki

13. In Vivo Cryotechniques for Preparation of Animal Tissues for Immunoelectron Microscopy 167
Shinichi Ohno, Nobuhiko Ohno, Nobuo Terada, Sei Saitoh, Yurika Saitoh, and Yasuhisa Fujii

14. Immunoelectron Microscopy of Cryofixed Freeze-Substituted Mammalian Tissue Culture Cells 181
Akira Sawaguchi

15. Immunoelectron Microscopy of Cryofixed Freeze-Substituted *Saccharomyces cerevisiae* 191
Jindriska Fiserova and Martin W. Goldberg

16. High-Resolution Molecular Localization by Freeze-Fracture Replica Labeling . . 205
Akikazu Fujita and Toyoshi Fujimoto

17. Pre-embedding Electron Microscopy Methods for Glycan Localization in Chemically Fixed Mammalian Tissue Using Horseradish Peroxidase-Conjugated Lectin 217
Yoshihiro Akimoto and Hayato Kawakami

18. Pre-embedding Nanogold Silver and Gold Intensification 225
Akitsugu Yamamoto and Ryuichi Masaki

19. The Post-embedding Method for Immunoelectron Microscopy of Mammalian Tissues: A Standardized Procedure Based on Heat-Induced Antigen Retrieval 237
Shuji Yamashita

20. Double-Label Immunoelectron Microscopy for Studying the Colocalization of Proteins in Cultured Cells 249
Haruo Hagiwara, Takeo Aoki, Takeshi Suzuki, and Kuniaki Takata

21. Serial Section Immunoelectron Microscopy of Algal Cells 259
Tetsuaki Osafune and Steven D. Schwartzbach

22. Freeze-Etch Electron Tomography for the Plasma Membrane Interface 275
Nobuhiro Morone

23. Localization of rDNA at Nucleolar Structural Components by Immunoelectron Microscopy 287
Seiichi Sato and Yasushi Sato

24. Immunogold Labelling for Scanning Electron Microscopy 297
Martin W. Goldberg and Jindriska Fiserova

25. Horseradish Peroxidase as a Reporter Gene and as a Cell-Organelle-Specific Marker in Correlative Light-Electron Microscopy 315
Thomas Schikorski

26. Monitoring Rapid Endocytosis in the Electron Microscope via Photoconversion of Vesicles Fluorescently Labeled with FM1-43 329
Thomas Schikorski

Subject Index 347

Contributors

YOSHIHIRO AKIMOTO • *Department of Anatomy, Kyorin University School of Medicine, Mitaka, Tokyo, Japan*

TAKEO AOKI • *Department of Anatomy and Cell Biology, Gunma University Graduate School of Medicine, Gunma, Japan*

STEVEN K. BACKUES • *Department of Biochemistry, University of Wisconsin, Madison, WI, USA*

SEBASTIAN Y. BEDNAREK • *Department of Biochemistry and Great Lakes Bioenergy Research Center, University of Wisconsin, Madison, WI, USA*

YOUNG-HEE CHO • *Department of Molecular Biology, Massachusetts General Hospital, Boston, MA, USA; Laboratory of Plant Signaling Systems Biology, Department of Biological Science, College of Natural Sciences, Sungkyunkwan University, Suwon, Korea*

JINDRISKA FISEROVA • *School of Biological and Biomedical Sciences, Durham University, Durham, UK*

YASUHISA FUJII • *Department of Anatomy and Molecular Histology, Interdisciplinary Graduate School of Medicine and Engineering, University of Yamanashi, Yamanashi, Japan*

TOYOSHI FUJIMOTO • *Department of Anatomy and Molecular Cell Biology, Nagoya University Graduate School of Medicine, Nagoya, Japan*

AKIKAZU FUJITA • *Department of Anatomy and Molecular Cell Biology, Nagoya University Graduate School of Medicine, Nagoya, Japan*

TAKUYA FURUICHI • *Research Institute for Bioresources, Okayama University, Okayama, Japan*

MARTIN W. GOLDBERG • *School of Biological and Biomedical Sciences, Durham University, Durham, UK*

SHALINI GUPTA • *Protein Research Laboratory, Research Resources Center, University of Illinois at Chicago, Chicago, IL, USA*

HARUO HAGIWARA • *Department of Occupational Therapy, School of Health Sciences, Faculty of Medicine, Gunma University, Gunma, Japan*

JIN-SHENG HUANG • *Protein Research Laboratory, Research Resources Center, University of Illinois at Chicago, Chicago, IL, USA*

G.D. LASANTHI P. JAYATHILAKA • *Protein Research Laboratory, Research Resources Center, University of Illinois at Chicago, Chicago, IL, USA*

HAYATO KAWAKAMI • *Department of Anatomy, Kyorin University School of Medicine, Mitaka, Tokyo, Japan*

SYED S. LATEEF • *Protein Research Laboratory, Research Resources Center, University of Illinois at Chicago, Chicago, IL, USA*

BAO-SHIANG LEE • *Protein Research Laboratory, Research Resources Center, University of Illinois at Chicago, Chicago, IL, USA*

JASON LOWERY • *Department of Cell Biology, University of Alabama at Birmingham, Birmingham, AL, USA*

RYUICHI MASAKI • *Department of Physiology, Kansai Medical University, Moriguchi, Osaka, Japan*
YOSHINOBU MINEYUKI • *Department of Life Science, Graduate School of Life Science, University of Hyogo, Hyogo, Japan*
KAZUTAKA MIYATAKE • *Department of Applied Biological Chemistry, Osaka Prefecture University, Osaka, Japan*
NOBUHIRO MORONE • *National Institute of Neuroscience, National Center of Neurology and Psychiatry, Tokyo, Japan; Institute for Integrated Cell-Material Sciences (iCeMS), Kyoto University, Kyoto, Japan*
MARI MUKUMOTO • *Department of Applied Biological Chemistry, Osaka Prefecture University, Osaka, Japan*
MASAMI NAKAZAWA • *Department of Applied Biological Chemistry, Osaka Prefecture University, Osaka, Japan*
NOBUHIKO OHNO • *Department of Anatomy and Molecular Histology, Interdisciplinary Graduate School of Medicine and Engineering, University of Yamanashi, Yamanashi, Japan*
SHINICHI OHNO • *Department of Anatomy and Molecular Histology, Interdisciplinary Graduate School of Medicine and Engineering, University of Yamanashi, Yamanashi, Japan*
TETSUAKI OSAFUNE • *Department of Life Science, Nippon Sport Science University, Yokohama, Japan*
DAVID M. RANCOUR • *Department of Biochemistry and Great Lakes Bioenergy Research Center, University of Wisconsin, Madison, WI, USA*
SEIICHI SATO • *Department of Biology, Faculty of Science, Ehime University, Matsuyama, Japan*
YASUSHI SATO • *Department of Biology, Faculty of Science, Ehime University, Matsuyama, Japan*
SEI SAITOH • *Department of Anatomy and Molecular Histology, Interdisciplinary Graduate School of Medicine and Engineering, University of Yamanashi, Yamanashi, Japan*
YURIKA SAITOH • *Department of Anatomy and Molecular Histology, Interdisciplinary Graduate School of Medicine and Engineering, University of Yamanashi, Yamanashi, Japan*
AKIRA SAWAGUCHI • *Department of Anatomy, Ultrastructural Cell Biology, Faculty of Medicine, University of Miyazaki, Miyazaki, Japan*
THOMAS SCHIKORSKI • *Neuroscience Department, Universidad Central del Caribe, Bayamon, Puerto Rico, USA*
STEVEN D. SCHWARTZBACH • *Department of Biology, University of Memphis, Memphis, TN, USA*
MELANIE L. STYERS • *Department of Biology, Chemistry, and Mathematics, University of Montevallo, Montevallo, AL, USA*
TAKESHI SUZUKI • *Department of Anatomy and Cell Biology, Gunma University Graduate School of Medicine, Gunma, Japan*
ELIZABETH SZTUL • *Department of Cell Biology, University of Alabama at Birmingham, Birmingham, AL, USA*
KEIJI TAKABE • *Division of Forest and Biomaterials Science, Graduate School of Agriculture, Kyoto University, Kyoto, Japan*
KUNIAKI TAKATA • *Department of Anatomy and Cell Biology, Gunma University Graduate School of Medicine, Gunma, Japan*

MIYUKI TAKEUCHI • *Department of Life Science, Graduate School of Life Science, University of Hyogo, Hyogo, Japan*
NOBUO TERADA • *Department of Anatomy and Molecular Histology, Interdisciplinary Graduate School of Medicine and Engineering, University of Yamanashi, Yamanashi, Japan*
AKITSUGU YAMAMOTO • *Department of Animal Bio-Science, Faculty of Bio-Science, Nagahama Institute of Bio-Science and Technology, Nagahama, Japan*
SHUJI YAMASHITA • *Electron Microscope Laboratory, School of Medicine, Keio University, Tokyo, Japan*
SADAKI YOKOTA • *Section of Functional Morphology, Faculty of Pharmaceutical Science, Nagasaki International University, Nagasaki, Japan*
SANG-DONG YOO • *Laboratory of Plant Signaling Systems Biology, Department of Biological Science, College of Natural Sciences, Sungkyunkwan University, Suwon, Korea*

Part I

Molecular Toolbox

Chapter 1

Protein Antigen Expression in *Escherichia coli* for Antibody Production

David M. Rancour, Steven K. Backues, and Sebastian Y. Bednarek

Abstract

Escherichia coli is a frequently used expression system for the generation of protein encoded by genes from diverse kingdoms and, thus, it is well suited for the production of protein antigens for antibody generation. It is a system of choice for many due to factors such as (1) the commercial availability of a vast array of reagents and materials needed for cloning, expression, and purification and (2) the potential high protein yields that can be acquired in a timely and cost-effective manner. This chapter will focus on (1) the general principles to keep in mind when choosing an antigen to express and (2) the use of a modified pGEX vector system (Rancour et al., *J. Biol. Chem.* 279:54264–54274, 2004) to use in its expression. Simplified protocols are provided for (1) assessing the expression of your protein, (2) testing whether your protein is or is not expressed as a soluble product, (3) performing bulk purifications of soluble or insoluble *E. coli*-expressed protein to acquire enough to be used for a complete immunization protocol, and (4) an optional procedure for epitope tag removal from your expressed protein of interest in order to avoid the unnecessary and sometimes unwanted production of antibodies against the fusion protein affinity chromatography tag. These four procedures have been used extensively and successfully in our lab as a basis for the production of recombinant protein and subsequent antibody production.

Key words: *Escherichia coli*, protein, expression, purification, glutathione-*S*-transferase, GST, epitope tag, antigen, tobacco etch virus (TEV) protease.

1. Introduction

Protein expression in *Escherichia coli* is a frequently used tool for the generation of protein encoded by genes from diverse kingdoms and, thus, it is well suited for the production of protein antigens for antibody generation. To facilitate the generation of useful antibodies against a target protein, there are several characteristics of the protein that need to be assessed prior to commencing with

S.D. Schwartzbach, T. Osafune (eds.), *Immunoelectron Microscopy*, Methods in Molecular Biology 657,
DOI 10.1007/978-1-60761-783-9_1, © Springer Science+Business Media, LLC 2010

expression. Taking the following details into consideration while determining the best antigen to express will greatly increase the chance that quality antibodies will be generated.

The first consideration is whether the protein of interest is known or predicted to be a soluble, a peripheral membrane, or an integral membrane protein. Soluble proteins or domains are much easier to work with in terms of expression, purification, and, in some case, immunization protocols. For peripheral and integral membrane proteins, select cDNA fragments predicted to encode hydrophilic soluble domains of the protein of interest (i.e., domains lacking long contiguous hydrophobic amino acid stretches or transmembrane segments as predicted by hydropathicity analysis using the Kyte–Doolittle (1) and/or Hopp–Woods (2) algorithms or utilizing newer methods of soluble protein prediction such as the method of Smialowski et al. (3)). Insoluble proteins can also be expressed and enriched as inclusion bodies in *E. coli* and then subsequently used for immunization. However, soluble proteins or protein domains most likely best represent the native folding organization of the protein in its cellular context and thus would be the optimal antigen to produce antibodies that would be used for immunocytochemistry. Additional drawbacks to using insoluble proteins include the purity of the protein not being high enough to generate specific anti-sera against the antigen, and affinity purification of specific antibodies becomes problematic (but not impossible) using the original antigen. Insoluble proteins can limit the use of standard liquid chromatography purification methods sans attempts at denaturation and renaturation procedures. Therefore, soluble proteins or domains should be the first choice for easy antigen production and purification to ensure a clean antigen to induce a specific immunological response.

The second consideration is whether the protein of interest is post-translationally modified. If so, does this modification influence its functionality and/or localization? Proteins from eukaryotes are commonly post-translationally modified. For example, in cases where these proteins are extensively decorated with carbohydrates, these modifications may adversely influence the reactivity of antibodies raised against an unmodified protein expressed in *E. coli*. Conversely, if you are interested in antibodies against antigens with specific modifications, you will need to keep in mind that most eukaryotic cell modifications do not take place in *E. coli* and thus you will need to seek other methods for antigen production.

The third consideration is whether the protein of interest is either a homo- or hetero-oligomer. Oligomerization may influence the ability to express the protein of interest as a soluble protein. A self-assembling soluble homo-oligomer [i.e., AtCDC48A (4)] is much easier to produce than a subunit of a multimeric, membrane-associated complex that needs specific chaperones for assembly. In the latter case, choosing a soluble domain fragment of the protein may be a better antigen choice.

The fourth consideration is whether the antigen is a product of a conserved gene family or relatively unique. To generate isoform-specific antibodies, expression of sequence divergent domains will be required. In cases where the divergent amino acid sequence is limited to short stretches (12–15 aa), synthetic peptides coupled to a carrier protein may be a better antigen choice. Alternatively, choosing a fragment of the protein that does not contain a highly conserve protein domain (e.g., Walker ATPase) could aid in minimizing cross-reactivity of your anti-serum.

Protein expression vectors for *E. coli* typically differ in the promoter/repressor system used for gene expression regulation and the type/position of epitope tags translationally fused to the protein product and are available from various commercial and academic sources. A modified pGEX4T plasmid expression system (4) encoding N-terminally fused glutathione-*S*-transferase (~26-kDa soluble protein from *Schistosoma japonicum*) is used in these protocols. The GST epitope tag has been shown to aid in increasing protein solubility, facilitating purification (5, 6), and allowing for higher yields (7). The Rosetta™ (DE3) pLysS *E. coli* strain (Novagen-EMD Biosciences) is used for protein expression in these protocols because the strain is protease deficient and contains a plasmid encoding six tRNAs underutilized by *E. coli* to alleviate codon bias which sometimes reduces protein expression of genes from divergent organismal origins.

The epitope tag is normally removed from an antigen to minimize the incidental production of antibodies to the tag in addition to the desired antigen. Removal of the epitope tag prior to use as an immunogen also facilitates subsequent affinity purification of antibodies to the protein of interest by allowing the original fusion protein to be used for affinity purification. A very useful feature of the modified pGEX4T-TEV vector system (4) used in these protocols is the inclusion of both the original thrombin and an added tobacco etch virus (TEV) sequence-specific cleavage sites between the GST tag and the protein insert. The specificity of TEV for its recognition sequence is more stringent than that of thrombin (8), thereby reducing unwanted cleavage of your protein of interest. In addition to its utility for tag removal, the use of TEV is quite cost effective because it can be easily produced in-house (9) from publicly available expression strains (10). TEV may also be purchased from several commercial sources including Promega, Eton Biosciences, and Invitrogen.

After the gene encoding the antigen is cloned into the modified pGEX4T-TEV vector, it should be first be transformed into the Rosetta™ (DE3) pLysS or other protein expression compatible *E. coli* strains. To produce a purified tag-free soluble bacterially expressed protein for use as an immunogen, the resulting clones must first be tested via small-scale induction to verify expression, and then subsequently by small-scale fractionation to determine the solubility of the protein product. Once the expression and solubility of the protein has been verified, large-scale expression and purification of the soluble protein by affinity chromatography using glutathione resin can be initiated. GST tag removal from purified soluble fusion proteins using sequence-specific proteases followed by protease removal by affinity chromatography is the final step in preparing a purified soluble bacterially expressed protein for use as an immunogen. For insoluble proteins, inclusion bodies are isolated and used for immunization. In this chapter, we present protocols for the preparation of tag-free soluble and insoluble bacterially expressed proteins for use as immunogens.

2. Materials

2.1. SDS-PAGE (Sodium Dodecylsulfate-Polyacrylamide Gel Electrophoresis)

1. Precast Ready Gel Tris–HCl SDS-PAGE gels, 4–15% acrylamide, Tris/glycine/SDS buffer (Bio-Rad Laboratories, Inc., Hercules, CA).
2. Small vertical electrophoresis unit (Hoefer, Inc., San Francisco, CA).
3. Power Supply (Bio-Rad Laboratories, Inc., Hercules, CA).
4. 10× SDS-PAGE running buffer: Mix 30.3 g Tris base, 144 g glycine, 10 g SDS, and distilled water (dH_2O) to a final volume of 1 L; dilute with dH_2O to 1× strength for use.
5. 5× SDS-PAGE sample buffer (5× SSB): Mix 3.9 mL 2 M Tris–HCl pH 6.8, 2.5 g SDS, 12.5 mL glycerol, 8 mg bromophenol blue, 6.25 mL β-mercaptoethanol, and double distilled water (ddH_2O) to a final volume of 25 mL. Use at 2× strength (2× SSB): 4 mL 5× SSB diluted with ddH_2O to final volume of 10 mL.
6. Protein Molecular Weight Markers: SDS-PAGE broad range standards (Bio-Rad Laboratories, Inc., Hercules, CA) diluted 1:40 in 2× SSB.
7. Coomassie stain for protein gels: 0.1% (w/v) Coomassie R-250 dissolved in fixative [40% (v/v) methanol, 10% (v/v) glacial acetic acid]; store capped at room temperature.
8. Coomassie destain solution for protein gels: 40% (v/v) methanol, 10% (v/v) glacial acetic acid; store capped at room temperature.
9. Kimwipes (Kimberly-Clark Corporation, Irving, TX).
10. Gel drying kit (Promega Corp., Madison, WI).

2.2. Analytical Scale Test of GST-Fusion Protein Expression

1. The modified pGEX4T plasmid engineered with a TEV protease cleavage site between endogenous Thrombin cleavage site and MCS (*see* **Note 1**) or pGEX4T plasmid (GE Healthcare Bio-Sciences Corp., Piscataway, NJ) containing the coding sequence for the protein/domain of interest in Rosetta™ (DE3) pLysS *E. coli* protein expression strain (Novagen-EMD Biosciences, Madison, WI).
2. LB broth: Mix 10.0 g Tryptone (BD Biosciences, San Jose, CA), 5.0 g yeast extract (BD Biosciences, San Jose, CA), 10.0 g sodium chloride, and ddH_2O to a final volume of 1 L; autoclave 20 min at 121°C; cool to 55°C before adding antibiotics.
3. Solid LB plates: Add 15.0 g/L Bactoagar (BD Biosciences, San Jose, CA) to freshly prepared LB broth prior to autoclaving. After autoclaving, cool broth to 55°C in a water bath, add antibiotic, mix by swirling, and aseptically pour a thin layer into 100-mm × 15-mm sterile plastic Petri dishes. Allow plates to cool to RT and store plates inverted in a plastic sleeve at 4°C.
4. Antibiotics: Carbenicillin (1,000×): 50.0 mg/mL in 50% (v/v) ethanol. Chloramphenicol (1,000×): 34.0 mg/mL in 95% (v/v) ethanol. Store both at −20°C.
5. Isopropyl-β-D-thiogalactopyranoside (IPTG) (Sigma-Aldrich, St. Louis, MO): 500 mM stock solution in ddH_2O, sterile filter, and store at −20°C.
6. Tris-buffered saline pH 7.4 (TBS pH 7.4; 10× strength): Mix 80.0 g NaCl, 0.2 g KCl, 30.0 g Tris base, and 800 mL of ddH_2O. With concentrated HCl, adjust pH to 7.4 and then bring to a final volume of 1 L with ddH_2O. Use at 1× strength and store at room temperature.
7. Spectrophotometer (600 nm capability) (GE Healthcare Bio-Sciences Corp., Piscataway, NJ).
8. Micro-tip sonicator (Branson Ultrasonics, Danbury, CT).

2.3. Protein Expression Test for Solubility and Capacity to Bind Glutathione–Sepharose

Items in **Section 2.2** are also included:

1. β-Mercaptoethanol (β-ME) (Sigma-Aldrich, St. Louis, MO).
2. NP-40 (Calbiochem-EMD Biosciences, San Diego, CA); stock solution of 20% (v/v) in dH_2O; store at room temperature.
3. Glutathione–Sepharose™ 4 Fast-Flow (GE Life-Sciences, Inc., Piscataway, NJ); working stock of 25% (v/v) in 1× TBS pH 7.4, supplemented with 0.1% (v/v) NP-40 and 0.02% (w/v) sodium azide; store at 4°C.

2.4. Large-Scale Expression and Purification

Items in **Section 2.3** are also included:

1. Phenylmethanesulfonyl fluoride (PMSF) (Sigma-Aldrich, St. Louis, MO) 200 mM stock in dry isopropanol (*see* **Note 2**).
2. Poly-Prep Chromatography Column (Bio-Rad Laboratories, Inc., Hercules, CA).
3. Adenosine 5'-triphosphate, disodium salt (Sigma-Aldrich, St. Louis, MO) made into 100 mM stock solution with dH_2O, stored at −20°C in small (~200 μL) aliquots to avoid freeze-thaw cycles.
4. $MgCl_2$: 1 M stock solution in dH_2O, autoclave, and store at room temperature.
5. L-Glutathione, reduced (Sigma-Aldrich, St. Louis, MO).
6. Tris-buffered saline pH 8.0 (TBS pH 8.0; 10× strength): Mix 8.0 g NaCl, 0.02 g KCl, 3.0 g Tris base, and dH_2O for a final volume of 100 mL adjusting the pH to 8.0 with concentrated HCl; use at 1× strength, store at room temperature.
7. Elution solution: 1× TBS pH 8.0 + 15 mM L-glutathione (reduced) (46 mg/10.0 mL). Prepare immediately before use.
8. Pierce 660 nm Protein Assay Kit (Thermo Scientific Pierce Protein Research Products, Rockford, IL).
9. Microplate spectrophotometer (660 nm capability) (Bio-Tek Instruments, Winooski, VT).
10. 96-Well, flat bottom microtiter plates (Greiner Bio-One North America, Monroe, NC).
11. Dialysis tubing (MWCO 12–14,000, 1 cm flat width, 0.3 mL/cm; Spectra/Por; Spectrum Laboratories, Inc., Rancho Dominguez, CA) and clips.
12. Glycerol.
13. Buffer A (need 50 mL): 25 mM HEPES–NaOH pH 7.5, 1 mM EDTA-NaOH pH 7.5.
14. Lysis buffer: To 50.0 mL buffer A add 50.0 μL NP-40 (100% stock), 5.0 mg lysozyme, 0.5 mg RNAse-A, 2.5 mg DNAse I, 100.0 μL $MgCl_2$ (1 M stock).
15. First wash buffer (need 150.0 mL): 50 mM HEPES–NaOH pH 7.5, 0.3 M NaCl, 1 mM EDTA–NaOH pH 7.5, 0.1% (v/v) NP-40.
16. Second wash buffer (need 25.0 mL): 50 mM HEPES–NaOH pH 7.5, 0.3 M NaCl, 1 mM EDTA–NaOH pH 7.5, 0.1% (v/v) NP-40, 10 mM β-ME.

17. Avanti J-E centrifuge with JA-20 and JA-14 rotors (Beckman-Coulter, Fullerton, CA).
18. 250-mL Polycarbonate centrifuge bottles and 50-mL polycarbonate centrifuge tubes.
19. Tabletop swinging bucket clinical centrifuge (Sorvall-Thermo Fisher Scientific, Waltham, MA).

2.5. Protease Removal of GST from Purified GST-Fusion Proteins

1. Purified GST-fusion protein (from **Section 3.4.2**).
2. Glutathione–Sepharose™ 4 Fast-Flow in TBS pH 7.4 at 25% (v/v).
3. Phosphate-buffered saline (PBS; 10× strength): Mix 80.0 g NaCl, 2.0 g KCl, 21.6 g Na_2HPO_4-$7H_2O$, 2.0 g KH_2PO_4, and dH_2O for a final volume of 1 L, sterile filter; use at 1× strength, sterile filter, and store at room temperature.
4. ProTEV Protease (Promega Corp., Madison, WI) (*see* **Note 3**).
5. Ni–Sepharose™ 6 Fast-Flow (GE Life-Sciences, Inc, Piscataway, NJ); washed and equilibrated in PBS.
6. Thrombin (Novagen-EMD Biosciences, Madison, WI).
7. *p*-Aminobenzamidine–agarose (Sigma-Aldrich, St. Louis, MO).

3. Methods

3.1. SDS-PAGE

1. Attach precast gel to electrophoresis apparatus. Fill upper and lower reservoirs with 1× strength running buffer.
2. Heat samples at 65°C for 15 min. Collect solution by centrifugation at 16,000×*g* for 5 min at room temperature.
3. Load 10 μL of each sample and protein molecular weight markers into individual wells.
4. Run gel according to manufacturer's suggested current/voltage.
5. Disassemble gel apparatus and transfer gel carefully to Coomassie stain in a clean plastic/glass dish and cover. Incubate gel for 30 min (minimum) with gentle orbital shaking.
6. Decant stain and save for later use. Add both destain solution and several wadded kimwipes to the gel containing dish and cover. Incubate with gentle orbital shaking until sufficient contrast is observed between protein bands and gel background. Frequently exchange used kimwipes for new ones to speed destaining.

7. Decant destain solution and save for later use. Rinse gel with dH_2O and then incubate in dH_2O for 10 min.
8. Setup gel in drying frame and dry according to manufacturer's instructions.

3.2. Analytical Scale Test of GST-Fusion Protein Expression

1. *Day 1*: Start individual 2.0 mL overnight cultures in LB plus appropriate antibiotic selection for each construct in an *E. coli* strain compatible with protein expression. Can stab individual colonies from primary transformation plate with sterile disposable pipette tip, streak on new solid LB plate with antibiotics (clone master plate), and eject tip into 2.0 mL culture tube with liquid LB supplemented with antibiotic. Grow overnight (16 h) at 37°C with vigorous orbital shaking (250 rpm).
2. *Day 2*: Prepare sufficient fresh LB supplemented with appropriate antibiotics for an appropriate number of 2.0 mL cultures and a single 1.0 mL spectrophotometer blank.
3. Use 0.1 mL saturated overnight culture to inoculate 2.0 mL cultures.
4. Grow new cultures 1.5 h at 37°C with vigorous orbital shaking (250 rpm).
5. Add IPTG to final concentration of 0.25 mM (add 1.0 μL of 500 mM IPTG stock solution) to all positive induction samples. Add sterile water for negative induction control.
6. Grow all cultures at 37°C with vigorous orbital shaking for 2 h.
7. Remove 1.0 mL from culture and collect cells by centrifugation in microcentrifuge at room temperature at maximum speed (~16,000×*g*) for 1 min. Remove medium by aspiration. Wash cells once with 0.5 mL ice-cold TBS pH 7.4. Recollect cells by centrifugation as before. Place cell pellets on ice.
8. With remaining culture perform $OD_{600\ nm}$ measurements and document.
9. To washed cell pellets from Step 7, add 2× SSB for a final concentration of 0.004 $OD_{600\ nm}$ per mL culture bacterial equivalents/1.0 μL 2× SSB and suspend by vortexing and sonication (*see* **Note 4**). Collect sample solution to tube bottom by pulse centrifugation in microfuge (*see* **Note 5**).
10. Run and stain SDS-PAGE gel (*see* **Section 3.1**).
11. Select the clone or clones that show the highest levels of induced protein (a dark band of the expected molecular weight in the induced but absent in the negative induction control sample) to proceed with the solubility test

(*see* **Section 3.3**). If no induced protein is observed, try (1) validation that IPTG and induction works with expression of GST from an empty vector, (2) changing the expression tag, or (3) choosing a different expression domain.

3.3. Protein Expression Test for Solubility and Capacity to Bind Glutathione–Sepharose

1. *Day 1:* Start individual 2.0 mL overnight cultures from the clone master plate (*see* **Section 3.2**, Step 1) in LB plus appropriate antibiotic selection for each construct/*E. coli* strain. Grow overnight (16 h) at 37°C.
2. *Day 2*: Prepare fresh LB supplemented with appropriate antibiotics for 2.5 mL cultures of each *E. coli* strain clone.
3. Use saturated overnight culture to inoculate new cultures at a 1:20 dilution (for 2.5 mL culture, use 0.125 mL of overnight culture).
4. Grow 1.5 h at 37°C with vigorous shaking.
5. Add IPTG to final concentration of 0.25 mM. Grow at 37°C with vigorous shaking for 2 h (*see* **Note 6**).
6. Collect cells by centrifugation in microfuge (room temp, full speed, 1 min) and remove cell medium completely by aspiration. In one 1.5 mL microfuge tube collect cells from 2.0 mL (2× 1.0 mL) of culture (i.e., collect cells from 1 mL culture by centrifugation, discard supernatant, add second 1 mL of culture to the same centrifuge tube). In a second tube generate a cell pellet for the "total induced protein sample" from 100 μL of culture.
7. Wash cells once with 0.5 mL of TBS pH 7.4 + 2 mM β-mercaptoethanol. Wash includes suspension of cells in solution with vortexing, recollection by centrifugation as above, and removal of wash solution by aspiration.
8. Suspend cells from 100 μL of culture ("second tube"; *see* Steps 6 and 7) in 30 μL 2× SSB (*see* **Note 5**). This is the total induced protein sample.
9. Suspend washed cells from 2.0 mL of induced culture (*see* Step 6–7) in 0.3 mL of TBS pH 7.4 + 2 mM β-mercaptoethanol and place on ice. Disrupt cold cells by sonication with micro-tip probe, a 5–10-s pulse two to three times at low power. Make sure to have cells on ice and prevent heating of sample (*see* **Note 7**).
10. Add 1.5 μL NP-40 detergent from a 20% stock to a final concentration of 0.1%. Mix by brief vortex and incubate on ice for 5 min.
11. Centrifuge samples 10 min at 10,000×*g* at 4°C.
12. Transfer the total supernatant to a new 1.5-mL microfuge tube. This is the soluble protein fraction.

13. To pellets, add 50 μL 2× SSB, suspend by vortexing (*see* **Note 5**). This is the insoluble protein fraction containing inclusion bodies and unbroken cells.
14. To the soluble protein fraction, add 25 μL of 25% (v/v) glutathione–Sepharose 4 Fast-Flow in TBS pH 7.4 + 0.1% NP-40 + 0.02% sodium azide (*see* **Note 8**). Mix by rotating at room temp for 10 min.
15. Collect beads gently by centrifugation (3,000×*g*, 1 min, room temp). Transfer supernatant to new 1.5-mL microfuge tube and store at −20°C for later analysis if needed. This is the unbound soluble protein fraction.
16. Wash beads three times with 1.0 mL TBS pH 7.4 + 0.1% NP-40 + 2 mM β-ME. Wash includes suspending beads gently by inverting tube by hand, collecting by centrifugation (3,000×*g*, 1 min, room temperature), and removal of wash solution by aspiration.
17. Wash beads once with 1 mL TBS pH 7.4 + 2 mM β-ME. Remove all wash solution using a 27-gauge needle on a 1.0-mL syringe by placing the beveled needle tip opening against the side of the microfuge tube.
18. Add 25 μL of 2× SSB to the beads (*see* **Note 5**). This solution contains the soluble protein bound to resin.
19. Analyze samples by SDS-PAGE (*see* **Section 3.1**). Samples to load on gel include (a) protein molecular weight markers, (b) total induced protein (*see* Step 8), (c) insoluble protein (*see* Step 13), (d) soluble protein bound to resin (*see* Step 18), and (e) optional BSA (0.5, 1, 2, and 5 μg) standards.
20. Verify the presence of the expressed protein bound to the glutathione–Sepharose beads and proceed with the large-scale expression and purification (*see* **Sections 3.4.1** and **3.4.2**). If no expressed protein is detected, try a different expression clone (*see* **Note 9**). If the expressed protein is seen only in the insoluble protein fraction, proceed instead with purification of *E. coli* inclusion bodies (*see* **Sections 3.4.1** and **3.4.3**), or attempt to regain solubility by lowering the induction temperature (*see* **Note 10**), changing the expression tag or choosing a different expression domain. If the expressed protein is observed in the total but not in either the soluble protein bound to resin or the insoluble fraction, analyze the unbound soluble protein (*see* Step 15) by protein precipitation (11) and SDS-PAGE (*see* **Section 3.1**). If the protein is expressed and soluble but does not bind to the resin, first increase the incubation time for bead binding. If protein still does not bind either (a) subtly

change the domain to be express through either addition or subtraction of residues to be expressed, (b) change the position of the epitope tag, or (c) choose a different expression tag.

3.4. Large-Scale Expression and Purification

3.4.1. Culture Growth and Protein Induction

1. *Day 1*: Start individual 50.0 mL overnight cultures in LB plus appropriate antibiotic selection for each construct/*E. coli* strain. Grow overnight at 37°C.
2. *Day 2*: Supplement with appropriate antibiotics 1.0 L autoclaved LB in 2.8 L Fernbach flask. Remove 1.0 mL for spectrophotometer measurement blank.
3. Use saturated overnight culture to make a freezer stock of the expression strain for future use. Either mix (a) 850 μL saturated culture with 150 μL sterile autoclaved glycerol or (b) 930 μL saturated culture with 70 μL DMSO in a cryovial, snap freeze in liquid nitrogen, and store at −80°C.
4. Use remaining saturated overnight culture to inoculate 1.0 L culture. Mix by swirling and check $OD_{600\ nm}$ of 1.0 mL.
5. Grow at 37°C with vigorous shaking to $OD_{600\ nm}$ = 0.4–0.6 (~1.5 h).
6. Induce culture by adding IPTG to a final concentration of 0.25 mM. Grow at 37°C with vigorous shaking for 2 h. Check $OD_{600\ nm}$.
7. Transfer induced cell culture to four 250 mL centrifuge bottles. Collect cells by centrifugation at 5,000×*g* for 10 min at 4°C. Decant used medium into bleach and then discard.
8. Suspend cell pellets in 20.0 mL/bottle of cold TBS pH 7.4. Pool into one bottle. Rinse empty bottles with TBS pH 7.4 and pool rinse into bottle with cells. Collect cells by centrifugation and discard wash solution after treating with bleach.
9. Suspend cell pellet in 20.0 mL of cold TBS pH 7.4. Transfer cell suspension to 50 mL centrifuge tube. Rinse bottles with 5.0 mL TBS pH 7.4 and pool rinse with cells. Collect cells by centrifugation at 5,000×*g* for 10 min at 4°C. Remove and discard wash solution after treating with bleach.
10. Snap freeze cell pellet in liquid nitrogen. Thaw for protein purification or store at −80°C for later purification.

3.4.2. Soluble Protein Purification

1. *Day 1*: Thaw cells in room temp water bath in the presence of 20 mL of cold TBS pH 7.4, 2 mM β-ME, and 1 mM PMSF (additional protease inhibitor cocktails can be added here if needed). Mix periodically until thawed. When thawed, place cell suspension on ice. Some cell breakages occurs during thawing and therefore solution should be highly viscous due to the release of bacterial chromosomal DNA.
2. Disrupt cold cells by sonication (25–30 s, 3–4 times on ice). Adequately broken cell suspension should become clear [less opaque = disrupted cells] and free-flowing [decreased viscosity = sheared DNA]) (*see* **Note 11**).
3. Add NP-40 detergent to a final concentration of 0.1% (add 110 μL of a 20% (v/v) stock). Vortex briefly and incubate on ice for 5 min. Remove 10 μL and add 40 μL 2× SSB for homogenate fraction (*see* **Note 5**).
4. Centrifuge samples for 10 min at 10,000×*g* at 4°C.
5. Transfer total supernatant to clean 50.0-mL conical tube. Remove 10 μL sample and add 40 μL 2× SSB for soluble fraction (*see* **Note 5**).
6. Suspend pellet from Step 4 with 20.0 mL TBS pH 7.4, take a 10 μL sample, and add 40 μL 2× SSB for insoluble inclusion bodies and unbroken cell fraction (*see* **Note 5**). To supernatant from Step 5, add 4.0 mL of 25% (v/v) glutathione–Sepharose 4 Fast-Flow in TBS pH 7.4 + 0.1% NP-40 + 0.02% sodium azide. Rotate mix at 4°C for 30 min.
7. Collect beads gently by centrifugation in a tabletop swinging-bucket rotor clinical centrifuge at 500×*g* for 5 min at 4°C. Transfer supernatant to clean 50.0 mL conical centrifuge tube. From supernatant, remove a 10 μL sample and add 40 μL 2× SSB for unbound soluble protein sample (*see* **Note 5**). Snap freeze remaining unbound fraction in liquid nitrogen and store at −80°C for later analysis if needed.
8. Wash beads three times in bulk with 50 mL TBS pH 7.4 + 0.1% NP-40 + 2 mM β-ME. Wash includes resuspending beads gently by inverting tube by hand, collecting by centrifugation (3,000×*g*, 1 min, room temperature), and removal of wash solution by aspiration. Transfer beads to pre-rinsed Bio-Rad Poly-Prep column. Allow the beads to settle then wash with 10 bed volumes TBS pH 7.4 + 0.1% NP-40 + 2 mM β-ME discarding the wash solution.
9. Wash column with 10.0 mL of TBS pH 7.4 + 0.1% NP40 + 2 mM β-ME + 1 mM ATP + 1 mM $MgCl_2$ to

remove any bacterial heat shock chaperone proteins associated with the GST-fusion protein.

10. Wash column with a minimum of 50 bed volumes (50.0 mL) of TBS pH 7.4 + 2 mM β-ME (no detergent).
11. Prepare 10.0 mL of fresh elution solution per column.
12. Perform a stepwise elution with elution solution: Fraction 1: 0.5 mL; Fractions 2–8: 1.0 mL each. Collect into individual 1.5 mL microfuge tubes and then place on ice (*see* **Note 12**).
13. Protein peak determination: In a microtiter plate, pipette 10 μL of each protein fraction into a well. To each protein sample add 150 μL Pierce 660 nm Protein Assay reagent, mix by pipetting, incubate 5 min at room temp, and measure $OD_{660\ nm}$ with a microplate reader. Remember to include a "no-protein" control sample.
14. Hydrate dialysis tubing by removing appropriate length from spool, place in beaker of dH_2O, and heat in microwave until hot. Incubate on bench until hydrated and pliable (~1 min). Keep wet once hydrated.
15. Pool fractions, usually fractions 2–4, containing protein peak ($OD_{660\ nm}$>0.05) and transfer to dialysis tubing with clip sealing one end. Seal other end with clip and ensure that clips are correctly fastened.
16. Dialyze against 2× 4 L of TBS pH 7.4 + 2 mM β-ME + 10% (v/v) glycerol at 4°C. Change first dialysis solution after 3–4 h then dialyze overnight.
17. Wash the used glutathione–Sepharose 4 Fast-Flow column with five column volumes of TBS pH 7.4. Store resin in TBS pH 7.4 + 0.02% NaN_3 at 4°C for future use.
18. *Day 2*: Transfer contents of dialysis tube to clean microfuge tube on ice.
19. Perform quantitative protein assay using BSA as a standard. Take 1–4 μg of pure GST-fusion protein and add 2× SSB keeping volume less than 15 μL. Snap freeze purified protein in aliquots and store at −80°C.
20. Perform SDS-PAGE analysis (*see* **Section 3.1**) of the following samples: (a) protein molecular weight markers, (b) homogenate (*see* Step 3), (c) soluble protein fraction (*see* Step 5), (d) insoluble inclusion bodies (*see* Step 6), (e) unbound protein (*see* Step 7), (f) residual resin bound (*see* **Note 13**), and (g) pure GST-fusion protein sample (*see* Step 19). Optional but recommended: run BSA standards (0.5, 1, 2, and 5 μg) on gel to provide a standard curve to quantify purified protein on gel.

21. Verify purity and yield. If overall yield is low, use more starting culture or adjust induction conditions to generate more soluble protein. If the purified protein shows a large amount of breakdown, try using more or different protease inhibitors, or performing the entire protocol at 4°C in a cold room.

3.4.3. Purification of E. coli Inclusion Bodies

1. Suspend/thaw cells in 50.0 mL lysis buffer; incubate thawed mixture 10 min on ice. Split evenly between two 50-mL centrifuge tubes.
2. Centrifuge for 10 min at 4°C and 17,400×*g*. Discard resulting supernatant.
3. Resuspend each cell pellet by vortexing in 25.0 mL first wash buffer. Incubate 5 min on ice.
4. Repeat centrifugation and washing twice.
5. Resuspend each pellet in 10.0 mL second wash buffer. Pool samples into one 50 mL centrifuge tube. Incubate 10 min on ice.
6. Centrifuge: 10 min at 4°C and 17,400×*g*. Discard resulting supernatant.
7. Resuspend pellet in 4.0 mL 2nd wash buffer supplemented with 5% (v/v) glycerol.
8. Take a 10 μL sample and add 40 μL 2× SSB. Make serial dilutions of protein in 2× SSB and analyze on SDS-PAGE gel (*see* **Section 3.1**) containing BSA standards.
9. Snap freeze remaining protein in liquid nitrogen and store at −80°C.

3.5. Protease Removal of GST from Purified GST-Fusion Proteins

1. In a rinsed poly-prep chromatography column (column 1), add 5.0 mg purified GST-fusion protein to 2.0 mL of 25% (v/v) glutathione–Sepharose™ 4 Fast-Flow in TBS pH 7.4 (*see* **Note 14**). Seal the column at both ends and incubate with end-over-end mixing at room temperature for 30 min.
2. Drain unbound material and wash resin with 10 bed volumes of PBS. Add PBS to a total solution/resin volume of 2.0 mL. If using thrombin, be sure to include 2.5 mM $CaCl_2$.
3. Add either ProTEV (200 units) or thrombin (20 units) to the column (column 1), seal, and incubate at room temperature with end-over-end mixing for 3 h at room temperature (*see* **Note 15**).
4. Drain column 1 solution directly into a second rinsed poly-prep chromatography column (column 2) (capped at the bottom) containing either (a) 50 μL of Ni–Sepahrose 6 Fast-Flow resin pre-washed with PBS (for the ProTEV) or (b) 50 μL of *p*-aminobenzamidine–agarose resin pre-washed with

PBS for the thrombin. Rinse column 1 resin with 1.0 mL of PBS and drain directly into column 2 (*see* **Note 16**).

5. Seal column 2 at both ends and incubate with end-over-end mixing for 30 min at room temperature.
6. Drain column 2 into a 5.0-mL polypropylene tube and transfer to ice.
7. Perform a quantitative protein assay on the solution drained from column 2 (*see* **Section 3.4.2**, Step 13) using BSA to generate a standard curve. This will allow determination of total protein content of the sample. Remove 1–4 μg of protein and analyze by SDS-PAGE (*see* **Section 3.1**) to verify the removal of the protease.
8. Make appropriate aliquot volumes (*see* **Note 17**), snap freeze in liquid nitrogen, and store at −80°C. You now have purified antigen, ready to use!

4. Notes

1. pGEX4T-TEV vector construction has been described (4). A straightforward method for its construction is to perform opposite direction "round-the-world" PCR of the desired pGEX4T vector resulting in a linearized vector with in-frame blunt ends directly between the sequence encoding the thrombin cleavage site and the *Bam*HI restriction site. Complementary 5′-phosphorylated hybridized oligos encoding the TEV recognition sequence can be ligated resulting in a circular plasmid for use in protein expression. Alternatively, the original plasmid from GE Healthcare can be used for expression with the GST removal accomplished through thrombin digestion.
2. PMSF is a toxic protease inhibitor so use caution when handling. In addition, PMSF has a short half-life in the presence of water (1 h at pH 7.5) so dissolving in a dry organic solvent and adding immediately before use to buffers is important for its efficacy.
3. A cost-effective alternative to commercially available TEV protease is to obtain an expression strain from the Center for Eukaryotic Structural Genomics-University of Wisconsin-Madison (10) and purify the recombinant His_6-tagged enzyme (i.e., according to (9)).
4. For calculating the volume of 2× SSB to add to induced cell pellets, take $OD_{600\ nm}$ value divide by 0.004. Resulting figure should be the μL volume of 2× SSB you add.

Processing the samples this way will normalize for culture growth rates. In most cases, the cellular DNA will make removal of sample difficult due to high viscosity, therefore pulse sonicate each sample (~5 s, low power) with a micro-tip sonicator to shear DNA.

5. Once protein sample buffer is added to protein samples, store samples at −20°C until ready to perform SDS-PAGE (*see* **Section 3.1**).
6. To find optimal expression conditions to maximize protein expression, try an induction time course (1–4 h) or an IPTG titration (0.1–1 mM).
7. Upon cell lysis by sonication, the cell suspension should become clear (less opaque = disrupted cells) and free flowing (decreased viscosity = sheared DNA).
8. Wear gloves when handling sodium azide since it is toxic!
9. Unfortunately, it is not uncommon for an expression clone to loose its ability to express the recombinant protein upon restreaking or storage. The frequency with which this happens varies greatly with different expression constructs. If a clone whose expression was verified no longer seems to give any expressed protein in a subsequent test (in a culture started from a restreaked plate), you must pick new colonies off of the original transformation plate. It may also be beneficial to perform a fresh transformation of the construct into the expression strain. If your construct is prone to loss of expression, we suggest making a freezer stock of each expression clone from the saturated culture (*see* **Section 3.2**, Step 1) instead of restreaking a plate. If even this freezer stock is later found to have lost its ability to express the recombinant protein, it may be necessary to use a colony directly from a fresh transformation for large-scale expression.
10. To facilitate solubility of the expressed protein during induction, the temperature at which the induction takes place can be lowered. For example, we use a temperature series of 28°C, room temperature (~20–22°C), and then, finally, 15°C to increase solubility. Do note that by decreasing the induction temperature, you also slow the growth of the culture and the protein expression rate and thus must account for this.
11. When using sonication for cell disruption, there are several points to keep in mind. First, try to prevent frothing of the solution. Frothing promotes protein denaturation. In addition, make sure that the sample does not heat up by keeping the sample on ice and allowing for time breaks in-between sonic applications. Very useful alternatives to cell

disruption by sonication are to (1) use a French Press at 20,000 psi or (2) use commercially available bacterial protein extraction reagents [i.e., BugBuster® Protein Extraction Reagent from Novagen (Novagen-EMD Biosciences, Madison, WI)].

12. Stepwise elution involves gently pipetting a specific volume of elution solution onto the resin, allowing the solution to flow into the bed while simultaneously collecting the eluate until the solvent meniscus reaches the top of the resin bed. Successive volumes of eluant are applied until no more protein elutes from the matrix. This procedure ensures that the eluted protein remains as concentrated as possible.

13. In some cases the GST-fusion protein does not elute efficiently and remains bound to the resin. Resin can be analyzed for the presence of residual protein by transferring 25 μL of a 25% (v/v) slurry of used beads to a new microfuge tube, removing the liquid and adding 30 μL of 2× SSB to solubilize protein for SDS-PAGE analysis. If inefficient elution is a problem, try (1) longer incubation times with the elution solution, (2) a higher pH for the elution solution, and/or (3) a higher glutathione concentration in the elution solution.

14. The starting mass amount of GST-fusion protein mixed with the beads may vary depending on your needs. If your protein of interest is smaller than GST, then the amount of pure, GST-free protein recovered from 5 mg GST-fusion protein will be less than 2.5 mg (assuming 100% recovery). You will need to keep in mind that a typical target protein amount for immunization of a single rabbit is approximately 1 mg.

15. The efficiency of GST removal from a GST-fusion protein with TEV or thrombin depends on several parameters and should be determined empirically on an analytical scale. Each GST-fusion behaves differently and therefore tag removal will need to be optimized for each. Common procedural factors to change include (1) the mass ratio of protease to target (more protease will result in more cleavage), (2) the time of the digestion reaction (1 h to overnight), and (3) the temperature at which the digestion reaction takes place (room temperature or 4°C). Keep in mind that you will need to optimize the digestion efficiency while considering the stability of your protein of interest. In addition, do not forget that thrombin needs Ca^{2+} ions for activity.

16. Verify that the resin has sufficient binding capacity to bind all of the protease.

17. You may need to stabilize your protein for other uses beyond use as an immunogen by adding glycerol to your solution to prevent precipitation.

Acknowledgments

This work was supported in part by the Department of Energy, Division of Energy Biosciences (Grant no. DE–FG02–ER20332 to S. Y. B.), and the DOE Great Lakes Bioenergy Research Center (www.greatlakesbioenergy.org), which is supported by the US Department of Energy, Office of Science, Office of Biological and Environmental Research, through Cooperative Agreement DE-FC02-07ER64494 between The Board of Regents of the University of Wisconsin System and the US Department of Energy.

References

1. Kyte, J. and Doolittle, R. F. (1982) A simple method for displaying the hydropathic character of a protein. *J. Mol. Biol.* **157**, 105–132.
2. Hopp, T. P. and Woods, K. R. (1981) Prediction of protein antigenic determinants from amino acid sequences. *Proc. Natl. Acad. Sci. USA* **78**, 3824–3828.
3. Smialowski, P., Martin-Galiano, A. J., Mikolajka, A., Girschick, T., Holak, T. A., and Frishman, D. (2007) Protein solubility: sequence based prediction and experimental verification. *Bioinformatics* **23**, 2536–2542.
4. Rancour, D. M., Park, S., Knight, S. D., and Bednarek, S. Y. (2004) Plant UBX domain-containing protein 1, PUX1, regulates the oligomeric structure and activity of arabidopsis CDC48. *J. Biol. Chem.* **279**, 54264–54274.
5. Dyson, M. R., Shadbolt, S. P., Vincent, K. J., Perera, R. L., and McCafferty, J. (2004) Production of soluble mammalian proteins in *Escherichia coli*: identification of protein features that correlate with successful expression. *BMC Biotechnol.* **4**, 32–49.
6. Shih, Y. P., Kung, W. M., Chen, J. C., Yeh, C. H., Wang, A. H., and Wang, T. F. (2002) High-throughput screening of soluble recombinant proteins. *Protein Sci.* **11**, 1714–1719.
7. Braun, P., Hu, Y., Shen, B., et al. (2002) Proteome-scale purification of human proteins from bacteria. *Proc. Natl. Acad. Sci. USA* **99**, 2654–2659.
8. Kapust, R. B., Tozser, J., Fox, J. D., et al. (2001) Tobacco etch virus protease: mechanism of autolysis and rational design of stable mutants with wild-type catalytic proficiency. *Protein Eng.***14**, 993–1000.
9. Blommel, P. G. and Fox, B. G. (2007) A combined approach to improving large-scale production of tobacco etch virus protease. *Protein Expr. Purif.* **55**, 53–68.
10. Blommel, P. G. and Fox, B. G. (2007) Structural genomics methods applied to production of TEV protease. In: Center for Eukaryotic Structural Genomics-Technology Dissemination Report 15: Center for Eukaryotic Structural Genomics, University of Wisconsin-Madison.
11. Wessel, D. and Flugge, U. I. (1984) A method for the quantitative recovery of protein in dilute solution in the presence of detergents and lipids. *Anal. Biochem.* **138**, 141–143.

Chapter 2

Expression of Epitope-Tagged Proteins in Plants

Takuya Furuichi

Abstract

Although immunoelectron microscopy is a powerful tool for visualizing the subcellular localization of target proteins, it is difficult to obtain and purify the specific antibodies required for this method. Instead of raising antibodies against individual target proteins, the use of transgenic plants expressing epitope-tagged proteins and commercially available antibodies simplifies the subcellular localization of target proteins. In this chapter, an improved method for producing transgenic plants that express epitope-tagged proteins and can be used for immunoelectron microscopic analysis is described.

Key words: Epitope tag, transformation, *Arabidopsis thaliana*, *Agrobacterium tumefaciens*.

1. Introduction

Agrobacterium tumifaciens is a bacterium that causes tumors (crown gall) in dicotyledonous plants by transferring a small segment of its own DNA ("transfer DNA" [T-DNA] in a Ti plasmid) into the plant's genome. It is therefore possible to transfer a gene of interest into the genomic DNA of a dicot by subcloning the gene into the T-DNA of *Agrobacterium* (between a pair of border sequences for gene transfer). Although this method of obtaining transgenic plants is applicable for various plant species including some monocots, it is complicated and requires careful handling for the establishment of the callus and the regeneration of the plant from transformed cells (1, 2). In order to transform *Arabidopsis thaliana*, a model plant used to study the development and stress responses of plants, a simplified,

S.D. Schwartzbach, T. Osafune (eds.), *Immunoelectron Microscopy*, Methods in Molecular Biology 657,
DOI 10.1007/978-1-60761-783-9_2,

Agrobacterium-mediated transformation method termed "floral dipping" (3) is commonly used.

In order to instantly identify the subcellular localization of proteins in plant cells, a fluorescent protein such as the green fluorescent protein (GFP) is fused to the N-terminus or C-terminus of the protein under study, and the fusion construct is then stably or transiently expressed in plant cells. The use of such fluorescent constructs enables direct subcellular localization of the GFP-tagged protein by confocal microscopy. Although GFP-based imaging is a simple and rapid tool, it has some limitations in terms of resolution and sensitivity. If the expression level of the mature target protein is low or if the protein is localized in small amounts in different regions, detection is difficult. In addition, GFP (27 kDa) has a tendency to localize at the nucleus.

In comparison, immunohistochemistry and immunoelectron microscopy, which involve the use of a specific antibody against native endogenous proteins, are more efficient techniques for studying the subcellular localization of proteins, even though their working procedures are more complex. For these techniques, recombinant proteins containing the entire or partial amino acid sequence of the target protein are purified, and antibodies (IgG) against these proteins are raised in animals such as mice or rabbits. A variety of monoclonal and polyclonal antibodies against both soluble and membrane proteins of human or animal origin have recently become commercially available. Unfortunately, such "antibody pools" are not available for research on plant cells.

Epitopes, such as FLAG, c-Myc, and polyhistidine, are supposed to be small enough (composed by 6–10 amino acids) to preserve the structure and the subcellular localization of epitope-tagged proteins resulting in the same localization as the corresponding endogenous proteins. Highly specific monoclonal antibodies against epitopes are commercially available, meaning that expression of epitope-tagged proteins is an alternative method for investigating the subcellular localization of target proteins by immunohistochemistry or immunoelectron microscopy without raising any specific antibodies. A further advantage of immunoelectron microscopy of transgenic cells expressing epitope-tagged proteins is the higher detection sensitivity based on the overexpression of the target protein, the high-affinity binding of antibodies against epitope tags, and the high-resolution localization by electron microscopy. This chapter provides a modified method to express epitope-tagged proteins in *A. thaliana* by *Agrobacterium*-mediated plant transformation.

2. Materials (*see* Note 1)

2.1. Cultivation of Arabidopsis for Agrobacterium-Mediated Transformation

1. Seeds of *A. thaliana* (ecotype Columbia-0 [Col-0]) (*see* **Note 2**)
2. Mixed soil (Potting soil:sand:pumice = 7:2:1 or potting soil:vermiculite:perlite = 4:3:2) (*see* **Note 3**)
3. Plastic pots for plant growth (70–80 mm diameter, 70–90 mm tall)
4. Spatula
5. Plant growth chamber with air conditioner and illumination controller (*see* **Note 4**)
6. Micropipettes and disposable tips (for 20–200 and 100–1,000 μL) (Eppendolf AG, Hamburg, Germany)
7. 10 mL of double-distilled H_2O (autoclaved)
8. 1.5-mL microfuge tube (Eppendolf AG, Hamburg, Germany)
9. Transparent film. (A plastic wrap for cooking is suitable)

2.2. Infection with A. tumefaciens

1. Binary plasmid with a cDNA encoding an epitope-tagged protein to be expressed and a plant selection marker such as hygromycin resistance (*see* **Note 5**)
2. Electro-competent cells of *A. tumefaciens* host strain (such as a strain of LBA4404 [Takara Bio Inc.; Otsu, Japan]) (*see* **Note 6**)
3. Gene Pulser II electroporation system (Bio-Rad; Hercules, CA, USA)
4. 0.2 cm gap width electroporation cuvettes (Bio-Rad; Hercules, CA, USA)
5. YEB medium: 5 g/L Bacto peptone (BD, Sparks, MD), 5 g/L beef extract, 1 g/L yeast extract (BD), 5 g/L sucrose, and 2 mM $MgSO_4$. Adjust the pH to 7.4 with 0.5 M NaOH, then sterilize by autoclaving at 121°C for 15 min.
6. LB (Luria Bertani) selection medium: 10 g/L Bacto tryptone, 5 g/L yeast extract, 5 g/L NaCl. Adjust the pH to 7.4 with 0.5 M NaOH, then sterilize by autoclaving at 121°C for 15 min. After the medium has cooled, add the corresponding antibiotics for transformant selection. (for hygromycin-resistant marker, use 50 μg/mL hygromycin for the selection of *Agrobacterium* colonies and transformed plants containing plasmid or T-DNA)

7. LB selection plates: Prepare LB medium and add 1% agar before autoclaving. After the medium has cooled, add antibiotic for the selection of transformants and pour the medium into plastic petri dishes (90 mm diameter; approximately 30 mL/dish) on a clean bench.
8. Infiltration medium: Mix 4.3 g Murashige and Skoog salt mixture (Sigma, St. Louis, MO), 0.12 g Gamborg's B5 vitamin mixture (Sigma, St. Louis, MO), 50 g sucrose, 0.5 g 2-(*N*-morpholino)ethanesulfonic acid (MES), and distilled water for a final volume of 1 L. Sterilize by autoclaving at 121°C for 15 min. Before using the medium for infection, add 10 μL/L 1 mg/mL benzylaminopurine in dimethyl sulfoxide [DMSO] and 10 μL/L Silwet L-77 (GE Silicones, WV, USA)
9. Tall beaker, 200 mL (60 mm diameter, 115 mm tall).
10. Paper towels
11. 100–200 mL of 70% ethanol
12. Disposable, steriled plastic petri dishes (90 mm diameter)
13. 15-mL test tubes (BD Falcon™ Conical Tubes, BD, Franklin Lakes, NJ)
14. Rotary shaker for culturing
15. Incubator for bacterial culture (at 30°C, keep inside dark)
16. Flat trays without drainage holes and clear plastic domes (Humi-dome, Hummert Intl., St. Louis, MO)
17. Vacuum pump
18. Vacuum desiccator
19. 100-mL and 1-L Erlenmeyer (conical) flasks
20. Centrifuge (up to 6,000×*g*) (Eppendorf AG, Hamburg, Germany)
21. 50-mL disposable tubes (BD Falcon™ Conical Tubes, BD, Franklin Lakes, NJ)

2.3. Seed Collection

1. Letter-size envelopes
2. Stainless steel tea filters (two to three pieces) to separate seeds from dried siliques and stems (diameter of mesh: 1–2 mm)

2.4. Selection of Putative Transformants

1. Selection medium: Mix 2.15 g Murashige and Skoog salt mixture, 100 mg myo-inositol, 10 g sucrose, 10 mL of 5% MES-KOH (pH 5.8), 3 g Gelrite (Wako, Japan) or 10 g agar, and distilled water for a final volume of 1 L. Sterilize by autoclaving at 121°C for 15 min. After the medium has cooled, add antibiotic for the selection of transformants and 100 μg/mL carbenicillin to sterilize the *Agrobacterium*.

Pour the medium into plastic petri dishes (90 mm diameter, 18 mm depth; approximately 30 mL/dish) on a clean bench.

2. 200 mL of double-distilled H_2O (autoclaved)
3. 10 mL of 0.1% sodium dodecyl sulfate (SDS)
4. 100 mL of bleach (sodium hypochlorite solution)
5. 0.1% sterile agarose solution: 0.1 g/L electrophoresis-grade agarose (Sigma, St. Louis, MO). Sterilize by autoclaving at 121°C for 15 min.

3. Methods

3.1. Cultivation of Arabidopsis for Agrobacterium-Mediated Transformation

1. Transfer 50–100 *Arabidopsis* seeds into a 1.5-mL microfuge tube using a spatula or by tapping.
2. Add 1 mL of water and tap the tube until the seeds separate, sink into the water, and swell. Store the tube in the dark at 4°C for 2 days (in a cold room or refrigerator) so that the seeds acquire the competence to germinate.
3. Pick up the seeds using a micropipette (100–1,000 μL) with a disposable tip and sow them in the moistened soil mixture in a plastic pot.
4. To prevent the seeds and surface of soil from drying out, cover the top of the pot with a petri dish or transparent film.
5. Grow the seedlings in a chamber with short-day conditions (8 h of daylight and 16 h of darkness). Normally, most of the seeds germinate within 3–5 days. After germination, remove the dish or film covering the top of the pot.
6. Two weeks after sowing, transplant individual seedlings into new pots.
7. Allow the seedlings to grow under short-day conditions until the rosette leaves are well developed (i.e., till they nearly cover the top of the pot **Fig. 2.1 A**). This may require 4–8 weeks. Usually, dipping of 6–8 plants yield an adequate number of transformants.
8. Transfer the pots into a chamber with long-day conditions (16 h of daylight and 8 h of darkness) to initiate the bolting process.
9. To obtain a greater number of floral buds per plant, cut the apical region of the inflorescence when bolting is complete in most of the plants. This process promotes the synchronized emergence of multiple secondary bolts (*see* **Note 7** and **Fig. 2.1 B**).

Fig. 2.1. *Arabidopsis* plants for *Agrobacterium*-mediated transformation. (**A**) A plant grown under short-day conditions. Rosette leaves (more than 20 pieces) are developed. Well-established roots support the soil from falling into the dipping bath when the plant is turned upside down. (**B**) A plant with multiple secondary bolts. *Arrowhead* indicates the clipped position of an inflorescent stem. (**C**) Upside down placement of a plant on a tall beaker (100 mL) for dipping. Diameter of the beaker is slightly smaller than the diameter of the pot to ensure that the pot is supported and does not fall into the beaker. *Arrowhead* indicates the approximate position of the surface of dipping solution (5–10 mm lower from the edge of beaker).

3.2. Infection with *A. tumefaciens*

1. Dissolve 40 μL of electro-competent *A. tumefaciens* cells in a microfuge tube on ice and add 1 μL (50–200 ng) of a binary plasmid construct.
2. Transfer the DNA cell mixture to an ice-cold electroporation cuvette for the electroporation. Electroporate the plasmids into the cells using a resistance of 200 Ω, a capacitance of 25 μF and a voltage of 2.5 V (*see* **Note 8**).
3. Immediately add 1 mL of YEB medium, mix by pipetting, and transfer to a 15 mL test tube. Incubate at 30°C, for 1 h on a rotary shaker rotating at 100–150 rpm.

4. Spread the electroporated cells onto a LB selection plate. To obtain a single colony of transformed *A. tumefaciens*, drop the inoculum of a differentiated portion (10–200 μm) onto 5–10 plates. After incubation at 30°C for 2 days in the dark, a number of colonies will be obtained.
5. When many unopened (immature) floral buds are observed, inoculate the *A. tumefaciens* transformed with the binary vector into 10 mL of LB selction medium in a 100-mL Erlenmeyer (conical) flask and incubate at 30°C overnight in the dark.
6. Dilute the overnight *Agrobacterium* culture with 300 mL of LB selection medium (1:100) in a 1-L Erlenmeyer flask. Culture at 30°C for 14–20 h until the cells reach the stationary phase (OD_{600} = 1.2–1.5).
7. Harvest the *A. tumifaciens* cells by centrifugation at 5,000–6,000×*g* for 10 min at room temperature and resuspend them in infiltration medium to obtain an OD_{600} of 0.8. The suspension termed "dipping solution" is used for the infection (*see* **Note 9**).
8. Pour 200–220 mL of the *Agrobacterium* suspension, dipping solution, into 200-mL-tall beakers. The beakers must have a small diameter to ensure that the pots are supported and do not fall into the beakers. Each "dipping bath" can be used up to five times.
9. Remove the flowers (opened buds) and siliques from the plants. Turn the plant upside down and gently submerge the aerial part (including the rosette leaves) into the dipping solution for 10–15 min (*see* **Note 10** and **Fig. 2.1C**).
10. Remove the plants from the dipping bath. Gently shake or wipe off the dipping solution remaining on the aerial part of the plant using a paper towel.
11. Place the plants in a plastic tray separating each plant from its neighbors to avoid contamination. Cover the plants with a plastic dome to maintain humidity because dipped plants are sensitive to drought owing to the hyperosmotic stress induced by the dipping solution. Place the plants under conditions of low light intensity to promote the breeding and infection of *Agrobacterium.*
12. Remove the plastic dome 2 days after infection. To avoid the withering of dipped plants, do not water them until the soil is relatively (not completely) dry. Usually, watering can be resumed within 1 week.
13. Allow the plants to grow for 3–5 weeks until the siliques developed from the dipped buds have matured (i.e., they appear brown and dry).

3.3. Seed Collection

1. Carefully harvest the aerial parts of the plants (except the rosette leaves) and transfer them into letter-size envelopes. Plants that have the same genotype and have been dipped in the solution containing the same cDNA construct can be put into a single envelope. Store the plants and seeds at room temperature until they are completely dry.
2. Release the seeds from the siliques by gently rubbing fingers over the envelope. Remove the stems and the outer portions of the siliques by filtering them through a stainless steel tea filter and collect the seeds on a clean piece of paper. Transfer the seeds into microfuge tubes and store them at 4°C under desiccation conditions for further use.

3.4. Selection of Putative Transformants

1. Transfer 2,000–3,000 seeds obtained from the dipped plants into a microfuge tube by using a spatula or by tapping. Add 600 μL of water and tap the tube until the seeds separate, sink into the water, and swell sufficiently.
2. Add 20 μL of 0.1% SDS and mix briefly by inverting the tube. Add 400 μL of bleach to sterilize the seed surface and remove any contaminating fungi and bacteria (*see* **Note 11**). Mix well for 10 min by inverting the tube or by gently shaking it on a rotary shaker.
3. Let the tube stand on a rack to allow the seeds to settle to the bottom. Remove the sterilization buffer and rinse the seeds five to seven times with 1 mL of sterilized double-distilled H_2O to remove the remaining sodium hypochlorite.
4. Suspend the seeds in 1 mL of 0.1% sterile agarose solution and plate this suspension on selection medium (500–600 seeds per plate). Place the plates in the dark at 4°C for 2 days (in a cold room or refrigerator) so that the seeds acquire the competence to germinate.
5. Transfer the plate to a growth chamber with long-day conditions. Within 2 weeks, the non-transformed plants (>90%) die (i.e., they turn white), because of the effect of the antibiotic corresponding to the selectable marker. Transformants appear green and have well-established roots. If some plates are crowded with too many transformants, transplant them to a new plate in order to eliminate false-positive clones which grow due to their proximity to antibiotic-resistant clones.
6. When the transformants have grown well on the plate, transplant them into a pot with moistened soil until they mature and produce transgenic seeds. The roots of the transformants should be rinsed well with water to remove the adherent agar, which may promote fungal growth and thus damage the plants. In order to investigate the

phenotype or function of the target protein, isolation of homozygous lines from the subsequent plant generation is recommended. It is recommended that the expression of the epitope-tagged protein be confirmed by western blotting before performing immunoelectron microscopic analysis.

4. Notes

1. Chemicals used in this chapter are purchased from Sigma (St. Louis, MO) when a specific supplier is not specified. Chemicals supplied by different suppliers may be substituted.
2. Ecotype Columbia-0 (Col-0) is suitable for use in floral dipping. The ecotypes Landsberg *erecta*, Wassilewskija, and Nossen can also be used, although they are less efficient in terms of the rate of transformation. In cases where mutant strains (knockout lines for specific genes) are used, the corresponding parent lines should be selected as controls.
3. Mix the different types of soil well in the ratios described with the help of a scoop. This specific soil composition aids aeration and retards fungus growth. The ratios of sand and pumice or vermiculite and perlite used can be modified depending on the texture and quality of the potting soil. Fill this soil mixture into plastic pots and place the pots in a tray. Before sowing the seeds, the soil must be moistened thoroughly with water.
4. Plant growth chamber with controlled temperature, humidity, and illumination (photoperiod and intensity). For optimal growth, set the temperature of the growth chamber at 25°C. A lower temperature is acceptable, but a higher temperature can be detrimental for the growth of *Arabidopsis* plants. In our laboratory, the plant growth chamber is set at 23°C to ensure that the temperature does not exceed 25°C when the fluorescent lamps and heating or cooling units are switched on and off. The optimal light intensity is 80–150 μmol m^{-2} s^{-1}. The light intensity can be adjusted by changing the capacity or number of daylight fluorescent lamps.
5. A variety of binary plasmids containing a pair of border sequences, a selection marker, and the origins of replication for *Escherichia coli*and *A. tumefaciens* can be obtained from the European Arabidopsis Stock Centre (http://arabidopsis.info/) and individual research groups. After choosing a binary vector, subclone the cDNA

generated by the method described above downstream of the promoter region. Use the Cauliflower mosaic virus 35S promoter region in order to obtain constitutive high expression. If the protein encoded by the cDNA is toxic and could damage the host plant when expressed at a high level, the use of an inducible promoter such as the dexamethasone-inducible promoter instead of the 35S promoter is recommended. To fuse the epitope tag sequence to the N-terminus of the target protein, the DNA sequence encoding the epitope should be inserted between the first codon (ATG = Met) and the second codon in a forward-specific primer for polymerase chain reaction (PCR). For C-terminal labeling, insert the epitope tag sequence immediately before the stop codon in a reverse-specific primer. The DNA sequences of the different epitope tags used are listed below.

c-Myc: GAA CAA AAA CTC ATC TCA GAA GAG GAT

FLAG: GAT TAC AAG GAT GAC GAC GAT AAG

Polyhistidine: CAT CAT CAC CAT CAC CAT

Repeating the epitope tag (two to three times) may enhance the detection sensitivity, although the cloning strategy becomes more complicated. Detection sensitivity of polyhistidine is less than the others, but polyhistidine-tagged proteins can be purified by affinity chromatography.

6. *A. tumefaciens* transformed with the binary vector construct is used as the gene carrier for transformation of the host plant. The strain LBA4404 is commonly used for this purpose and competent cells for electroporation are commercially available from various companies. However, the C58C1RifR and EHA101 strains exhibit higher transformation efficiency than the LBA4404 strain. If LBA4404-mediated transformation is unsuccessful, we recommend the use of one of these strains.
7. Plants germinated and grown under long-day conditions can also be used. In long-day conditions, plants can be dipped 4–6 weeks after sowing onto the soil. In this case, sow 2–4 seeds in a pot covered with nylon window screen secured with a rubber band to prevent the soil from falling into the dipping bath. These plants have less rosette leaves and thin bolts and wither easier than the plants grown under our conditions.
8. Transformation efficiency is strongly dependent upon the electro-pulse and gap width of the cuvette. Suitable combinations should be empirically determined. The conditions given are optimal for transformation of *A. tumefaciens*.

9. The concentration of *Agrobacterium* in dipping solution could be reduced to an OD_{600} of 0.4 if most of the dipped plants wither due to *Agrobacterium* overgrowth.
10. To improve the transformation efficiency, vacuum infiltration (4) can be used. Because vacuum-infiltrated plants wither easily, caution should be exercised during the subsequent steps involving shade control and watering of the plants.
11. The addition of bleach may cause the formation of small air bubbles on the seed surface. These bubbles should be released by tapping or vortexing for a short period.

References

1. Valvekens, D., Montague, M. V., and Lijsbettens, M. V. (1988) *Agrobacterium tumefaciens*-mediated transformation of *Arabidopsis thaliana*root explants by using kanamycin selection. *Proc. Natl. Acad. Sci. USA* **85**, 5536–5540.
2. Weising, K., Schell, J., and Kahl, G. (1998) Foreign genes in plants: transfer, structure, expression and application. *Annu. Rev. Genet.* **22**, 421–477.
3. Clough, S. J. and Bent, A. F. (1998) Floral dip: a simplified method for *Agrobacterium*-mediated transformation of *Arabidopsis* thaliana. *Plant J.* **16**, 735–743.
4. Bechtold, N., Ellis, J., and Pelletier, G. (1993) *In planta Agrobacterium* mediated gene transfer by infiltration of adult *Arabidopsis thaliana* plants. *C. R. Acad. Sci. Paris, Life Sci.* **316**, 1194–1199.

Chapter 3

Expression of Epitope-Tagged Proteins in *Arabidopsis* Leaf Mesophyll Protoplasts

Young-Hee Cho and Sang-Dong Yoo

Abstract

Advances in genomic and proteomic platforms enable high-throughput studies for regulatory factors and interactors involved in signaling network at a molecular level. However, it has never been trivial to verify the omics data in vivo or functionally integrate the data in a cell signaling context. For plants, genetic approaches using knockout mutants and transgenic lines have been mainly used to characterize functions of gene products in vivo. In general, such approaches demand a longer time and a higher cost and have difficulties in understanding gene functions comprehensively in a high-throughput manner. Transient gene expression is a method of choice to examine the cellular functions of genetic components in vivo. The leaf mesophyll protoplasts (LMP) provide a perfect system to transiently express a gene encoding an epitope-tagged protein of interest and quickly and reliably trace subcellular locales of the protein in a near high-throughput manner. Here, a simple and straightforward method for isolating leaf meshophyll protoplasts from *Arabidopsis* has been described in detail to help beginners initiate their first cell-based functional genomic analysis.

Key words: Leaf mesophyll protoplasts, transient expression, epitope tag, high-throughput, *Arabidopsis*.

1. Introduction

Plant protoplasts are a powerful and easily manipulated system for quickly and efficiently exploring molecular and cellular functions of plant genes (1, 2). Protoplasts can be obtained from a variety of different plant tissue types through enzymatic digestion of their cell walls (3–5). Leaf mesophyll protoplasts (LMP) are the most widely used cell types due to their uniformity and relatively easy isolation. In addition, the procedure for isolating LMP appears to

S.D. Schwartzbach, T. Osafune (eds.), *Immunoelectron Microscopy*, Methods in Molecular Biology 657,
DOI 10.1007/978-1-60761-783-9_3, © Springer Science+Business Media, LLC 2010

be rather universal and applicable with only very minor modifications to various plant species including *Arabidopsis*, maize, rice, barley, alfalfa, and faba bean (3–6).

Currently, many LMP-based analyses have been applied for the model plant *Arabidopsis* which is one of the most widely studied plants due to its low genetic complexity, short life cycle, and ease of transformation. There are also enriched genomic and genetic resources available for *Arabidopsis* including a well-curated whole genome sequence and a range of mutant lines (The Arabidopsis Information Resource; www.arabidopsis.org). With these resources, molecular epistatic analysis using transient expression of LMP isolated from a mutant in the same pathway can complement traditional genetic studies and be a very powerful tool to discover in vivo gene functions and relationships of a genome product in a cell signaling network.

Information about a protein locale within cells of tissue sections can provide a strong clue in understanding protein behaviors and functions. Protein distribution and localization can be traced at a subcellular level by exploring the principle of antibodies binding specifically to antigens in biological tissues. This technique is also utilized widely in the diagnosis of abnormal cells such as those found in cancerous tumors in animals, because certain classes of molecular markers are specific to particular cellular events like cell proliferation or cell death. For such immunolocalization, a high-titer antibody is required to clearly and unambiguously determine protein locales by reducing non-specific background. Unfortunately, achieving such specific antibody to a protein of interest is often expensive, laborious, and time consuming. Above all, its success is not guaranteed. Transient expression in the LMP system is a valuable alternative to identify the localization of proteins of interest within cells quickly and comprehensively (7–11). A protein sequence can be fused with an epitope or several epitopes recognized by a commercially available highly specific antibody or tagged with green fluorescent protein (GFP) and its derivatives (12). Small epitope tags which are short peptides do not normally interfere with protein localization, but the bulky GFP tag sometimes causes mislocalization (**Table 3.1**). In certain cases, epitope tags have a clear advantage over a GFP tag in protein visualization. For instance, GFP-tagged proteins localized in plant vacuoles are often degraded in the acidic pH environment in the light (13). This instability of GFP hampers observation of vacuole-targeted proteins in higher plants. In such cases, epitope tags can replace GFP to comprehensively characterize protein behavior in subcellular organelles.

Several epitopes are available for vectors expressing in LMP (**Table 3.1**). All of these epitopes are derived from animal peptides, so backgrounds from cross-reactions of the antibodies are diminutive in plant cells. The immunolocalization of epitope-

Table 3.1
Useful epitope tags for transient protein expression in protoplasts

Name	Epitope tag	Size	MW
HA	YPYDVPDYA	9 aa	1.1 kDa
Myc	EQKLISEEDL	10 aa	1.2 Da
Flg	DYKDDDDK	8 aa	1.0 kDa
GFP	*	238 aa	ca. 26.9 kDa

*GenBank: ABL09837.1

tagged proteins in the LMP system can be conducted in a high-throughput manner becoming an indispensable tool to explore the genomic and genetic information available in *Arabidopsis.*

To isolate LMP from mature leaves, leaf stripes are infiltrated with a mixture of fungal cell wall digesting enzymes. Intact cells are released from the cell wall matrix and then transfected with plasmid DNA containing a gene of interest using a calcium–polyethylene glycol (PEG) method. The transfection efficiency of LMP is dependent upon the purity of plasmid DNA. Because a high LMP transfection rate facilitates statistical analysis of protein visualization data, it is desirable to reach a technical status of reasonable transfection efficiency. Plasmid DNA can be prepared as closed circular plasmids by centrifugation to equilibrium in CsCl–ethidium bromide gradients. Despite the fact that the CsCl procedure is more labor intensive and time consuming compared to various commercial DNA preparation kits, it gives the highest yield of pure DNA resulting in the highest transfection efficiencies. LMP are cultured after transfection to allow the gene to be expressed and subjected to further analysis. Methods for large-scale preparation of high-quality plasmid DNA, plant growth and LMP isolation, and calcium–PEG transfection of LMP are described in this chapter.

2. Materials

2.1. Large-Scale Preparation of High-Quality Plasmid DNA

1. Culture solution: Terrific broth containing 0.5% (v/v) glycerol
2. Orbital rotary shaker (Thermo-Fisher Scientific, Waltham, MA)
3. 250 mL polypropylene copolymer centrifuge tubes (Nalgene, Rochester, NY)

4. Midspeed centrifuge with Sx4750 rotor (Beckman Coulter, Fullerton, CA)
5. 70% (v/v) ethanol
6. 95% (v/v) ethanol
7. Isopropanol
8. Miracloth (Calbiochem, San Diego, CA)
9. 15 mL plastic centrifuge tube (Corning, Corning, NY)
10. Cesium chloride
11. Ethidium bromide solution (10 mg/mL)
12. Quickseal tube (Beckman Coulter, Fullerton, CA)
13. Ultracentrifuge with NVT90 rotor (Beckman Coulter, Fullerton, CA)
14. 1-Butanol
15. Syringe with a 18 g hypodermic needle
16. Lysis solution 1 (LS1): 10 mM EDTA–NaOH, pH 8.0. Prepare 0.5 M EDTA adjusted to pH 8.0 with 2 M NaOH as a 50× stock solution. Mix 1 volume with 49 volumes of distilled water
17. Lysis solution 2 (LS2): 0.1 M NaOH containing 1% (w/v) SDS. Prepare 2 M NaOH and 20% (w/v) SDS as 20× stock solutions. Mix 1 volume of 2 M NaOH and 1 volume of 20% (w/v) SDS with 18 volumes of distilled water
18. Neutralization solution (NS): 2.5 M potassium acetate–acetic acid, pH 4.8

2.2. Plant Growth and LMP Isolation

1. *Arabidopsis* seeds: Col-0 and *L*er accession (Arabidopsis Biological Resource Center; http://abrc.osu.edu/)
2. Plant growth chamber with light and humidity control (Conviron, Pembina, ND)
3. Metro-Mix 360 (Sun Gro, Belleve, WA)
4. Bench-top centrifuge (Krackeler Scientific, Albany, NY)
5. 30 mL round bottom centrifuge tube
6. 10 × 25 mm petri dish (Thermo-Fisher Scientific, Waltham, MA)
7. Sterile 0.45 μm syringe filters (GE Health care, Piscataway, NJ)
8. Generic razor blades
9. Bell-shaped desiccators (Thermo-Fisher Scientific, Waltham, MA)
10. Vacuum pump (Thermo-Fisher Scientific, Waltham, MA)
11. 70 μm nylon mesh (Carolina Biological Supplies, Burlington, NC)

12. Miracloth (Calbiochem, San Diego, CA)
13. Kimwipe (Kimberly-Clark, Dallas, TX)
14. Neubauer 0.1 mm deep hemocytometer (Hausser Scientific. Horsham, PA)
15. Cell wall digesting enzyme mix: In a petri dish (100 × 25 mm for 10 mL enzyme solution), dissolve 1–1.5% (w/v) cellulase R10 (Yakult Pharmaceutical Ind. Co., Tokyo) and 0.2–0.4% (w/v) macerozyme R10 (Yakult Pharmaceutical Ind. Co., Tokyo) in preheated 20 mM MES–KOH, pH 5.7, 0.4 M mannitol, 20 mM KCl (*see* **Note 1**). Incubate the enzyme solution at 55°C for 10 min to inactivate biochemical activities other than cell wall degrading. After cooling on ice, add $CaCl_2$ to a final concentration of 10 mM from a 1 M stock solution and BSA to a final concentration of 0.1% (w/v) from a 10% stock solution. Prepare the enzyme mix just before starting LMP isolation and sterilize by filtration through a 0.45 μm syringe filters
16. W5 solution: 2 mM MES–KOH, pH 5.7, 154 mM NaCl, 125 mM $CaCl_2$, 5 mM KCl
17. MMg solution: 4 mM MES–KOH, pH 5.7, 0.4 M D-mannitol, 15 mM $MgCl_2$

2.3. Calcium–PEG Transfection of LMP

1. WI solution: 4 mM MES–KOH, pH 5.7, 0.5 M D-mannitol, 20 mM KCl.
2. Calcium–PEG solution: Prepare a solution containing 40% (w/v) PEG4000 (Fluka, USA), 0.2 M D-mannitol, 100 mM $CaCl_2$ at least 2 h before use.
3. 2-mL round bottom microfuge tubes
4. 6-well cell culture plate (Thermo-Fisher Scientific, Waltham, MA)

3. Methods

3.1. Large-Scale Preparation of High-Quality Plasmid DNA

1. Grow bacteria transformed with a plasmid encoding an epitope-tagged protein by inoculating 5-mL terrific broth containing appropriate antibiotics for plasmid selection with a single *E. coli* colony. Incubate overnight at 37°C on an orbital rotary shaker at 220 rpm.
2. Transfer 1 mL of overnight culture to 200 mL of terrific broth containing appropriate antibiotics for plasmid selection and incubate on an orbital rotary shaker at 220 rpm for 10–13 h at 37°C (*see* **Note 2**). Pour the culture into a

centrifuge bottle and recover the bacterial cells by centrifugation at 4,105 × *g* for 15 min at 4°C.

3. Decant the culture broth and resuspend cells in 40 mL of LS1 making sure that visible cell aggregates are not present in the suspension.
4. Add 80 mL of LS2 to the cell suspension in the centrifuge bottle and mix gently by inverting the bottle several times. Keep the mixture at room temperature approximately 5 min or until it turns clear.
5. Add 30 mL of NS and mix gently by inverting the tube several times. The solution has to be completely mixed but avoid vortexing which could shear the plasmid DNA.
6. Remove the cell debris by centrifugation at 4,105 × *g* for 5 min at 4°C.
7. Transfer to a fresh 250 mL centrifuge bottle by pouring through a layer of Miracloth (*see* **Note 3**). Add an equal volume of isopropanol to the 250 mL centrifuge bottle. Invert the bottles several times to fully mix and the DNA should immediately precipitate.
8. Recover the precipitated DNA by centrifugation at 4,105 × *g* for 5 min at 4°C.
9. Decant the supernatant, wash the DNA pellet with approximately 5 mL of 95% (v/v) ethanol, and then air-dry the pellet for approximately 10 min (*see* **Note 4**).
10. Resuspend the DNA pellet completely in 3.5 mL of LS1 and place the tube on an orbital rotary shaker. Shake at 80 rpm for at least 5 min.
11. Transfer the DNA solution to a 15 mL plastic centrifuge tube and adjust the final volume to 4.7 mL with LS1 using a disposable pipette.
12. Dissolve 5.5 g of CsCl in the DNA solution by gently inverting the tube (*see* **Note 5**). Add 200 μL of ethidium bromide, mix quickly, and then remove insoluble particles by centrifugation at 2,095 × *g* for 5 min at 4°C.
13. Transfer the DNA solution to a 5.5 mL quick seal tube and seal the tube.
14. Place the tubes in the NVT90 rotor and centrifuge to generate a CsCl gradient and separate the plasmid DNA from cellular DNA. Centrifuge at 286,000 × *g* (60,000 rpm) for 12 h or 645,000 × *g* (90,000 rpm) for 3 h at 20°C (*see* **Note 6**).
15. Cut off the cap of the quick seal tube. Insert a No. 18 g hypodermic needle into the side of the tube just below the red ethidium bromide-stained plasmid DNA band

and collect the DNA in a 15 mL plastic centrifuge tube (*see* **Note 7**).

16. To remove the ethidium bromide from the DNA, add 7 mL 1-butanol saturated with 1 M NaCl and mix the two phases by shaking vigorously. Let the phases separate and slowly add 1–2 mL of water to form a layer between the dense DNA CsCl solution at the bottom of the tube and the red ethidium bromide-1-butanol layer at the top of the tube.
17. Remove the top layer of 1-butanol containing ethidium bromide and a half of the water layer without disturbing the DNA CsCl layer.
18. Repeat the 1-butanol extraction two to three times until the DNA layer is clear.
19. Precipitate the DNA by adding 3 volumes of 95% (v/v) ethanol (*see* **Note 8**).
20. Wash DNA aggregates or pellets recovered by centrifugation with 70% (v/v) ethanol and air-dry.
21. Resuspend the DNA in about 300 μL of sterile water and determine the DNA concentration (*see* **Note 9**). Add sterile water for a final DNA concentration of 2 μg/μL DNA
22. Store the DNA at – 20°C.

3.2. Plant Growth and LMP Isolation

1. Grow *Arabidopsis* on soil in a growth chamber with a 23°C/13 h light and 20°C/11 h dark cycle. Adjust the light intensity to 75–80 μE and the relative humidity to 50–65% (*see* **Note 10**). If the cyclic control of temperature is not available, you can use a constant temperature of 22°C (*see* **Note 11**). These growth conditions maximize leaf expansion prior to bolting.
2. Detach and collect expanded leaves 3–5 from 21- to 28-day-old plants that have not yet bolted for flowering (*see* **Note 12**). Slice the leaves to a thickness of approximately 1 mm with a sharp razor blade. Chang the blade when the cutting edge becomes dull (*see* **Note 13**).
3. Submerge the leaf pieces completely in the enzyme mixture prepared in a petri dish (10 mL mixture per 20 leaves).
4. Place the petri dish containing the leaf pieces in a bell-shaped desiccator and vacuum infiltrate the enzyme mixture into the leaf pieces in the dark for 30 min at room temperature. After releasing the vacuum, digest the cell walls in the dark for another 2–3 h at room temperature (*see* **Note 14**). Place the petri dish on an orbital rotary shaker at 60–80 rpm for 5–10 min at room temperature to separate LMP from partially digested cell walls.

5. Filter the enzyme–LMP mixture through a layer of 70 μm nylon mesh into a 30 mL round bottom centrifuge tube. Pellet the LMP by gentle centrifugation at 60×*g* for 1–2 min at room temperature.
6. Remove the enzyme solution and resuspend the LMP in 5 mL of cold W5 solution.
7. Stand the LMP solution on ice for 30 min allowing the LMP to settle by gravity.
8. Carefully remove the W5 solution using a pipette and resuspend the loose green LMP pellet in 3 mL MMg solution. Count the cell number under a microscope using a hemocytometer and adjust the cell density to 1–2 × 10^5 cells/mL in MMg.
9. Keep LMP in MMg solution at room temperature until transfected (*see* **Note 15**).

3.3. Calcium–PEG Transfection of LMP

1. Aliquot 20 μL of DNA (2 μg/μL) into a 2 mL round bottom microfuge tube.
2. Add 200 μL of LMP (2–4 × 10^4 cells) and mix by gently tapping two to three times.
3. Add 220 μL of PEG solution, and mix the contents immediately and completely by gently tapping several times.
4. Place the microfuge tube at room temperate for 3–30 min. This process is stressful to LMP, so the shorter time will produce the higher transfection rate (*see* **Note 16**).
5. Add 900 μL of W5 solution and mix well by inverting the microfuge tube two to three times.
6. Recover the transfected LMP by centrifugation of the transfection tube at 60×*g* for 1 min at room temperature.
7. Remove the supernatant and resuspend the transfected LMP in 1 mL of WI and transfer into a well of a 6-well cell culture plate.
8. Incubate the transfected LMP at room temperature under dim light (30–40 μE) (*see* **Note 17**). Harvest transfected LMP by centrifugation at 100×*g* and subject to further analysis.

4. Notes

1. Concentrations of enzymes should be determined empirically for each genetic background and tissues type of *Arabidopsis*.

2. Bacteria growth to too high a density often results in a lower yield of plasmid DNA.
3. If miracloth is not available, you can use two layers of KimWipes instead.
4. Avoid vacuum drying because it could make the pellet hard to resuspend.
5. The final concentration of CsCl is approximately 1 g/mL.
6. Alternatively, you can use the MLA-80 rotor (Beckman coulter, Fullerton, CA) with 6.5 mL thick-wall polycarbonate tubes by centrifuging at 440,000×*g* (80,000 rpm) for approximately 12 h.
7. If a thick-wall polycarbonate tube is used instead of a Quickseal tube, you can simply remove the DNA using a pipette.
8. If the DNA preparation is successful, you will see a cloudy solution with large DNA aggregates that settle to the bottom of the tube allowing the solution to be decanted. If the aggregates do not form, the DNA needs to be recovered by centrifugation at 2,095×*g* for 5 min at 4°C.
9. One unit of O.D. at 260 nm equals 50 μg/μL DNA. The typical yield is 2–3 mg of DNA.
10. Growth optimization is necessary for each growth facility. However, *Arabidopsis* requires little maintenance, once optimal growth conditions have been established even in artificial indoor settings.
11. LMP with high transfection efficiency can be reproducibly obtained by preparing the protoplasts from healthy leaves of mature plants. We cannot emphasize enough the need to take good care of the plants to harvest healthy LMP.
12. Younger leaves often have smaller cells. Older leaves have bigger cells with larger vacuoles which have a tendency to be easily broken during LMP isolation making them unsuitable. If you need more cells, use more of the younger leaves rather than the older ones. Remember, cell size and uniformity are important factors that affect the reproducibility and reliability of the LMP transfection.
13. During leaf slicing, the sharper the blade, the less tissue is crushed or wounded. A consistent slicing action from sample to sample with sharp blades guarantees a consistently high yield of LMP from each experiment.
14. We generally avoid an extended digestion since this process can be stressful to LMP. However, leaves in different genetic backgrounds require different digestion times, so this has to be optimized for each genotype as well as each mutant.

15. You can keep LMP at room temperature for at least next 1 h without compromising the transfection rate.
16. Determine the transfection time empirically for each genotype. You can transfect a sample containing GFP to estimate transfection rate.
17. Incubation time should be determined for each gene product. Typically protein accumulation is detectable after 6–9 h.

Acknowledgments

We thank Dr. Donald Hunter for his help in the preparation of this chapter. This work was supported by National Research Foundation of Korea Grant 2010-0007068 to Y.-H. Cho and grants 2009-0085565 and 2010-0016989 to S.-D. Yoo.

References

1. Sheen, J. (2001) Signal transduction in maize and Arabidopsis mesophyll protoplasts. *Plant Physiol.* **127**, 1466–1475.
2. Yoo, S. D., Cho, Y. H., and Sheen, J. (2007) A versatile experimental cell system, Arabidopsis leaf mesophyll protoplasts. *Nat. Protoc.* **2**, 1565–1572.
3. Kovtun, Y., Chiu, W. L., Tena, G., and Sheen, J. (2000) Functional analysis of oxidative stress-activated mitogen-activated protein kinase cascade in plants. *Proc. Natl. Acad. Sci. USA* **97**, 2940–2945.
4. Sheen, J. (1993) Protein phosphatase activity is required for light-inducible gene expression in maize. *EMBO J.* **12**, 3497–3505.
5. Jacobsen, J. V. and Beach, L. R. (1985) Control of transcription of alpha-amylase and rRNA genes in barley aleurone protoplasts by gibberellin and abscisic acid. *Nature* **316**, 275–277.
6. Assmann, S. M., Simoncini, L., and Schroeder, J. I. (1985) Blue light activates electrogenic ion pumping in guard cell protoplasts of *Vicia faba*. *Nature* **318**, 285–287.
7. Lee, Y., Kim, D., Kim, Y., and Hwang, I. (2001) Identification of a signal that distinguishes between the chloroplast outer envelope membrane and the endomembrane system in vivo. *Plant Cell* **13**, 2175–2190.
8. Jin, J., Kim, Y., Kim, S., Lee, S., Kim, D., Cheong, G., and Hwang, I. (2001) A new dynamin-like protein, ADL6, is involved in trafficking from the *trans*-golgi network to the central vacuole in Arabidopsis. *Plant Cell* **13**, 1511–1526.
9. Sohn, E., Kim, E., Zhao, M., Kim, S., Kim, H., Kim, Y., Lee, Y., Hillmer, S., Sohn, U., Jiang, L., and Hwang, I. (2003) Rha1, an Arabidopsis Rab5 homolog, plays a critical role in the vacuolar trafficking of soluble cargo proteins. *Plant Cell* **15**, 1057–1070.
10. Kim, H., Park, M., Kim, S., and Hwang, I. (2005) Actin filaments play a critical role in vacuolar trafficking at the golgi complex in plant cells. *Plant Cell* **17**, 888–902.
11. Cho, Y. H., Yoo S. D., and Sheen, J. (2006) Regulatory functions of nuclear hexokinase1 complex in glucose signalling. *Cell* **127**, 579–589.
12. Chiu, W.-L., Niwa, Y., Zeng, W., Hirano, T., Kobayashi, H., and Sheen, J. (1996) Engineered GFP as a vital reporter in plants. *Curr.Biol.* **6**, 325–330.
13. Tamura, K., Shimada, T., Ono, E., Tanaka, Y., Nagatani, A., Higashi, S., Watanabe, M., Nishimura, M., and Hara-Nishimura, I. (2003) Why green fluorescent fusion proteins have not been observed in the vacuoles of higher plants. *Plant J.* **35**, 545–555.

Chapter 4

Transient Expression of Epitope-Tagged Proteins in Mammalian Cells

Melanie L. Styers, Jason Lowery, and Elizabeth Sztul

Abstract

Before the advent of molecular methods to tag proteins, the visualization of proteins within cells by immunoelectron microscopy required the use of highly specific antibodies directed against the protein of interest. Thus, only proteins for which antibodies were available could be visualized. Current technologies allow the detection of proteins for which specific antibodies are not available. This procedure involves the generation of DNA constructs that express the protein of interest tagged with an epitope that is recognized by a well-characterized commercially available antibody. Proteins can be tagged with a wide variety of epitopes, small and large, using commercially available vectors that allow expression in mammalian cells. Epitope-tagged proteins are easily transfected into many mammalian cell lines and, in most cases, tightly mimic the distribution of the endogenous protein. Prior to immunoelectron microscopy, expression and localization of tagged proteins can be assessed by Western blotting and immunofluorescence. Furthermore, specialized fluorescent tags, such as the green fluorescent protein (GFP), can be used to rapidly screen for transfection efficiency and localization. The use of epitope-tagged protein expression has increased the versatility of immunoelectron microscopy to explore the function of uncharacterized proteins for which highly specific antibodies are not available.

Key words: Immunoelectron microscopy, transfection, Western blot, immunofluorescence microscopy, epitope tag, green fluorescent protein (GFP).

1. Introduction

In the past, immunoelectron microscopic approaches to study protein localization within cells required the availability of highly specific antibodies for each protein. This requirement necessitated the purification of often rare cellular proteins and the generation of either monoclonal or polyclonal antibodies to those

S.D. Schwartzbach, T. Osafune (eds.), *Immunoelectron Microscopy*, Methods in Molecular Biology 657,
DOI 10.1007/978-1-60761-783-9_4,

antigens. In addition to their specificity, these antibodies also had to recognize the desired antigen subsequent to the fixation conditions necessary to preserve cells and tissues for immunoelectron microscopy. However, with the advent of molecular cloning and the mapping of the human genome, scientists can now introduce tags into any protein of interest to facilitate visualization of proteins for which no antibodies are currently available.

A wide variety of DNA vectors for tagging proteins are commercially available. Tags range from small peptide epitopes, such as the Hemagglutinin (HA), FLAG, and myc tags, to small proteins, such as the green fluorescent protein (GFP). A wide variety of well-characterized monoclonal and polyclonal antibodies to all common tags, useful for electron microscopy, are readily available from commercial sources. Thus, the wide variety of tags and the spectrum of available anti-tag antibodies allow for the visualization of single or multiple proteins within the cell by immunoelectron microscopy.

A number of key considerations are essential for assessing the cellular distribution of tagged proteins by immunoelectron microscopy. First, it is essential that the tagged protein is expressed and that the expressed protein is the correct size. Second, the anti-tag antibodies must be specific and not recognize other cellular components. Third, the tagged protein must localize to the correct cellular compartment, based on current knowledge regarding its distribution and/or function. Fourth, it is essential that the tagged protein is expressed at levels sufficient for detection at the ultrastructural level. These four criteria should be tested prior to electron microscopy by using Western immunoblotting and immunofluorescence localization.

Western immunoblotting allows the determination of the level of expression of the epitope-tagged protein and ensures that a protein of the desired size is produced. This procedure is also used to assess the specificity of the anti-tag antibodies. If more than one protein or proteins of the wrong size are recognized, this signifies that the antibodies recognize additional cellular proteins. In this case, either a different, more specific antibody should be used or the protein to be studied must be tagged with a different tag that is recognized by antibodies that do not react with other cellular components.

Immunofluorescence is an essential preliminary to immunoelectron microscopy because it establishes the gross distribution of the tagged protein and provides a measure of the strength of the antibody signal. Some tagged proteins do not express well in mammalian cells, and immunofluorescence is the method of choice to decide whether immunoelectron microscopic detection will be possible. In general, a strong localized fluorescence signal must be detected when the anti-tag antibodies are used at a 1:50 dilution to make immunoelectron microscopy feasible. Light-level analysis of the cellular localization of a tagged protein after expression is essential to ensure that the protein shows a specific localization. Immunofluorescence can provide information regarding the association of the tagged protein with specific subcellular compartments. This is important because it greatly facilitates ultrastructural analysis by focusing imaging on those specific subcellular organelles on grid sections. Immunofluorescence also suggests specific markers to be used in double label analyses to refine the distribution of the tagged protein. For example, immunofluorescence localization of a tagged protein to the Golgi complex can be then followed by double label immunoelectron microscopy using antibodies to proteins known to reside in different cisternae of the Golgi. Finally, immunofluorescence can also be used to assess and optimize the transfection efficiency of the tagged construct.

Although in most cases epitope-tagged proteins faithfully mimic the localization and function of their endogenous counterparts, the attachment of a peptide or protein tag can interfere with endogenous bonding interactions required for protein folding. Misfolding of the tagged protein can in some cases lead to its aggregation or mislocalization. Aggregation can be easily detected by immunofluorescence, with the aggregated proteins accumulating in large cytoplasmic aggresomes (1, 2). Misfolding can also cause mislocalization of the tagged proteins and thus, whenever possible, the cellular distribution of the tagged protein should be compared to the known localization and/or function of the protein. Luckily, since a wide variety of tags are available, varying the size and/or type of epitope can often eliminate misfolding problems. There is no a priori method to determine how a specific protein will be affected by a specific tag, and it requires empirical trial and error to identify the optimal tag. In general, problems with protein folding or mislocalization can often be minimized through the use of smaller tags, such as HA or myc. In addition, varying whether the tag is fused to the N- or the C-terminus of the protein of interest often alleviates folding or localization problems. Tagging at the N-terminus can be very important for proteins that are lipid-modified at the C-terminus, such as the Rab members of the Ras superfamily. These proteins must be tagged at the N-terminus in order not to interfere with lipid modification. In contrast, the ARF members of the Ras superfamily are often tagged at the C-terminus because they are myristoylated within their N-terminal domains. Similarly, it is best not to place tags close to known functional domains within proteins. For example, secreted proteins in which the N-terminal signal sequences are cleaved require tagging at the C-terminus.

Western immunoblotting and immunofluorescence localization are useful for assessing the level of expression of an epitope-tagged protein. Most of the commonly used vectors for expression of proteins in mammalian cells drive expression using the strong cytomegalovirus (CMV) promoter. Some examples include the pCMV-myc and pCMV-HA vectors from Clontech. Although these vectors are useful for most applications, at times excessive overexpression can mask the true localization of a protein. Thus, it is advisable to perform preliminary immunoblotting and immunofluorescence experiments to monitor the expression and the localization of the tagged protein beginning approximately 6 h after transfection and continuing until 48 h post-transfection to assess the level of expression and to note any changes in localization as the protein levels increase. For proteins affected by high overexpression, tetracycline inducible or repressible systems are commercially available and allow for tight control of protein expression levels. These systems are most often used in stably transfected cells and require drug-mediated selection of stable cell clones expressing gene products from two distinct vectors. The first vector constitutively expresses either a tetracycline transactivator ("tet on") or repressor ("tet off"), and the second vector expresses the epitope-tagged protein of interest under the control of a promoter containing a tetracycline response element. Level of expression can be tightly controlled by varying the concentration of tetracycline present in the media. Commonly used tetracycline-regulated systems include the "Tet On" and "Tet Off" expression systems available from Clontech.

With careful control, expression of epitope-tagged constructs in mammalian cells can be a highly useful technique for immunoelectron microscopy-based studies. In general, expression of epitope-tagged constructs in mammalian cells can allow unprecedented insight into cellular architecture and physiology. In this chapter, we will discuss approaches and provide examples detailing how to express epitope-tagged proteins in mammalian cells and how to assess expression, transfection efficiency, and preliminary localization of the tagged proteins by Western blotting and immunofluorescence. Techniques to be discussed include mammalian tissue culture on glass and plastic, transfection of DNA constructs into mammalian cells, Western blotting, and immunofluorescence. A list of the materials required will be followed by a detailed discussion of the protocols for these techniques.

2. Materials

Solutions may be stored at room temperature unless noted otherwise.

2.1. Cell Culture and Transfection

1. HeLa cell stock stored at −80°C (*see* **Note 1**)

2. PBS: Sterile 1X phosphate-buffered saline (Mediatech, Inc., Manassas, VA)
3. 10% FBS MEM: Supplement Minimum Essential Media with glucose and L-glutamine (MEM, Mediatech, Inc., Manassas, VA) with 10% fetal bovine serum (FBS, Life Technologies, Grand Island, NY), 1 mM sodium pyruvate, 0.075% sodium bicarbonate, 100 units/mL penicillin, and 100 μg/mL streptomycin using a concentrated sterile penicillin–streptomycin solution (Mediatech, Inc., Manassas, VA). Store at 4°C.
4. Trypsin/EDTA: Sterile solution of 0.25% trypsin and 1 mM ethylenediaminetetraacetic acid (EDTA) (Mediatech, Inc., Manassas, VA). Store at 4°C.
5. 35 and 100 mm tissue culture dishes (Fisher Scientific, Pittsburgh, PA)
6. 24-well tissue culture plates (Fisher Scientific, Pittsburgh, PA)
7. 12 mm circular coverslips (Fisher Scientific, Pittsburgh, PA)
8. 1.5 mL sterile, nuclease-free microcentrifuge tubes (Fisher Scientific, Pittsburgh, PA)
9. Serum-free RPMI (Mediatech, Inc., Manassas, VA). Store at 4°C.
10. TransIT-LT1 polyamine transfection reagent (Mirus Corporation, Madison, WI). Store at 4°C (*see* **Note 2**).
11. Miniprep of transformation vector encoding epitope-tagged protein in sterile nuclease-free water. The DNA concentration should be at least 50 μg/mL for transfection (*see* **Notes 3** and **4**). Store at −20°C.
12. Miniprep of control transformation vector that does not contain an insert; empty vector. The DNA concentration should be at least 50 μg/mL for transfection (*see* **Notes 3** and **4**). Store at −20°C.
13. 37°C incubator containing 5% CO_2 (Fisher Scientific, Pittsburgh, PA)

2.2. Cell Lysis

1. 10X PBS (phosphate-buffered saline, not sterile): 137 mM NaCl, 27 mM KCl, 43 mM KH_2PO_4, 14 mM Na_2HPO_4, pH 7.4. Dilute to 1X PBS by combining 100 mL 10× PBS with 900 mL H_2O for final use.
2. 50× Complete protease inhibitors (Roche Diagnostics, Indianapolis, IN) stock solution: Prepare a 50× stock solution of protease inhibitors according to the manufacturer's instructions by diluting one pellet of Complete protease inhibitors in 1 mL of deionized water. Store stock solution at −20°C.

3. RIPA buffer: 150 mM NaCl, 1% NP-40, 0.5% sodium deoxycholate, 0.1% SDS, 50 mM Tris–HCl, pH 8.0. Store at 4°C. Just prior to use, add protease inhibitors to a final concentration of 1× from the 50× Complete protease inhibitors stock solution.
4. Teflon cell scrapers (Fisher Scientific, Pittsburgh, PA)
5. 1.5 mL microcentrifuge tubes (Fisher Scientific, Pittsburgh, PA)
6. Microcentrifuge (Fisher Scientific, Pittsburgh, PA)
7. Pierce BCA (bicinchoninic acid) Protein Assay kit (Thermo Scientific, Rockford, IL)
8. Visible wavelength spectrophotometer (Fisher Scientific, Pittsburgh, PA)
9. Cuvettes (Fisher Scientific, Pittsburgh, PA)
10. 4× Laemmli sample buffer: 8% sodium dodecyl sulfate (SDS) (*see* **Note 5**), 40% glycerol, 20% 2-mercaptoethanol, 0.008% bromophenol blue, 0.25 M Tris–HCl, pH 6.8. Store at 4°C.
11. Deionized water

2.3. SDS-Polyacrylamide Gel Electrophoresis and Western Blotting

1. 30% acrylamide/bisacrylamide solution 29:1 (Bio-Rad, Hercules, CA) (*see* **Note 6**)
2. 1.5 M Tris–HCl, pH 8.8
3. 1.0 M Tris–HCl, pH 6.8
4. 10% SDS: Prepare a 10% solution in H_2O (*see* **Note 5**).
5. 10% APS: 10% ammonium persulfate (Bio-Rad, Hercules, CA) in H_2O. Prepare solution fresh every 2–4 weeks and store at 4°C in order to promote rapid gel polymerization.
6. TEMED: *N*,*N*,*N*,*N′*-tetramethylethylenediamine (Bio-Rad, Hercules, CA)
7. Isopropanol
8. Kimwipes (Kimberly Clarke, Neenah, WI)
9. Tris/glycine/SDS running buffer (10×): 250 mM Tris base, 1.9 M glycine, 1% SDS, pH 8.3 (*see* **Note 5**). Dilute to 1X Tris/glycine/SDS running buffer by combining 100 mL 10× Tris/glycine/SDS running buffer with 900 mL H_2O for final use.
10. Kaleidoscope prestained molecular weight markers (Bio-Rad, Hercules, CA)
11. Gel loading tips (Fisher Scientific, Pittsburgh, PA)
12. Mini-PROTEAN gel electrophoresis system (Bio-Rad, Hercules, CA)

13. Nitropure nitrocellulose membrane (Cole Parmer, Vernon Hills, IL)
14. Whatman paper (Fisher Scientific, Pittsburgh, PA)
15. Sponge (Bio-Rad, Hercules, CA): Fits transfer cassette.
16. Methanol (Sigma Aldrich, St. Louis, MO)
17. 10× Tris–glycine Transfer Buffer: 20 mM Tris base, 150 mM glycine. For 1× solution, combine 100 mL of 10× buffer, 800 mL of H_2O, and 100 mL of methanol. Prepare 1× buffer just prior to use to prevent evaporation of methanol.
18. Ice container (Bio-Rad, Hercules, CA): Fits transfer apparatus.
19. Tupperware or other small plastic container
20. 10× TBS (Tris-buffered saline): 1.5 M NaCl, 0.1 M Tris–HCl pH 8.0. Prepare 1X solution by combining 100 mL of 10X TBS and 900 mL of H_2O.
21. TBST (Tris-buffered saline containing 0.1% Tween-20): Prepare 1 L of 1× TBS and add 1 mL of Tween-20.
22. Blocking solution: 5% carnation non-fat evaporated milk (can be obtained from your local grocer) in TBST. Store at 4°C for up to 1 week.
23. Primary antibody solution: 5% bovine serum albumin (BSA, Sigma Aldrich) in TBST containing the primary antibody to the epitope tag diluted at the manufacturer's recommended concentration for Western blotting. 0.05% sodium azide (NaN_3, Sigma Aldrich, St. Louis, MO) may be added to primary antibody solution and the solution stored at 4°C. This allows the solution to be reused for 3–4 months.
24. Secondary antibody solution: Dilute HRP-conjugated secondary antibody (Invitrogen, Carlsbad, CA) into blocking solution at the manufacturer's recommended dilution for Western blotting (*see* **Note 7**).
25. Super Signal West Pico Chemiluminescent substrate (Thermo Scientific, Rockford, IL). Store solution at 4°C and use according to manufacturer's instructions.
26. Mini-Trans Blot tank transfer apparatus (Bio-Rad, Hercules, CA).

2.4. Immunofluorescence

1. Millipore 0.22 μm mixed cellulose ester membranes for filtration (Fisher Scientific, Pittsburgh, PA). Should be sized appropriately to fit Büchner funnel below.
2. Filter apparatus: Side-arm Erlenmeyer flask capped with a Büchner funnel with glass frit (Sigma Aldrich, St. Louis,

MO). Place Millipore 0.22 μm mixed cellulose ester membrane in funnel prior to filtration. Connect apparatus to a vacuum line for efficient filtration.

3. 10× PBS_f (phosphate-buffered saline): 137 mM NaCl, 27 mM KCl, 43 mM KH_2PO_4, 14 mM Na_2HPO_4, pH 7.4. Vacuum filter solution using filter apparatus and 0.22 μm mixed cellulose ester membrane to remove particulate matter. Dilute to 1× PBS_f by combining 100 mL 10× PBS_f with 900 mL H_2O for final use.
4. 3% PFA: 3% paraformaldehyde in 1× PBS_f. Dissolve paraformaldehyde (Sigma Aldrich, St. Louis, MO) in 1× PBS_f by heating at 60°C for 20–30 min. The final solution will remain slightly cloudy and must be vacuum filtered as described above. Aliquot filtered solution in approximately 10 mL aliquots and store at −20°C (*see* **Note 8**).
5. 10 mM NH_4Cl (Sigma Aldrich, St. Louis, MO) in 1× PBS_f. Vacuum filter solution as described above to remove particulate matter. Store solution at −20°C.
6. 0.01% Triton X-100 in 1× PBS_f: Prepare solution immediately prior to use.
7. PBST (phosphate-buffered saline containing 0.01% Tween-20): 1× PBS_f + 0.01% Tween-20. Vacuum filter solution as described above to remove particulate matter.
8. 2.5% Goat serum (Sigma Aldrich, St. Louis, MO) in PBST: Vacuum filter solution as described above to remove particulate matter, aliquot, and store at −20°C.
9. 0.4% fish skin gelatin in PBST: 0.4% Gelatin from cold water fish skin (Sigma Aldrich, St. Louis, MO) in PBST. Vacuum filter solution as described above to remove particulate matter, aliquot, and store at −20°C.
10. Secondary antibody: Goat α-mouse or Goat α-rabbit Alexa 488 (Molecular Probes/Invitrogen, Eugene, OR) (*see* **Note 9**).
11. Hoescht 33258 (Molecular Probes/Invitrogen, Eugene, OR) solution: Dilute stock solution (1 mg/mL in H_2O, store at 4°C) in PBS to a final concentration of 0.1 μg/mL just prior to use.
12. Mounting media: 20 mM Tris–HCl, pH 8.0, 0.5% *N*-propyl gallate (Sigma Aldrich, St. Louis, MO), 50% glycerol. Dissolve *N*-propyl gallate in Tris–HCl, pH 8.0, prior to the addition of glycerol. Mix solution by gentle rocking overnight at 4°C to minimize introduction of bubbles into the media. Aliquot mounting media in 50 μL aliquots and store at −80°C.

13. Glass slides (Fisher Scientific, Pittsburgh, PA).
14. Clear nail polish.

3. Methods

3.1. Cell Culture and Transfection

1. Rapidly thaw a vial containing 1 mL of HeLa cells preserved in liquid nitrogen in a water bath at 37°C. Immediately following thawing, dilute cells into 10 mL of 10% FBS MEM in order to prevent toxicity from concentrated DMSO present in the freezing media. Pellet cells at $800 \times g$ for 5 min in a clinical centrifuge. Aspirate the media and resuspend in 10 mL of fresh 10% FBS MEM. Add cell suspension to a 100 mm tissue culture dish. After cells have attached to the substratum (approximately 6–8 h), change the media using 10 mL of fresh 10% FBS MEM.
2. 100 mm stock culture dishes of HeLa cells should be fed every 2–3 days by changing the media to 10 mL of fresh 10% FBS MEM. Passage cells when they reach confluence (approximately 5 days after a 1:7 dilution – *see* **Note 10**). Passage HeLa cells from a confluent 100 mm stock culture dishes by first washing cells with 10 mL of sterile PBS and then incubating cells in 1 mL trypsin/EDTA for approximately 5 min at room temperature. Rinse cells off of the plate using 6 mL of 10% FBS MEM and add 1 mL of cell suspension to a 100 mm plate containing 9 mL of 10% FBS MEM.
3. One day before transfection (*see* **Note 11**), seed 3 mL of cell suspension (diluted 1:7 as described above) to each of two 35 mm dishes or 1 mL of cell suspension to each of two wells of a 24-well plate containing a glass coverslip sterilized by autoclaving or by flame (*see* **Note 12**). Duplicate dishes/coverslips allow for transfection of both the epitope-tagged DNA construct and the control empty vector (*see* **Note 13**). A 1:7 split of cells will generally provide experimental cultures appropriate for transfection the following day, in addition to a stock culture in a 100 mm dish.
4. For each DNA construct to be transfected (empty vector and vector containing the epitope-tagged protein), add 125 μL of serum-free RPMI to two 1.5 mL sterile, nuclease-free microcentrifuge tubes (50 μL for 24-well plate – *see* **Note 14**).
5. For each construct, to one of the tubes containing RPMI, add 3 μL of TransIT LT1 (1.5 μL for 24-well plate)

(*see* **Note 15**). To the second tube, add 1 μg of plasmid DNA (0.5 μg for a 24-well plate). The total volume of the plasmid DNA should be less than 20 μL (10 μL for a 24-well plate) (*see* **Notes 3** and **4**).

6. Incubate the two solutions for 5 min at room temperature.
7. Following the incubation, combine the two solutions and mix by pipetting or inversion. Incubate mixture for 20 min at room temperature to allow the DNA and transfection reagent to complex. The mixed solutions should not be incubated for longer than 45 min before addition to cells.
8. Add each transfection mixture (containing either the empty vector or the vector containing the epitope-tagged protein) to a single well of cells containing 2.25 mL of fresh media (0.9 mL for 24-well plate).
9. Incubate cells in the transfection media for 4–6 h in a standard incubator in 5% CO_2 at 37°C. At this time, change the media to fresh 10% FBS MEM. This step decreases toxicity caused by the transfection reagent. Cell lysis or fixation for immunoelectron microscopy or immunofluorescence should be performed 24–48 h after transfection, dependent upon the protein being expressed.

3.2. Cell Lysis

1. Wash transfected cells three times with 2 mL of PBS in order to remove media and cellular debris. After aspirating the final wash, add 0.5 mL of RIPA buffer containing protease inhibitors to the cells.
2. Lift cells from the tissue culture dish by scraping the bottom of the dish with a Teflon cell scraper. Collect cell lysates in a 1.5 mL microcentrifuge tube and incubate for 30 min at 4°C to ensure efficient lysis (*see* **Note 16**).
3. Remove cellular debris and unbroken cells by centrifugation at 16,100×*g* for 30 min at 4°C in a standard microcentrifuge. Transfer soluble lysates to a fresh tube to be maintained at 4°C. At this point, lysates may be stored at −80°C for later use.
4. Quantify the protein concentrations of each cell lysate using the Pierce BCA assay kit (*see* **Note 17**) according to the manufacturer's instructions. Formation of the colored product is assessed using a standard visible wavelength spectrophotometer and cuvettes.
5. To equal amounts of protein from each lysate (15–25 μg), add 10 μL of 4X Laemmli sample buffer and sufficient deionized water to make a final volume of 40 μL. Store samples at 4°C while SDS-PAGE gel is being prepared. Samples may also be stored at −20°C for extended storage.

3.3. SDS-PAGE Electrophoresis and Western Blotting

1. Assemble the gel using glass plates and 1.5 mM spacers freshly cleaned with 70% ethanol.
2. Combine 4.0 mL H_2O, 3.3 mL 30% acrylamide/bisacrylamide solution 29:1 (*see* **Notes 6** and **18**), 2.5 mL 1.5 Tris–HCl, pH 8.8, 0.1 mL 10% SDS, 0.1 mL 10% APS, and 0.004 mL of TEMED to make the resolving (lower) gel solution. Immediately, pipette the lower gel solution into the apparatus using disposable pipettes to a level approximately 5 mm below where the bottom of the sample wells will be located. Pipette a small volume (1–2 mL) of isopropanol on top of the resolving gel solution in order to form a flat surface. Polymerization generally occurs within 30 min.
3. After the gel has polymerized, pour off the isopropanol and remove the excess using a Kimwipe. Prepare the stacking (upper) gel solution by combining 3.4 mL H_2O, 0.83 mL 30% acrylamide/bisacrylamide solution 29:1 (*see* **Note 6**), 0.63 mL 1.0 M Tris–HCl, pH 6.8, 0.05 mL 10% SDS, 0.05 mL 10% APS, and 0.005 mL TEMED. Immediately pipette the stacking solution into the apparatus and insert the comb slowly, making sure to avoid bubbles. Polymerization will occur in less than 30 min.
4. After polymerization, transfer the gel to the running setup and fill the tank with 1× Tris/glycine/SDS running buffer.
5. Prior to loading, boil samples prepared as described in **Section 3.2**, step 5 for 5 min to denature proteins.
6. Load 10 μL of Kaleidoscope prestained molecular weight markers in the first well using gel loading tips. (Note that molecular weight markers should not be boiled prior to loading.) Boiled control and experimental samples should be loaded in adjacent wells (*see* **Note 19**).
7. Run the gel at approximately 80 V constant voltage until the proteins exit the stacking gel and enter the resolving gel. The gel can then be run at approximately 180 V until the dye front nears the bottom of the gel.
8. Turn off the power, disassemble the electrophoresis apparatus, and carefully cut off the stacking gel using a spacer.
9. Cut the Whatman paper and the Nitropure nitrocellulose membrane (*see* **Note 20**) to match the size of the sponges and the gel, respectively. Presoak the Whatman paper in transfer buffer and the Nitropure nitrocellulose membrane in water for 2 min prior to assembly of the transfer cassette.
10. Assemble the transfer cassette in the following order: sponge, Whatman paper, gel, Nitropure nitrocellulose

membrane, and Whatman paper (*see* **Note 21**). Insert the cassette into the buffer-filled transfer apparatus with the membrane facing the positive pole. An ice container can be added to reduce heating. Transfer the proteins to the nitrocellulose membrane by running the apparatus at low constant voltage (10 V) overnight or higher constant voltage (80 V) for 4 h at 4°C.

11. Disassemble the transfer apparatus and place the nitrocellulose membrane in a small Tupperware container using forceps. To inhibit non-specific antibody binding, incubate membrane in blocking solution for 45 min at room temperature. All incubation steps for Western blotting should be performed with agitation at room temperature unless otherwise specified.
12. Incubate the membrane in 10 mL of the primary antibody solution in a small plastic container for 2–4 h at room temperature (or overnight at 4°C). Sealed plastic bags can be used for smaller volumes.
13. Wash the membrane in 15 mL TBST three times for 10 min each to remove unbound primary antibody.
14. Incubate the membrane in 10 mL of secondary antibody solution for 1–2 h.
15. Wash the membrane in 15 mL TBST three times for 10 min each to remove unbound secondary antibodies.
16. Detect bound antibodies using Super Signal West Pico Chemiluminescence substrate. Combine 1 mL of luminol/enhancer solution and 1 mL of stable peroxide buffer solution. Distribute the mix uniformly over the surface of the membrane, wrap the moist membrane in plastic wrap, and incubate for 5 min. Transfer the membrane to a new piece of plastic wrap and wick off the excess solution using a Kimwipe. Expose the blot to film. Exposure times will vary depending upon antibody affinity and level of expression (*see* **Note 22**). **Figure 4.1A,B** show Western blots of HeLa cells either mock transfected (Control) or transfected with epitope-tagged GBF1 (Golgi-specific Brefeldin A Resistant Guanine Nucleotide Exchange Factor 1) constructs. In **Fig. 4.1A**, GBF1 was tagged using the small myc epitope, and the blot was probed with anti-myc antibodies. In **Fig. 4.2A**, GBF1 was tagged with GFP, and the blot was probed with anti-GFP antibodies. Note that bands corresponding to the expected molecular weights of myc- and GFP-tagged GBF1 were evident in transfected cell lysates but not in controls. This result indicates the expression of the tagged protein and the specificity of the antibody that recognizes only the tagged protein. Blots for

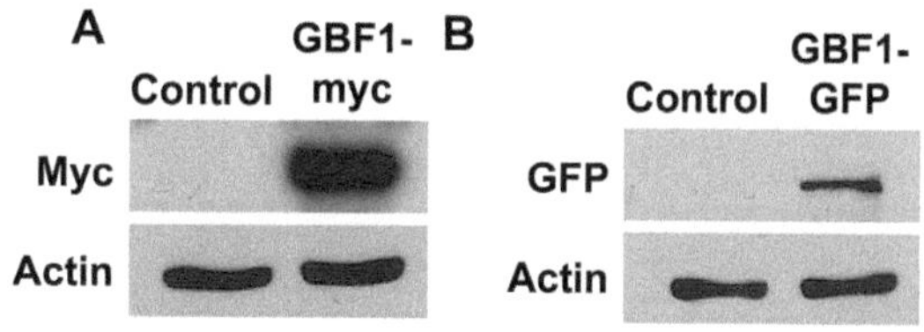

Fig. 4.1. Western blotting can be used to test expression of epitope-tagged proteins. HeLa cells were either mock transfected (Control), transfected with myc-tagged GBF1 (GBF1-myc), or transfected with GFP-tagged GBF1 (GBF1-GFP). Cells were lysed 24 h later. Equal amounts of each lysate were processed for SDS-PAGE. Gels were transferred to nitrocellulose, and the nitrocellulose filters were cut at a molecular weight of 90 kDa according to the migration of molecular markers. The upper parts of the blots were Western blotted using either (**A**) anti-myc antibodies (Covance, Princeton, NJ) or (**B**) anti-GFP antibodies (Santa Cruz Biotechnology, Inc., Santa Cruz, CA). The lower part of each blot was probed using an anti-actin antibody (Sigma Aldrich) to demonstrate equal loading of cell lysates in each lane. Bands corresponding to the expected molecular weights of myc-tagged GBF1 (~204 kDa) and GFP-tagged GBF1 (~230 kDa) were evident in transfected cell lysates, but not in controls.

the cellular housekeeping protein actin demonstrate equal protein loading of control and experimental samples.

3.4. Immunofluorescence

1. Wash cells growing on glass coverslips in 24-well dishes three times with 0.5 mL 1× PBS_f (each) to remove media and cellular debris (*see* **Note 23**).
2. Fix cells in 0.5 mL of 3% PFA for 10–15 min. (*see* **Note 24**).
3. In order to stop the fixation reaction, aspirate the 3% PFA, and quench coverslips by adding 0.5 mL of 10 mM NH_4Cl in 1× PBS_f for 10 min.
4. Wash coverslips with 0.5 mL 1× PBS_f (each) three times (*see* **Note 25**).
5. Permeabilize cells by incubating in 0.5 mL 0.1% Triton-X100 in 1× PBS_f for 7–10 min.
6. Wash cells in 0.5 mL 1× PBS_f three times for 2 min each. At this point, coverslips can be stored in 1 mL 1× PBS_f at 4°C for up to 2 weeks.
7. To block remaining reactive sites, incubate coverslips in 0.5 mL 2.5% Goat serum in PBST for 5 min.
8. Incubate coverslips in primary antibody diluted as suggested by the manufacturer in 0.5 mL 0.4% fish skin gelatin in PBST for 45 min to 1 h at 37°C. Note that smaller volumes of primary antibody can be used by transferring the coverslips to a humidified chamber (*see* **Note 26**). For colocalization studies, two different antibodies (from different species or with different isotypes) may be utilized to visualize the tagged protein and the organelle of interest.

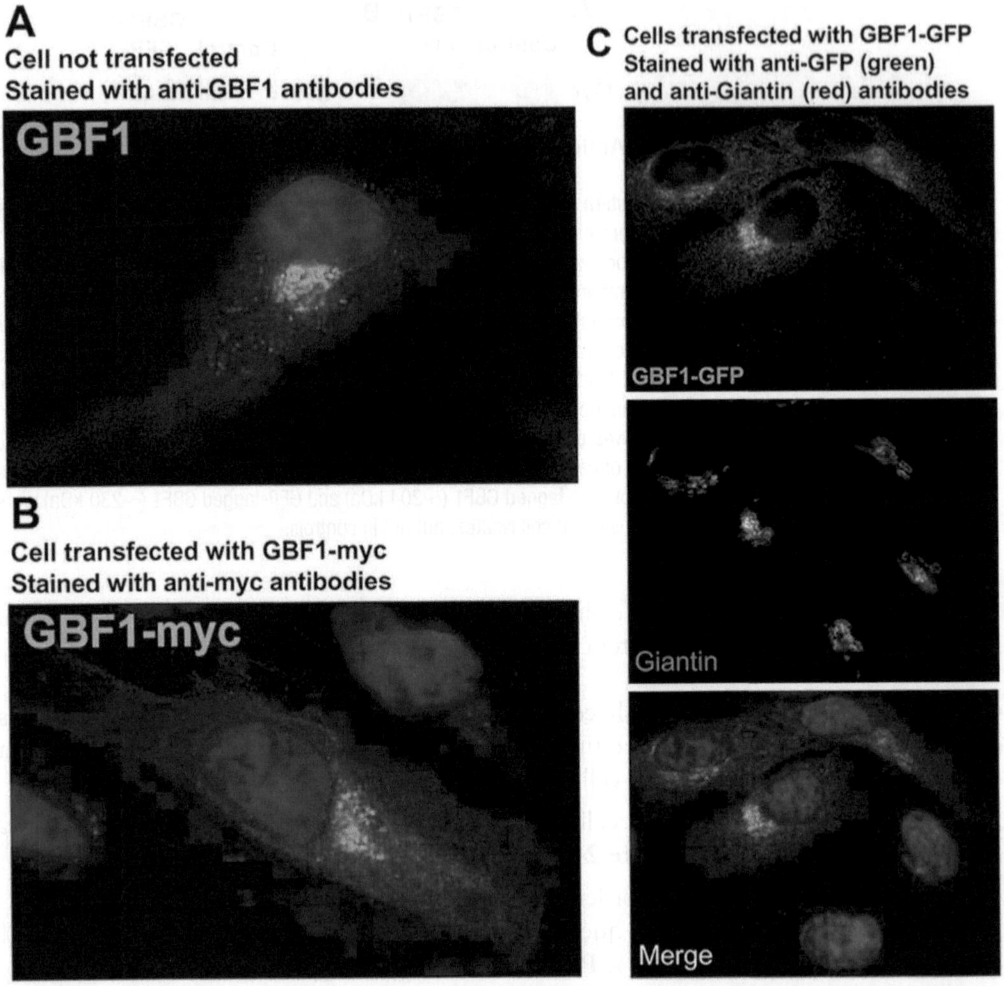

Fig. 4.2. Immunofluorescence can be used to assess the transfection efficiency and localization of tagged proteins. (**A**) HeLa cells were stained with antibodies against endogenous GBF1. GBF1 localizes to the perinuclear region of the cell, and this staining pattern is characteristic of Golgi staining. (**B**) HeLa cells were transfected with a construct expressing myc-tagged GBF1. Cells were then stained with antibodies directed against myc. The localization of the myc-tagged GBF1 mimics the localization of endogenous GBF1. (**C**) HeLa cells were transfected with a construct expressing GFP-tagged GBF1. Cells were then stained with antibodies directed against GFP to detect GFP–GBF1 (*top panel*) and against Giantin (*middle panel*). Giantin is a known Golgi protein and is used here as a Golgi marker. GFP-tagged GBF1, like Giantin, localizes to the Golgi (see merged image in *lower panel*). Nuclei were stained using Hoescht 33258 solution. The myc- and GFP-tagged GBF1 localize correctly and express at detectable levels. This indicates that tagged GBF1 can be used for ultrastructural analyses to further dissect the intra-Golgi distribution of GBF1.

9. Wash coverslips five times with 0.5 mL PBST for 5 min each.
10. Block remaining reactive sites by incubating in 0.5 mL 0.4% Fish Skin Gelatin in PBST for 5 min.
11. Incubate coverslips in species-specific secondary antibody conjugated to Alexa 488 diluted 1:1,000 into 0.5 mL 2.5% Goat Serum in PBST for 30–45 min (*see* **Note 9**). For

colocalization studies, two secondary antibodies that correspond to the primary antibodies may be utilized. Each of the antibodies must be conjugated to a different fluor for independent visualization.

12. Incubate coverslips in Hoescht 33258 solution for 5 min to stain nuclei.
13. Wash coverslips five times with 0.5 mL PBST for 5 min each.
14. Mount coverslips on glass slides by gently placing coverslip cells facing down onto a small (5 μL) drop of mounting media. Aspirate excess media and seal the edges of the coverslip with clear nail polish.
15. Visualize cells by epifluorescence or confocal microscopy immediately after the nail polish dries. **Figure 4.2** shows images of HeLa cells stained with antibodies directed against GBF1 (Golgi-specific Brefeldin A Resistant Guanine Nucleotide Exchange Factor 1) or images of cells expressing myc- or GFP-tagged GBF1 and stained with anti-tag antibodies. Endogenous GBF1 localizes to the perinuclear Golgi apparatus (**Fig. 4.2A**). Staining of cells transfected with myc-tagged GBF1 shows that the myc-tagged protein exhibits a localization similar to the endogenous GBF1 (**Fig. 4.2B**). Colocalization of GFP-tagged GBF1 with the Golgi marker Giantin shows that the tagged protein does in fact localize to the Golgi (**Fig. 4.2C**). Myc- and GFP-tagged GBF1 localize correctly and express at detectable levels, indicating that tagged GBF1 can be used for immunoelectron microscopy analyses.

4. Notes

1. Epitope-tagged proteins can be expressed in many different types of tissue culture cells; however, expression varies significantly among different cell types. Two of the most commonly used cell lines are HeLa cells (ATCC number CCL-2) and Cos-7 cells (ATCC number CRL-1651), both easily transfectable and highly resilient cell models. However, other cell types, such as HEK-293T or MDCK cells, can also be transfected to high efficiency.
2. Commonly used transfection reagents include salt-mediated reagents, such as $CaCl_2$, and newer cationic lipid-based reagents, such as Mirus TransIT LT1 (the reagent utilized in this protocol), Lipofectamine 2000 (Invitrogen), and Fugene 6 (Roche). In all cases, the reagent is added to the DNA to form complexes which are

then added to cells. Salt-based complexes are taken into the cells by endocytosis, while the lipid–DNA complexes can enter the cell either through endocytosis or through fusion with the plasma membrane. However, in the event that the cells utilized are not easily transfected, electroporation, which utilizes electric shock to force DNA plasmids into cells, can also be utilized to produce high levels of expression of epitope-tagged proteins.

3. DNA may be prepared from 1 to 5 mL of an overnight bacterial culture using any standard miniprep kit, such as the QIAprep spin miniprep kit (Qiagen, Valencia, CA), according to the manufacturer's instructions. DNA should be resuspended in sterile nuclease-free water.
4. DNA concentration can be assessed either by determining the absorbance of the DNA solution at a wavelength of 260 nm using a standard spectrophotometer (1 absorbance unit = 50 μg/mL DNA) or by comparing 1 μL of plasmid DNA to quantitative DNA standards on an agarose gel. For most plasmids, a standard miniprep resuspended in 50 μL of sterile, nuclease-free water will yield approximately 200 ng DNA/μL. If a sufficient concentration for transfection is not achieved, the DNA may be concentrated by ethanol or isopropanol precipitation, followed by resuspension in a smaller volume of sterile nuclease-free water.
5. A mask should be worn when weighing SDS, as it aerosolizes easily and can be an irritant.
6. Note that unpolymerized acrylamide is a potent neurotoxin and should be handled with care, using gloves and standard precautions.
7. A wide variety of HRP- and gold-conjugated secondary antibodies are commercially available from many companies, including Invitrogen (Carlsbad, CA). The secondary antibody should be specific for the species from which the primary antibody was isolated. Generally anti-mouse and anti-rabbit secondary antibodies are significantly more specific than anti-goat, anti-sheep, or anti-chicken secondary antibodies. In addition, isotype-specific secondary antibodies are also commercially available if the isotype of the primary antibody is known.
8. Paraformaldehyde is highly toxic. One should always wear a mask while weighing dry paraformaldehyde, and aerosolization should be avoided.
9. For use with other wavelengths of light or with other species of primary antibodies, secondary antibodies of varying species specificities attached to Alexa Fluors that

fluoresce at different wavelengths are available from Molecular Probes/Invitrogen (Eugene, OR).

10. The notation "1:7" indicates that the cells from a single 100 mm tissue culture plate are diluted in such a way that they can be used to generate seven new 100 mm plates.
11. Cells should be transfected at least 1 day after seeding to achieve optimal survival. The confluency of the cells should be between 65 and 90% for optimal transfection efficiency.
12. Some cell types do not adhere well when glass is used as a substratum. In this case, glass coverslips can be coated with fetal bovine serum, collagen, or fibronectin overnight at 37°C to improve adherence.
13. Transfections are routinely performed by transfecting both an empty vector (control) and the same vector containing an insert encoding the tagged protein (experimental) in parallel. This method provides a negative control for Western blotting and immunofluorescence.
14. Volumes for transfections are described for 35 mm dishes, with volumes for 24-well plates following in parentheses.
15. Transfection efficiency varies greatly dependent upon the cell type and the protein being expressed. It may be necessary to optimize transfection efficiency by using varying amounts of transfection reagent and DNA to obtain optimal efficiency. Transfection efficiency can reach up to 80%, but generally varies between 10% and 50%. For some primary or difficult to transfect cell lines, other techniques, such as electroporation, in which a short electric shock is used to allow DNA penetration into cells, may be necessary to achieve sufficient levels of transfection.
16. In the case of inefficient lysis or for solubilization of membrane-bound proteins, brief sonication or mechanical disruption using a dounce homogenizer can increase solubilization. However, care should be taken to avoid the formation of bubbles, as this can lead to protein precipitation.
17. It is imperative when using RIPA buffer as the lysis buffer to utilize the BCA assay, as the detergents present in RIPA buffer may interfere with the Bradford assay.
18. This protocol specifies a 10% acrylamide gel, useful for most applications; however, the percentage of acrylamide may be adjusted for better separation of very large or very small proteins.
19. Western blotting is routinely performed on lysates from cells transfected with an empty vector (control) and with the same vector containing an insert encoding the tagged

protein (experimental) loaded in adjacent lanes on SDS-PAGE. The lysates in the control lane contain only the tag, and the antibodies should detect either no bands (for small tags) or a single band corresponding to the molecular weight of the tag alone (for large tags) in that lane. The lysates in the experimental lanes should contain a single band of the molecular weight appropriate for the tagged protein. Large tags such as GFP (molecular weight of 27 kDa) add significantly to the molecular weight of a protein, and this must be taken into consideration when calculating the expected molecular weight. If bands other than the expected tagged protein are detected in both lanes, this indicates that the antibodies recognize additional cellular proteins. In this case, a more specific antibody or different protein tag should be used.

20. The Nitropure nitrocellulose membrane should be manipulated with forceps to avoid transfer of keratins and other proteins to the membrane prior to use.
21. Care should be taken to remove any air bubbles present between the gel and the membrane. Air bubbles block transfer of proteins to the membrane. Air bubbles can be removed by gently rolling a glass test tube or glass pipette over the assembled stack.
22. For low-affinity antibodies or low expressing proteins, stronger ECL detection kits can be utilized. The West Femto Maximum Sensitivity Chemiluminescent Substrate kit from Thermo Scientific is highly sensitive and can be used to detect low signals.
23. All steps should be performed at room temperature unless stated otherwise. Solutions should be removed from coverslips by aspiration, taking care not to aspirate directly from the center of the coverslip. Touching the aspirator directly to the coverslip may result in shearing of cells. Coverslips should never be aspirated to the point of being completely dry, as this will lead to alterations in cellular morphology.
24. Some antibodies may require fixation in methanol. In these cases, cells should be incubated in methanol at −20°C for 5 min, followed by – five to six washes with PBS. Methanol causes both fixation and cell permeabilization, so steps 5 and 6 may be eliminated when using this method.
25. GFP-tagged proteins can be visualized without the use of antibodies, although antibodies may enhance signal. For GFP-tagged proteins, fluorescence may be visualized directly in live cells or after fixation and washing.

26. Volumes used for primary antibody incubation can be minimized by placing the coverslips in a chamber containing damp Whatman paper covered with parafilm. The coverslips can be placed face up on the parafilm and a 50 μL drop of primary antibody solution can be placed on top of the coverslip. However, care should be taken to minimize evaporation and prevent drying of the coverslips.

References

1. Garcia-Mata, R., Bebök, Z., Sorscher, E. J., and Sztul, E. S. (1999) Characterization and dynamics of aggresome formation by a cytosolic GFP-chimera. *J. Cell Biol.* **146**(6), 1239–1254.
2. Garcia-Mata, R., Gao, Y. S., and Sztul, E. (2002) Hassles with taking out the garbage: aggravating aggresomes. *Traffic* **3**(6), 388–396.

Chapter 5

Production and Purification of Polyclonal Antibodies

Masami Nakazawa, Mari Mukumoto, and Kazutaka Miyatake

Abstract

Polyclonal antibodies are derived from multiple B-cell clones that have differentiated into antibody-producing plasma cells in response to an immunogen. Polyclonal antibodies raised against a single molecular species of antigen recognize multiple epitopes on a target molecule resulting in signal amplification in indirect immunoassays, including immuno-electron microscopy. In this chapter, we present a basic procedure to generate polyclonal antibodies in rabbits.

Key words: Polyclonal antibody, rabbit, ammonium sulfate precipitation, cardiac puncture, subcutaneous injection.

1. Introduction

The original meaning of the term "polyclonal antibody" is the total population of antibodies present in an animal's serum. Polyclonal antibodies are derived from multiple B-cell clones that have differentiated into antibody-producing plasma cells in response to an immunogen. In contrast to a monoclonal antibody which recognizes a single epitope, a polyclonal antibody against a single molecular species of antigen recognizes more than one epitope on the target molecule. This allows multiple antibodies to bind through the antigen amplifying the signal in indirect immunoassays, including immuno-electron microscopy. Polyclonal antibodies with low nonspecific binding are a powerful tool for immuno-electron microscopy as they will produce a stronger signal than monoclonal antibodies.

Polyclonal antibodies can be produced in all animals that have an immune response. The choice of host is based on several

S.D. Schwartzbach, T. Osafune (eds.), *Immunoelectron Microscopy*, Methods in Molecular Biology 657,
DOI 10.1007/978-1-60761-783-9_5, © Springer Science+Business Media, LLC 2010

factors, the most important of which is the intended amount of polyclonal antibody. For most work in which small volumes of antiserum are required (<100 mL), the rabbit is the most common species for polyclonal antibody production. In this chapter, we present a basic procedure to generate polyclonal antibody in rabbits.

2. Materials

2.1. Immunization of Rabbits

1. Institutional Animal Care and Use Committee approval
2. Sterile phosphate-buffered saline (PBS): prepare 10X PBS by mixing 87.7 g NaCl, 30 g $Na_2HPO_4{\cdot}12H_2O$, 2.9 g $NaH_2PO_4{\cdot}2H_2O$, and distilled water for a final volume of 1 L. Dilute the 10X PBS tenfold with distilled water to prepare PBS as needed. Sterilize by autoclaving at 121°C for 20 min
3. Antigen solution: resuspend antigen in sterile PBS for a final concentration of 1 mg/mL. A minimum of 5 mg antigen is required to immunize two rabbits
4. New Zealand White female rabbits, 2-kg weight, specific pathogen free
5. Freund's complete adjuvant (Sigma-Aldrich, St. Louis, MO; *see* **Note 1**)
6. Freund's incomplete adjuvant (Sigma-Aldrich, St. Louis, MO)
7. 5-mL Luer-lock, all-plastic-type syringes for preparing emulsion (*see* **Note 2**)
8. 22-gauge 1 1/2″ syringe needles.
9. 18-gauge double hub micro-emulsifying needle with reinforcing bar (Popper & Sons, New Hyde Park, NY)
10. Alcohol swabs
11. A rabbit restrainer
12. Sterile 1.5-mL plastic sample tubes
13. Sterile 1-mL plastic syringes

2.2. Indirect Enzyme-Linked Immunosorbent Assay (Indirect ELISA)

1. PBS: prepare 10X PBS by mixing 87.7 g NaCl, 30 g $Na_2HPO_4{\cdot}12H_2O$, 2.9 g $NaH_2PO_4{\cdot}2H_2O$, and distilled water for a final volume of 1 L. Dilute the 10X PBS tenfold with distilled water to prepare PBS as needed
2. Antigen solution: dilute antigen with PBS for a final concentration of 5 μg/mL

3. PBS-T: PBS containing 0.05% Tween 20
4. 1% BlockAce (DS Pharma Biomedical, Osaka, Japan) in water
5. Microtiter plate washer (Bio-Rad Laboratories, Hercules, CA)
6. Microtiter plate reader with 490-nm filter (Bio-Rad Laboratories, Hercules, CA)
7. Test antisera
8. Secondary antibody: horseradish peroxidase-conjugated goat anti-rabbit IgG-(H+L) (Bio-Rad Laboratories, Hercules, CA)
9. 10 mg *o*-phenylenediamine dihydrochloride tablets (Sigma-Aldrich, St. Louis, MO)
10. 30% hydrogen peroxide
11. 0.5 M H_2SO_4
12. Multichannel pipet and disposable pipet tips
13. 96-well ELISA plate (Asahi Techno Glass, Chiba, Japan)
14. Citrate phosphate buffer: prepare 10X citrate phosphate buffer stock solution by mixing 36.8 g $Na_2HPO_4{\cdot}12H_2O$, 10.2 g citric acid and distilled water for a final volume of 100 mL. Store stock solution at 4°C. Dilute 10X citrate phosphate buffer stock solution tenfold as needed (*see* **Note 3**).

2.3. Preparation of Blood Sample by Cardiac Puncture and Preparation of Serum from Blood

1. Sterile 50-mL plastic syringes
2. Sterile 2.5-mL plastic syringes
3. 26-gauge 1/2″ syringe needles
4. 18-gauge 1 1/2″ syringe needles
5. A rabbit restrainer
6. Weight scales
7. Sterile 50-mL plastic centrifuge tubes with conical bottom
8. Sterile PBS: prepare 10X PBS by mixing 87.7 g NaCl, 30 g $Na_2HPO_4{\cdot}12H_2O$, 2.9 g $NaH_2PO_4{\cdot}2H_2O$, and distilled water for a final volume of 1 L. Dilute the 10X PBS tenfold with distilled water to prepare PBS as needed. Sterilize by autoclaving at 121°C for 20 min
9. 5% (w/v) sodium pentobarbital solution: dissolve sodium pentobarbital (Sigma-Aldrich, St. Louis, MO) in sterile PBS.
10. Alcohol swabs
11. A rabbit restraint board

2.4. Precipitation of Immunoglobulin G Fraction with Ammonium Sulfate

1. Saturated ammonium sulfate solution: add 800 g ammonium sulfate to 1 L of distilled water. Heat until the salt dissolves completely and let cool to room temperature. Adjust the pH to 7.4 with ammonium hydroxide. Allow undissolved ammonium sulfate to settle to the bottom of the container and be careful not to remove the ammonium sulfate crystals when using the solution.
2. Antisera
3. Magnetic stirrer
4. Cold room or cold chamber (4°C)
5. PBS: prepare 10X PBS by mixing 87.7 g NaCl, 30 g $Na_2HPO_4{\cdot}12H_2O$, 2.9 g $NaH_2PO_4{\cdot}2H_2O$, and distilled water for a final volume of 1 L. Dilute the 10X PBS tenfold with distilled water to prepare PBS as needed
6. Dialysis tubing, MWCO 14,000 (Viskase companies Inc., Darien, IL)
7. 200 or 300-mL glass beaker

2.5. SDS Gel Electrophoresis and Western Blotting

1. 29.2% acrylamide/0.8% bisacrylamide stock solution: mix 29.2 g acrylamide (Nacalai tesque, Kyoto, Japan) and 0.8 g bisacrylamide (Nacalai tesque, Kyoto, Japan) with distilled water for a final volume of 100 mL. Store at 4°C in the dark for up to 3 weeks. Unpolymerized acrylamide is a neurotoxin and care should be taken to avoid inhalation and skin contact.
2. Separating gel solution: 1.875 M Tris–HCl, pH 8.8
3. Stacking gel solution: 0.6 M Tris–HCl, pH 6.8
4. 10% ammonium persulfate (Nacalai tesque, Kyoto, Japan) in sterile water prepared immediately before use
5. *N,N,N′,N′*-Tetramethylethylenediamine (TEMED) (Nacalai tesque, Kyoto, Japan)
6. 10% (w/v) SDS: prepared in distilled water
7. SDS-PAGE running buffer: 25 mM Tris, 192 mM glycine, 0.1% SDS
8. 3X sample loading buffer: 180 mM Tris–HCl, pH 6.8, 30% glycerol, 6% SDS, 0.003% bromophenol blue, 5% β-mercaptoethanol. Store in aliquots at −20°C
9. Electrophoresis apparatus: Dual slab PAGE system AE-6500 (mini-slab) (ATTO corporation, Tokyo, Japan) 1-mm spacers, 1-mm Teflon combs with 14 teeth.
10. Blotting apparatus: Trans-Blot SD Semi-Dry Electrophoretic transfer cell (Bio-Rad, Hercules, CA)
11. Blotting buffer: prepare 10X blotting buffer by mixing 58.14 g Tris and 29.27 g glycine and distilled water for a

final volume of 1 L. Dilute the 10X blotting buffer tenfold with distilled water to prepare blotting buffer as needed. Store diluted blotting buffer at 4°C

12. 0.45 μm immobilon-P membrane (Millipore, Billerica, MA): cut into 6×9 cm pieces, must be wetted in 100% methanol for 1 min, and then soaked in blotting buffer with gentle shaking before use.
13. Filter paper 3 MM sheets (No.514A, Advantec, Tokyo, Japan) cut into 6×9 cm pieces
14. PBS: prepare 10X PBS by mixing 87.7 g NaCl, 30 g $Na_2HPO_4{\cdot}12H_2O$, 2.9 g $NaH_2PO_4{\cdot}2H_2O$, and distilled water for a final volume of 1 L. Dilute the 10X PBS tenfold with distilled water to prepare PBS as needed
15. Blocking buffer: 5% powdered skim milk in PBS
16. Washing buffer: PBS
17. Impulse heat sealer (Consolidated plastics, Stow, OH)
18. Plastic boxes
19. Heat-sealable plastic bags
20. Primary antibody diluted 1,000 times in 0.1% bovine serum albumin in PBS (*see* **Note 4**)
21. Secondary antibody: Goat anti-rabbit horseradish peroxidase-conjugated antibody (Bio-Rad, Hercules, CA) diluted 10,000 times in 0.1% bovine serum albumin in PBS
22. Chemiluminescent substrate: SuperSignal West Pico chemiluminescent substrate (Thermo Fisher Scientific, Rockford, IL)
23. Luminescent image analyzer LAS-4000 (Fujifilm, Tokyo, Japan)
24. Saran wrap

3. Methods

3.1. Immunization of Rabbits

1. Collect a preimmune blood sample from the rabbit's marginal ear vein with a sterile 1-mL syringe using a 22-gauge needle (*see* **Note 5**). Transfer the blood to a sterile 1.5-mL plastic sample tube. Keep at 37°C for 1 h to allow clots to form. Transfer to 4°C and leave overnight. Centrifuge at 1,000×*g* for 20 min at 4°C. Collect the supernatant fraction as preimune serum. Aliquot and store at −80°C.
2. Prepare an emulsion (1.2 mL/rabbit) using 0.6 mL sterile PBS containing antigen (1 mg/mL) and 0.6 mL Freund's

complete adjuvant. Use a 22-gauge needle to load one syringe with antigen solution and the other with adjuvant. Link the two syringes with a double-ended Luer-Lock connector. Press the syringe barrels back and forth until a stable emulsion is produced (for 5–10 min till thickened). Let a drop of the emulsion fall into water to check the endpoint of emulsification. A stable emulsion will not disperse.

3. Immobilize a rabbit with a rabbit restrainer or by help from an assistant and wipe the back of the rabbit with alcohol swabs to sterilize. Inject the rabbit subcutaneously with 1 mL of the antigen emulsion delivered to 4–10 injection sites using a 22-gauge needle (*see* **Note 6**). At least two rabbits should be immunized for each antigen.
4. Boost rabbits 2 weeks later by subcutaneously injecting an emulsion (1 mL/rabbit) of equal volumes of sterile PBS containing 0.5 mg antigen/rabbit and Freund's incomplete adjuvant. The emulsion is prepared and injected as in steps 2 and 3.
5. Collect blood 1 week after the booster injection to determine the antibody titer. Place rabbits in a rabbit restrainer and collect the blood from the marginal ear vein using a sterile 1-mL plastic syringe with a 22-gauge needle. Transfer the blood to a sterile 1.5-mL plastic sample tube. Keep the blood at 37°C for 1 h and centrifuge at 1,000×*g* for 20 min to separate the serum sample (supernatant fraction).
6. Determine the antibody titer in the serum by indirect ELISA (**Section 3.2**, *see* **Note 3**).
7. Boost rabbits as in step 4 every 2 weeks until a sufficient titer (a significant ELISA signal at <1/10,000 dilution) is achieved.
8. When the antibody titer is sufficient, boost rabbits by injecting 0.5 mg antigen in sterile PBS without adjuvant into a marginal ear vein (500 μL) using a sterile 1-mL plastic syringe with a 22-gauge needle.
9. Three days after the final immunization (step 8), collect all of the blood from the rabbits by cardiac puncture (*see* **Section 3.3**).

3.2. Indirect Enzyme-Linked Immunosorbent Assay (Indirect ELISA)

1. Coat plate(s) with antigen by dispensing 100 μL of antigen solution (5 μg/mL in PBS) into each well of an ELISA plate using a multichannel pipet and tips. Close the lid and incubate at 37°C for 1 h or 4°C overnight to allow the antigen to coat the wells.
2. Rinse the coated plate once with PBS-T using a microplate washer apparatus.

3. Add 200 μL of 1% BlockAce solution to each well and incubate at 37°C for 1 h or 4°C overnight.
4. Rinse the plate once with PBS-T using a microplate washer apparatus.
5. Add 100 μL of serum diluted in PBS to each of the coated wells, close the lid, and incubate at 37°C for 1 h (*see* **Note 7**).
6. Wash the plate five times with PBS-T using a microplate washer and remove residual liquid completely by vigorously patting on a paper towel.
7. Add 100 μL of diluted secondary antibody diluted 1:10,000 in PBS-T or as specified by the supplier to each well, close the lid, and incubate at 37°C for 1 h.
8. Wash the plate five times with PBS-T using a microplate washer and remove residual liquid completely by vigorously patting on a paper towel.
9. Prepare the substrate solution by dissolving one tablet of *o*-phenylenediamine dichloride in 25 mL of citrate phosphate buffer in the dark and then add 5 μL of 30% H_2O_2.
10. Add 200 μL of substrate solution to each well and incubate at 37°C for 30 min in the dark.
11. Add 50 μL of 0.5 M H_2SO_4 to each well to stop the peroxidase reaction.
12. Read absorbance at 490 nm with a microtiter plate reader. Generally, a signal which shows both OD>0.7 in <1/1,000 dilution and higher absorbance compared to the same dilution of preimmune serum is considered as a positive signal.

3.3. Preparation of Blood Sample by Cardiac Puncture and Preparation of Serum from Blood

1. Stop feeding the rabbit 18 h before bleeding. Continue to supply water as usual.
2. Weigh the rabbit on a scale. Transfer and keep the rabbit in the restrainer. Anesthetize the rabbit with 0.5 mL/kg weight of 5% sodium pentobarbital solution injected into the marginal ear vein using a 2.5-mL plastic syringe with a 26-gauge needle.
3. When the rabbit is well anesthetized, lay the rabbit on its back using the restraint board. Secure all limbs using cotton ties (*see* **Note 8**).
4. Wipe an area between the fifth and sixth rib on the rabbit's left side at the point where the heartbeat is the strongest with an alcohol swab and insert a 1 1/2″×18-gauge needle attached to a sterile 50-mL plastic syringe.
5. Advance the needle slowly until blood begins to fill the syringe and then draw the plunger back slowly.

6. After filling the syringe, remove it from the needle leaving the needle in the heart. Attach a new syringe to the needle. Continue bleeding until the heart beat stops. A total of 100–150 mL of blood can be obtained.
7. Aliquot 30 mL blood/sterile 50-mL plastic tube (*see* **Note 9**). Keep at 37°C for 1 h with a little tilting to allow clots to form. Transfer the tubes to 4°C and leave overnight.
8. Centrifuge at 1,000×*g* for 20 min at 4°C. Collect the supernatant fraction which is the antiserum. A total of 40–70 mL of antisera is usually obtained.

3.4. Precipitation of Immunoglobulin G Fraction with Ammonium Sulfate

1. Centrifuge the antiserum at 3,000×*g* for 30 min at 4°C. Measure the volume of the supernatant and transfer it to a pre-chilled beaker. The following steps should be performed at 4°C.
2. Dilute the antiserum by adding an equal volume of ice-cold PBS to the beaker and mix with a magnetic stirrer.
3. Add saturated ammonium sulfate solution dropwise with gentle stirring to the diluted antiserum to produce a 40% saturated solution (e.g., 40 mL saturated ammonium sulfate solution to 60 mL diluted antiserum).
4. After finishing the ammonium sulfate addition, continue mixing constantly for 2–4 h at 4°C to allow a precipitate to form.
5. Centrifuge at 10,000×*g* for 30 min at 4°C.
6. Dissolve the pellets in a minimal volume of PBS.
7. Dialyze for 3 h at 4°C against 1 L of PBS and then overnight at 4°C against 1 L of fresh PBS.
8. Freeze 1 mL aliquots of antibody at −80°C.

3.5. SDS Gel Electrophoresis and Western Blotting

1. Assemble the glass plates and spacers locking the assembled glass sandwich to the gel casting stand.
2. Prepare a 10% polyacrylamide gel by using a magnetic stirrer to mix 1.5 mL separating gel solution, 2.5 mL 29.2% acrylamide/bisacrylamide, 3.38 mL distilled water. Degas using a vacuum supplied by an aspirator. Rapidly add 75 μL 10% SDS, 45 μL 10% ammonium persulfate, and 7 μL TEMED and mix using a magnetic stirrer. Gently pour the running gel solution between the glass plates prepared in step 1.
3. Overlay the gel with 200 μL distilled water. Allow the gel to polymerize for 30 min.
4. Pour off the water on the top of the separating gel.

5. Prepare the stacking gel solution by using a magnetic stirrer to mix 3.6 mL distilled water, 0.5 mL stacking gel solution, and 0.83 mL 29.2% acrylamide/bisacrylamide. Degas using a vacuum supplied by an aspirator. Rapidly add 50 μL 10% SDS, 80 μL freshly prepared 10% ammonium persulfate, and 4.5 μL TEMED and mix using a magnetic stirrer. Immediately pour the stacking gel solution and position the sample comb. Let the gel polymerize for 10–20 min.
6. Pour SDS-PAGE running buffer into the lower chamber (anode) and put the gel sandwich on the lower chamber. Hold the gel with stopper and pour the SDS-PAGE running buffer into the upper chamber (cathode). Remove the comb and wash the sample wells with SDS-PAGE running buffer.
7. Dilute 2 volumes sample with 1 volume 3X loading buffer, mix, and denature the samples by incubation at 100°C for 3 min with boiling water.
8. Load 20 μL of denatured sample containing 20 μg protein into each well. Load prestained molecular weight markers into the first lane.
9. Run the gel at 20 mA/gel constant current until the bromophenol blue tracking dye reaches the bottom of the gel (about 1 h).
10. Gloves should be worn for all subsequent steps. Prewet the Immobilon-P membrane in 100% methanol for 1 min and then soak in blotting buffer with gentle shaking before use. Soak six filter paper sheets slightly larger than the gel in blotting buffer.
11. Disassemble the gel and cut off the stacking gel. Make a diagonal cut on the left-hand side of the gel for orientation. Equilibrate the gel in blotting buffer for at least 1 min.
12. Assemble the blotting stacks onto the platinum anode (+) in the following order: three sheets of pre-wetted filter paper, the pre-wetted Immobilon-P membrane, the equilibrated gel, and three sheets of pre-wetted filter paper. Roll a pipet over each layer to remove air bubbles.
13. Carefully place the cathode onto the stack and transfer at 20 mA/cm^2 membrane for 1 h.
14. Disassemble the stacks, mark the position of the prestained markers on the membrane with a pencil, and place the membrane in a plastic box containing blocking buffer. Incubate for 1 h at room temperature with constant shaking.
15. Wash the membrane with washing buffer once and place the membrane in a heat-sealable plastic bag.

16. Pour enough diluted primary antibody to cover the membrane into the plastic bag and seal. Incubate for 1 h at room temperature.
17. Remove the primary antibody and place the membrane in a clean plastic box containing washing buffer. Wash the membrane three times by incubation with washing buffer for 10 min at room temperature with constant shaking.
18. Place the membrane in a clean plastic box with enough secondary antibody solution to cover the membrane diluted according to the manufacturer's instructions and incubate for 30 min at room temperature with constant shaking.
19. Remove the secondary antibody and place the membrane in a clean plastic box containing washing buffer. Wash the

Fig. 5.1. Western blot of whole cell extracts of *Euglena gracilis* probed with antibody prepared in rabbits against pyruvate:NADP$^+$ oxidoreductase. Rabbits were immunized with purified *E. gracilis* pyruvate:NADP$^+$ oxidoreductase and the antibody was purified by ammonium sulfate precipitation. Thirty micrograms of *Euglena* whole cell protein was electrophoresed on a 10% polyacrylamide cell, transferred to an immobilone P membrane, and incubated with a 1,000-fold dilution of anti-pyruvate:NADP$^+$ oxidoreductase followed by a 10,000-fold dilution of anti-rabbit IgG horse radish peroxidase secondary antibody. The blot was incubated with SuperSignal West Pico chemiluminescent substrate for 5 min and detected with a Fujifilm LAS-4000 imaging system.

membrane three times by incubation with washing buffer for 10 min at room temperature with constant shaking.

20. Place the membrane on a plastic bag or saran wrap. Prepare the chemiluminescent substrate and place over the membrane surface ensuring a uniform distribution across the membrane according to the manufacturer's instructions. Incubate 5 min.
21. Drain the excess substrate from the membrane and capture the signal directly with a LAS-4000 imaging system. **Figure 5.1** shows that an antibody prepared in rabbits against *Euglena gracilis* pyruvate:NADP$^+$ oxidoreductase recognizes a single protein in whole cell extracts of *E. gracilis.*

4. Notes

1. Freund's complete adjuvant (FCA) elicits a strong response to most antigens. However, it may cause inflammation, induration, pain, and necrosis at the injection site, and its use is therefore ethically *controversial.* It is also a risk to workers, as it can cause localized soft tissue damage following accidental needlestick injuries. Due to the harmful responses produced by this adjuvant, the ability of other adjuvants such as Titermax Gold (Titermax USA, Norcross, GA) to elicit the desired immune response should be considered.
2. Syringes should be all-plastic and Luer-lock type to prepare antigen-containing adjuvant. Plastic syringes with rubber pistons contain a lubricant that fails in the presence of adjuvant and causes the syringes to stick.
3. The ELISA assay is a rapid, sensitive, and quantitative assay making it the most popular immunoassay for antibody selection. However, ELISA-positive antibodies may not be suitable for immuno-electron microscopy. Polyclonal antibodies to be used for immuno-electron microscopy should be identified using both ELISA and Western blotting which identifies antibodies that have a high affinity for denatured proteins. Western blotting will also detect the nonspecific binding to proteins other than the target antigen.
4. The antibody dilution required for western blotting is determined by cutting the membrane into individual lanes and incubating each lane with a different dilution of the primary antibody over the range of 1/100 to 1/5,000. The optimal

antibody dilution is the one producing the strongest signal with the least background.

5. The marginal ear vein is found on the inner edge of the ear. If the vein is not visible, warm the ear with a hair drier and keep the rabbit warm. Apply gentle pressure to the injection site with an alcohol swab to stop the blood flow after the bleeding.
6. The use of multiple injection sites provides for an enhanced interaction of the immunogen with the immune system because more lymph nodes are exposed to the preparation. Repeated injection on the same site should be avoided as it can cause skin ulceration.
7. The antibody titer is determined by performing tenfold serial dilutions of antibody from 1/100 to 1/10,000 in PBS with 0.1% BSA. Add 100 μL of each dilution to a well.
8. Prepare extra anesthetic agents (1–2 mL) to use if the animal recovers from the anesthesia.
9. Both plastic and glass tubes can be used. Clots form a little faster in glass tubes.

Chapter 6

Production and Purification of Monoclonal Antibodies

Masami Nakazawa, Mari Mukumoto, and Kazutaka Miyatake

Abstract

Monoclonal antibodies (mAb's) have been used extensively in biochemical and biomedical studies, including immunoelectron microscopy. Production of mAb's consists of four steps: immunizing the animal usually a mouse, obtaining immune cells from the spleen of the immunized mouse, fusing the spleen cells with myeloma cells to obtain hybridomas, and selecting the hybridoma cell line producing the desired mAb. In this chapter, we present a general method for monoclonal antibody production.

Key words: Monoclonal antibody, myeloma cell, hybridoma, cell fusion, mouse ascites method, protein A.

1. Introduction

Monoclonal antibodies (mAb's) are used extensively in basic biomedical and biochemical research, including immunoelectron microscopy. In general, mAb shows lower nonspecific binding than do polyclonal antibodies because they are a homogeneous antibody recognizing a single epitope.

Construction of cell lines producing a desired mAb requires the following four steps: immunizing an animal, usually a mouse; obtaining immune cells from its spleen, fusing the spleen cells with myeloma cells to make them immortal, and selecting the hybridoma cell lines producing the desired mAb. The hybrid antibody-producing cell resulting from the fusion of an antibody-secreting cell and an immortal cell (i.e., myeloma) is called a hybridoma. The development of the immortal hybridoma requires the use of animals.

S.D. Schwartzbach, T. Osafune (eds.), *Immunoelectron Microscopy*, Methods in Molecular Biology 657,
DOI 10.1007/978-1-60761-783-9_6,

Producing excellent mAb's needs antigens which are suitable for the experiment and a good selection system. In this chapter, we also describe the indirect ELISA antibody assay method, which is one of the most popular and convenient assay systems used in the selection of mAb's.

2. Materials

2.1. Immunization of Mice

1. Institutional Animal Care and Use Committee approval.
2. Sterile phosphate-buffered saline (PBS): prepare 10× PBS by mixing 87.7 g NaCl, 30 g $Na_2HPO_4{\cdot}12H_2O$, 2.9 g $NaH_2PO_4{\cdot}2H_2O$, and distilled water for a final volume of 1 L. Dilute the 10× PBS tenfold with distilled water to prepare PBS as needed. Sterilize by autoclaving at 121°C for 20 min.
3. Antigen solution: resuspend antigen in sterile PBS for a final concentration of 500 μg/mL. A minimum of 2 mg antigen is required to immunize five mice.
4. Balb/c female mice, 7 weeks old (*see* **Note 1**).
5. Freund's complete adjuvant (Sigma-Aldrich, St. Louis, MO; *see* **Note 2**).
6. Freund's incomplete adjuvant (Sigma-Aldrich, St. Louis, MO).
7. Luer-lock, all-plastic type syringes for preparing emulsion (*see* **Note 3**).
8. 22-Gauge 1.5-in. syringe needles.
9. 26-Gauge 0.5-in. syringe needles.
10. 18-Gauge double hub micro-emulsifying needle with reinforcing bar (Popper & Sons, New Hyde Park, NY).

2.2. Indirect Enzyme-Linked Immunosorbent Assay (Indirect ELISA)

1. PBS: prepare 10× PBS by mixing 87.7 g NaCl, 30 g $Na_2HPO_4{\cdot}12H_2O$, 2.9 g $NaH_2PO_4{\cdot}2H_2O$, and distilled water for a final volume of 1 L. Dilute the 10× PBS tenfold with distilled water to prepare PBS as needed.
2. Antigen solution: dilute antigen with PBS for a final concentration of 5 μg/mL.
3. PBS-T: PBS containing 0.05% Tween 20.
4. Blocking buffer: 1% BlockAce (DS Pharma Biomedical, Osaka, Japan) in water.
5. Microtiter plate washer (Bio-Rad Laboratories, Hercules, CA).

6. Microtiter plate reader with 490-nm filter (Bio-Rad Laboratories, Hercules, CA).
7. Test antisera or test antibodies.
8. Horseradish peroxidase-conjugated goat anti-mouse IgG-(γ chain) (SouthernBiotech, Birmingham, AL).
9. 10 mg *o*-Phenylenediamine dihydrochloride tablets (Sigma-Aldrich, St. Louis, MO).
10. 30% Hydrogen peroxide.
11. 0.5 M H_2SO_4.
12. Multichannel pipette and disposable pipette tips.
13. 96-Well ELISA plate (Asahi Techno Glass, Chiba, Japan).
14. Citrate phosphate buffer: prepare 10× citrate phosphate buffer stock solution by mixing 36.8 g $Na_2HPO_4 \cdot 12H_2O$, 10.2 g citric acid, and distilled water for a final volume of 100 mL. Store stock solution at 4°C. Dilute 10× citrate phosphate buffer stock solution tenfold as needed.

2.3. Preparation of Myeloma Cells

1. Frozen stock of P3X63-Ag8.653 murine myeloma cell line (Japanese Collection of Research Bioresources, Tokyo, Japan).
2. DMEM: sterile serum-free Dulbecco's Modified Eagle's Medium containing 4,500 mg/L D-glucose, 584 mg/L L-glutamine, 110 mg/L sodium pyruvate (supplied as sterile liquid medium; Life Technologies, Carlsbad, CA) supplemented with 100 U/mL penicillin and 100 μg/mL streptomycin (use Penicillin-Streptomycin, liquid, Life Technologies, Carlsbad, CA).
3. 10% FBS DMEM: DMEM supplemented with 10% fetal bovine serum (FBS; supplied as sterile liquid; Thermo Trace, Sydney, Australia; *see* **Note 4**).
4. 8-Azaguanine myeloma medium: 10% FBS DMEM supplemented with 130 μM 8-azaguanine (Sigma-Aldrich, St. Louis, MO; *see* **Note 5**).
5. Sterile 100-mm tissue culture dishes (Asahi Techno Glass, Chiba, Japan).
6. Sterile 15-mL plastic centrifuge tubes with conical bottom.
7. Humidified 37°C, 5% CO_2 tissue culture incubator (HIRASAWA WORKS, Tokyo, Japan).
8. Inverted phase contrast microscope.
9. Alcohol swabs.
10. 0.4% Trypan blue solution (Life Technologies, Carlsbad, CA).
11. Hemocytometer.

2.4. Fusion of Murine Myeloma Cells with Murine Spleen Cells

1. DMEM: sterile serum-free Dulbecco's Modified Eagle's Medium containing 4,500 mg/L D-glucose, 584 mg/L L-glutamine, 110 mg/L sodium pyruvate (supplied as sterile liquid medium; Life Technologies, Carlsbad, CA) supplemented with 100 U/mL penicillin and 100 μg/mL streptomycin (use penicillin–streptomycin, liquid; Life Technologies, Carlsbad, CA).
2. 15% FBS DMEM: DMEM supplemented with 15% fetal bovine serum (FBS; supplied as sterile liquid; Thermo Trace, Sydney, Australia; *see* **Note 4**).
3. HAT (hypoxanthine/aminopterin/thymidine) medium: prepare 500 mL 15% FBS DMEM containing 5% BriClone (Archport, Dublin, Ireland; *see* **Note 6**). Take 10 mL of this medium and reconstitute one vial of HAT Media Supplement (50×) Hybri-Max (Sigma-Aldrich, St. Louis, MO). Add the contents of the vial to the 490 mL of 15% FBS DMEM containing 5% BriClone for a final concentration of 100 μM hypoxanthine, 0.4 μM aminopterin, and 16 μM thymidine
4. HT (hypoxanthine/thymidine) medium: prepare 500 mL 15% FBS DMEM containing 5% BriClone (Archport, Dublin, Ireland; *see* **Note 6**). Take 10 mL of this medium and reconstitute one vial of HT Media Supplement (50×) Hybri-Max (Sigma-Aldrich, St. Louis, MO). Add the contents of the vial to the 490 mL of 15% FBS DMEM containing 5% BriClone for a final concentration of 100 μM hypoxanthine, 16 μM thymidine.
5. Red blood cell lysis buffer: mix 180 mL of 0.16 M NH_4Cl (8.3 g/L) and 20 mL of 0.17 M Tris–HCl, pH 7.2. Autoclave at 121°C for 15 min and store at 4°C.
6. DMSO: sterile-filtered, hybridoma tested dimethyl sulfoxide (Hybri-Max; Sigma-Aldrich, St. Louis, MO).
7. PEG solution: autoclave 1.5 g of polyethyleneglycol 4000 (gas chromatography grade; Merck Chemicals, Nottingham, UK) at 121°C for 20 min in a glass bottle. Add 1.5 mL of DMEM and 0.3 mL of DMSO (hybridoma tested grade, Sigma-Aldrich, St. Louis, MO) before cooling in a sterile hood (*see* **Note 7**).
8. 50-mL Sterile plastic centrifuge tubes with conical bottom.
9. Dissecting board or plastic tray.
10. Sterile scissors: two general operating scissors with sharp/blunt points (14.5 cm) and one general operating scissors with sharp/sharp points (11.5 cm).
11. Sterile forceps: one forceps with teeth to hold the skin of the mouse tightly (15 cm), two forceps without teeth (15 cm), and one forceps without teeth (11 cm).

12. Sterile 60-mm plastic culture dishes (Asahi Techno Glass, Chiba, Japan).
13. Sterile 100-mm plastic culture dishes (Asahi Techno Glass, Chiba, Japan).
14. Sterile 70 μm mesh cell strainer (BD Bioscience, San Jose, CA).
15. Sterile 10-mL plastic syringe with 22-gauge needle.
16. Sterile polystyrene 96-well microtiter plates (Asahi Techno Glass, Chiba, Japan).
17. Stopwatch
18. Humidified 37°C, 5% CO_2 tissue culture incubator (HIRASAWA WORKS, Tokyo, Japan).

2.5. Cloning of Hybridoma Cell Lines

1. HT medium: prepare 500 mL 15% FBS DMEM containing 5% BriClone (Archport, Dublin, Ireland; *see* **Note 6**). Take 10 mL of this medium and reconstitute one vial of HT Media Supplement (50×) Hybri-Max (Sigma-Aldrich, St. Louis, MO). Add the contents of the vial to the 490 mL of 15% FBS DMEM containing 5% BriClone for a final concentration of 100 μM hypoxanthine, 16 μM thymidine.
2. Sterile polystyrene 96-well microtiter plates (Asahi Techno Glass, Chiba, Japan).
3. Sterile polystyrene 24-well microtiter plates (Asahi Techno Glass, Chiba, Japan).
4. Sterile 15-mL plastic centrifuge tubes with conical bottom.
5. Humidified 37°C, 5% CO_2 tissue culture incubator (HIRASAWA WORKS, Tokyo, Japan).
6. Inverted phase contrast microscope.
7. 0.4% Trypan blue solution (Life Technologies, Carlsbad, CA).
8. Hemocytometer.
9. Additional reagents and equipment for indirect ELISA (*see* **Section 2.2**).

2.6. Freezing and Recovery of Hybridoma Cell Lines

1. DMEM: sterile serum-free Dulbecco's Modified Eagle's Medium containing 4,500 mg/L D-glucose, 584 mg/L L-glutamine, 110 mg/L sodium pyruvate (supplied as sterile liquid medium; Life Technologies, Carlsbad, CA) supplemented with 100 U/mL penicillin and 100 μg/mL streptomycin (use Penicillin–Streptomycin, liquid, Life Technologies, Carlsbad, CA).
2. Sterile 15-mL plastic centrifuge tubes with conical bottom.
3. Sterile 50-mL plastic centrifuge tubes with conical bottom.
4. CELLBANKER 1 (Nippon zenyaku kogyo, Fukushima, Japan).

5. Sterile 2-mL cryotubes with internal thread (Asahi Techno Glass, Chiba, Japan).
6. Inverted phase contrast microscope.
7. 0.4% Trypan blue solution (Life Technologies, Carlsbad, CA).
8. Hemocytometer.
9. 15% FBS DMEM: DMEM supplemented with 15% fetal bovine serum (FBS; supplied as sterile liquid; Thermo Trace, Sydney, Australia; *see* **Note 4**).
10. Alcohol swabs.
11. Pasteur pipettes.
12. 37°C water bath.
13. Humidified 37°C, 5% CO_2 tissue culture incubator (HIRASAWA WORKS, Tokyo, Japan).
14. −80°C freezer (SANYO Electric, Tokyo, Japan).
15. Liquid nitrogen freezer (TAIYO NIPPON SANSO CORPORATION, Tokyo, Japan).

2.7. Production of mAb in the Murine Ascites Fluid

1. Cloned hybridoma cell line producing desired antibody.
2. Male Balb/c mice (syngenic host) older than 8 weeks.
3. Pristane (2,6,10,14-tetramethylpentadecane; Sigma-Aldrich, St. Louis, MO; *see* **Note 8**).
4. DMEM: sterile serum-free Dulbecco's Modified Eagle's Medium containing 4,500 mg/L D-glucose, 584 mg/L L-glutamine, 110 mg/L sodium pyruvate (supplied as sterile liquid medium; Life Technologies, Carlsbad, CA) supplemented with 100 U/mL penicillin and 100 μg/mL streptomycin (use penicillin–streptomycin, liquid; Life Technologies, Carlsbad, CA).
5. 15% FBS DMEM: DMEM supplemented with 15% fetal bovine serum (FBS; Thermo Trace, Sydney, Australia; *see* **Note 4**).
6. Pastuer pipettes.
7. Scissors with sharp/blunt points (14.5 cm).
8. Forceps, 15 cm, with teeth.
9. Sterile normal saline solution: 0.9% NaCl.
10. Sterile 5-mL luer-lock all-plastic type syringes for pristane injections (*see* **Note 3**).
11. Sterile 5-mL plastic syringes.
12. 22-Gauge needle.
13. 26-Gauge needle.

14. 0.4% Trypan blue solution (Life Technologies, Carlsbad, CA).
15. Hemocytometer.
16. Inverted phase contrast microscope.
17. No. 2 filter paper.
18. Sterile 25 mm mixed cellulose acetate 0.45-μm syringe filter (Toyo roshi kaisya, Tokyo, Japan).

2.8. Purification of mAb Using Affinity Chromatography

1. IsoStrip Mouse Monoclonal Antibody Isotyping Kit (Roche diagnostics, Mannheim, Germany).
2. HiTrap Protein A HP (1 mL) column (GE Healthcare UK, Buckinghamshire, England).
3. 25 mm Mixed cellulose acetate 0.45-μm syringe filter, not sterile (Toyo roshi kaisya, Tokyo, Japan).
4. 47-mm 0.45-μm Mixed cellulose acetate membrane filter.
5. Binding buffer:
   a. For IgG_1 subclass (1.5 M glycine-NaOH, pH 8.9, 3 M NaCl): dissolve 56.3 g of glycine and 87.7 g of NaCl in 350 mL of Milli-Q water. Adjust the pH to 8.9 with NaOH and add Milli-Q water for a final volume of 500 mL. Filter with a 47-mm 0.45-μm mixed cellulose acetate membrane filter and degas under vacuum supplied by an aspirator.
   b. For IgG_{2a}, IgG_{2b}, and IgG_3 subclasses (20 mM sodium phosphate buffer, pH 7.0): Dissolve 1.56 g of $NaH_2PO_4{\cdot}2H_2O$ in 350 mL of Milli-Q water. Adjust the pH to 7.0 with NaOH and add Milli-Q water for a final volume of 500 mL. Filter with a 47-mm 0.45-μm mixed cellulose acetate membrane filter and degas under vacuum supplied by an aspirator.
6. Elution buffer:
   a. For IgG_1, IgG_{2a}, and IgG_3 subclasses (100 mM sodium citrate buffer, pH 4.5): dissolve 10.5 g of citric acid monohydrate in 350 mL of Milli-Q water. Adjust the pH to 4.5 with NaOH and add Milli-Q water for a final volume of 500 mL. Filter with a 47-mm 0.45-μm mixed cellulose acetate membrane filter and degas under vacuum supplied by an aspirator.
   b. For IgG_{2b} subclass (100 mM sodium citrate buffer, pH 3.5): dissolve 10.5 g of citric acid monohydrate in 350 mL of Milli-Q water. Adjust the pH to 3.5 with NaOH and add Milli-Q water for a final volume of 500 mL. Filter with a 47-mm 0.45-μm mixed cellulose acetate membrane filter and degas under vacuum supplied by an aspirator.

7. Ascites fluid containing desired mAb.
8. 1 M Tris–HCl, pH 9.0.
9. 2.5-mL Plastic syringes.
10. 10-mL Plastic syringes.

3. Methods

3.1. Immunization of Mice

1. Prepare an emulsion (200 μL/mouse) using equal volumes of sterile PBS containing 50 μg antigen/mouse and Freund's complete adjuvant, by linking two locking syringes, one loaded with antigen solution (500 μg/mL of antigen in sterile PBS) and the other loaded with adjuvant (using 22-gauge needle), using a double ended Luar-Lock connector. Press the syringe barrels back and forth until a stable emulsion is produced (for 5–10 min till thickened). Let a drop of the emulsion fall into water to check the endpoint of emulsification. A stable emulsion will not disperse.
2. Inject a mouse intraperitoneally with 200 μL of the antigen emulsion using a 26-gauge needle. Five mice are immunized for each antigen.
3. Boost mice 2 weeks later by intraperitoneally injecting an emulsion (200 μL/mouse) of equal volumes of sterile PBS containing 50 μg antigen/mouse and Freund's incomplete adjuvant. The emulsion is prepared and injected as in Steps 1 and 2.
4. Collect blood 1 week after the booster injection to determine the antibody titer. Place mice in a retaining device and use a 26-gauge needle to make a small incision in the mouse's tail vein. Immediately collect the blood (<20 μL) in a microcentrifuge tube using a 200-μL pipette or glass capillary. Keep the sample blood at 37°C for 1 h and centrifuge at 5,000×*g* for 10 min to separate the serum sample (supernatant fraction).
5. Determine the antibody titer in the serum by indirect ELISA (*see* **Section 3.2, Note 9**).
6. If the antibody titer is too low for cell fusion, boost mice as in Step 3 every 2 weeks until a sufficient titer is achieved (e.g., a significant signal at <1/10,000 dilution).
7. When the antibody titer is sufficient, boost mice by injecting 50 μg antigen in sterile PBS without adjuvant intraperitoneally (100 μL).

8. Four days after the final immunization (Step 7), fuse murine myeloma cells with spleen cells from the immunized mice (*see* **Section 3.4**).

3.2. Indirect Enzyme-Linked Immunosorbent Assay (Indirect ELISA)

1. Coat plate(s) with antigen by dispensing 100 μL of antigen solution (5 μg/mL in PBS) into each well of an ELISA plate using a multichannel pipette and tips. Close the lid and incubate at 37°C for 1 h or 4°C overnight.
2. Rinse the coated plate once with PBS-T using microplate washer apparatus.
3. Add 200 μL of 1% BlockAce solution to each well and incubate at 37°C for 1 h or 4°C overnight.
4. Rinse plate once with PBS-T using microplate washer apparatus.
5. Add 100 μL of antibody samples diluted in PBS to each of the coated wells, close the lid, and incubate at 37°C for 1 h.
6. Wash plate five times with PBS-T using microplate washer and remove residual liquid completely by vigorously patting on a paper towel.
7. Add 100 μL of diluted antibody-enzyme conjugate (Goat anti-mouse IgG (γ chain)-horseradish peroxidase conjugated; 1:3,000 dilution in PBS-T) to each well (*see* **Note 10**), close the lid, and incubate at 37°C for 1 h.
8. Wash plate five times with PBS-T using microplate washer and remove residual liquid completely by vigorously patting on a paper towel.
9. Prepare the substrate solution by dissolving one tablet of *o*-phenylenediamine dichloride in 25 mL of citrate phosphate buffer in the dark and then add 5 μL of 30% H_2O_2.
10. Add 200 μL of substrate solution to each well and incubate at 37°C for 30 min in the dark.
11. Add 50 μL of 0.5 M H_2SO_4 to each well to stop the peroxidase reaction.
12. Read the absorbance at 490 nm with a microtiter plate reader (*see* **Note 9**).

3.3. Preparation of Myeloma Cells

3.3.1. Culture of Myeloma Cells

1. At least 7 days before the cell fusion experiment, start culturing myeloma cells. Place a sterile 15-mL centrifuge tube containing 5 mL of DMEM prewarmed in a 37°C water bath and a sterile 100-mm culture dish in the sterile hood.

2. Recover a cryotube containing frozen myeloma cells from the liquid nitrogen storage tank.
3. Put the cryotube in 37°C water bath for 2–5 s to thaw the surface of frozen cells and move it to a sterile hood.
4. Wipe the surface and top of the cryotube with an alcohol swab and transfer the cells to a 15-mL centrifuge tube. Steps 1–4 should be completed within 1 min in order to maintain maximum cellular viability.
5. Centrifuge at 200×*g* for 5 min. The cells will thaw while centrifuging.
6. Aspirate the supernatant.
7. Resuspend the cell pellet in 10 mL of 10% FBS DMEM.
8. Transfer the resuspended cells to a 100-mm culture dish and spread the cells.
9. Incubate the culture overnight in a CO_2 incubator at 37°C.
10. Aspirate the culture medium and add 10 mL of new culture medium (10% FBS DMEM).
11. Incubate the culture in a CO_2 incubator at 37°C until the culture becomes confluent. Check the condition of the cells with an inverted microscope.
12. Prepare four new culture dishes containing 8 mL of 10% FBS DMEM.
13. Completely resuspend the cell culture with a Pasteur pipette and subculture 2 mL of cell suspension in the new dishes.
14. Continue to subculture as in Steps 12–13 when the cells become almost confluent (*see* **Note 11**).

3.3.2. Cell Viability Test by Trypan Blue Exclusion

1. Suspend cell culture with a Pasteur pipette.
2. Mix 10 μL of trypan blue solution and 10 μL of cell suspension.
3. Place a clean coverslip on the hemocytometer. Introduce the trypan blue cell mixture (Step 2) into the counting chamber from the edge of the coverslip. Allow the suspension to be drawn into the chamber by capillary action. Fill the chamber on the opposite side in the same manner.
4. Count the unstained (viable) and stained (dead) cells separately in the hemocytometer. Each counting chamber consists of nine 1-mm squares divided into smaller squares. One of the 1-mm squares represents a volume of 10^{-4} mL. Count all the cells within each of the four corner squares. In your counting, count the cells that lie on the top and the left-hand lines of each square but not those on the bottom

or the right-hand lines. Count the chamber on the opposite side in the same manner.

5. Calculate the mean number of cells in each of the 1-mm squares. Multiply by 10^4 and the dilution factor (2× in this case) to obtain the cell concentration in cells/mL culture.
6. Calculate the percentage of viable cells from the number of total cells (viable + dead) and unstained cells (viable).

3.4. Fusion of Murine Myeloma Cells with Murine Spleen Cells

3.4.1. Isolation of Splenocytes

1. Euthanize a mouse by cervical dislocation and fully immerse the mouse in a beaker containing 70% ethanol (*see* **Note 12**).
2. Place the mouse with its right side down on a dissecting board or plastic tray (head-side left).
3. Lift the skin over the thorax using sterile forceps and cut the skin away with sterile scissors to expose the left side of the rib cage.
4. Place the mouse in a sterile hood.
5. Using another set of small sterile forceps and scissors, open the muscle layer. Remove the spleen from the left upper abdomen of the mouse. The spleen is a small dark red organ.
6. Place the spleen in a sterile 60-mm culture dish containing 5 mL of DMEM and dissect away surface fat and other adhering tissue using a sterile forceps and scissors.
7. Transfer the cleaned spleen to a sterile 100-mm culture dish.
8. Set a cell strainer on a sterile 50-mL centrifuge tube.
9. Fill a 10-mL sterile syringe with a 22-gauge needle with 10 mL of DMEM. Inject DMEM into the spleen at several sites to push immune cells out of the spleen. Transfer the cell suspension to the cell strainer with a serological pipette. Repeat three times each time transferring the cell suspension to the cell strainer (*see* **Note 13**).
10. Centrifuge the combined cell suspension at 200×*g* for 5 min at room temperature.
11. Remove the supernatant and resuspend the splenocytes in 10 mL of ice-cold red blood cell lysis buffer and let the suspension stand for 10 min on ice to ensure erythrocyte hemolysis.
12. Centrifuge the tube containing the spleen cells at 200×*g* for 5 min at room temperature.

13. Remove the supernatant. Wash the spleen cells once with 10 mL of DMEM followed by centrifugation at 200×*g* for 5 min and discard the supernatant.

3.4.2. Fusion of Myeloma Cells and Spleen Cells

1. While the spleen cells are incubating on ice in the Red blood cell lysis buffer (**Section 3.4.1**, Step 11), transfer the myeloma cells (almost confluent) to a 50-mL centrifuge tube. Count the myeloma cells and check their viability using the trypan blue exclusion method (*see* **Section 3.3.2**). Incubate the centrifuge tube at 37°C in a CO_2 incubator until cell fusion.
2. Resuspend the washed spleen cells in 10 mL of DMEM in a sterile 50-mL centrifuge tube (*see* **Section 3.4.1**, Step 13) and add 4×10^7 viable myeloma cells. Recover the mixed cells by centrifugation at 200×*g* for 5 min (*see* **Note 13**).
3. Wash the mixed cells twice by resuspension in 10 mL of DMEM and centrifuge at 200×*g* for 5 min at room temperature.
4. After the second wash, discard the supernatant and loosen the mixed cell pellet by gently flicking the tip of the tube.
5. Initiate fusion of the spleen and myeloma cells in the 50-mL centrifuge tube by adding 1 mL of PEG solution at a rate of 0.1 mL/6 s by micropipette with gentle mixing (*see* **Note 14**).
6. Mix the PEG cell solution gently for 1 min.
7. Add 5 mL of DMEM solution at a rate of 0.1 mL/6 s with gentle mixing.
8. Add 4 mL of DMEM and 2 mL of FBS and mix gently.
9. Centrifuge the cells at 100×*g* for 7 min.
10. Pour off the supernatant and wash the cells twice with 15% FBS DMEM by centrifugation for 10 min at 150×*g*.
11. Resuspend the cell pellet in 15 mL of 15% FBS DMEM. Transfer the cell suspension to a sterile 100 mm culture dish and incubate for 3–15 h at 37°C in a CO_2 incubator.
12. Resuspend the cells from the 100-mm culture dish in 160 mL of HAT medium placing 40 mL of cell suspension in each of four 50-mL centrifuge tubes.
13. Add 200 μL of cell suspension to each of the wells of eight 96-well culture plates using a multichannel pipette.
14. To select for hybrid cells, incubate the plates at 37°C in a CO_2 incubator (*see* **Note 15**).
15. After 24 h, remove 100 μL of the culture medium from each well and add 100 μL fresh HAT medium.

16. Grow the hybrids in HAT medium for 2 weeks by removing 100 μL of the culture medium from each well and adding 100 μL fresh HAT medium every 3 days.
17. Check for antibody production by indirect ELISA assay using 100 μL of culture supernatant.

3.5. Cloning of Hybridoma Cell Lines

1. From each well containing antigen-specific antibody in the hybridoma supernatant (positive signal in the indirect ELISA), transfer all of the cells into a separate well of a 24-well culture plate that has been preincubated with 0.5 mL of HT medium and culture 2 or 3 days in a CO_2 incubator (*see* **Note 16**).
2. Check the antibody titer by indirect ELISA. Select the higher titer and/or faster growing hybridoma cell lines.
3. Perform a trypan blue cell viability test (*see* **Section 3.3.2**) of the ELISA-positive hybridoma cell lines.
4. Transfer 200 viable hybridoma cells of a selected hybridoma cell line to a sterile 15-mL centrifuge tube containing 5 mL of HT medium producing a 40 cells/mL hybridoma cell suspension. Mix well.
5. Using sterile 15-mL centrifuge tubes containing 2.5 mL of HT medium, perform three serial dilutions of the 40 cells/mL hybridoma suspension by mixing 2.5 mL of hybridoma suspension with 2.5 mL HT medium producing cell suspensions containing 20, 10, and 5 cells/mL.
6. Fill 24 wells of a 96-well plate with 100 μL of each dilution for 4, 2, 1, and 0.5 cells/well.
7. Incubate the plate(s) at 37°C in a CO_2 incubator for 5 days. Monitor cell growth with an inverted microscope and mark the wells containing monoclonal cells. Monoclonal cells grew from a single cell in each well.
8. Feed the culture by adding 100 μL of HT medium to every well and continue incubation in a CO_2 incubator until cell growth in the bottom of the wells is 10–50% confluent.
9. Check the antibody titer by indirect ELISA. Cells in the positive wells are transferred to a new 24-well culture plate.
10. Repeat the cloning procedure (Steps 4–9) to establish a stable and monoclonal hybridoma cell line.
11. Established hybridomas are maintained in 15% FBS DMEM.

3.6. Freezing and Recovery of Hybridoma Cell Lines

1. For each cryotube, grow two cultures in 100-mm dishes to 80–100% confluency.
2. Resuspend the cell culture in the culture media with a Pasteur pipette and transfer to a sterile 50-mL centrifuge tube.

3. Perform a trypan blue cell viability test (*see* **Section 3.3.2**). Cells should be nearly 100% viable.
4. Centrifuge at 200×*g* for 5 min at room temperature.
5. Aspirate the supernatant and resuspend the cell pellet in 10 mL of DMEM.
6. Centrifuge at 200×*g* for 5 min.
7. Resuspend the cell pellet in CELLBANKER 1 to give a cell density of 1×10^7 cells/mL.
8. Aliquot 1 mL/cryotube.
9. Label the cryotubes with the date and information concerning the hybridoma line.
10. Freeze in a −80°C freezer overnight.
11. Transfer the cryotubes to a liquid nitrogen freezer.
12. To start a hybridoma culture from frozen cells, remove a cryotube from the freezer, and initiate a culture as was done for the frozen myeloma cells but use 15% FBS DMEM (*see* **Section 3.3.1**).
13. Propagate the cells in 15% FBS DMEM by incubating at 37°C in a CO_2 incubator. Subculture the cells when the cells are almost confluent. Check the condition of the cells with an inverted microscope.

3.7. Production of mAb in the Murine Ascites Fluid

1. Ten days and 3 days prior to inoculation with hybridomas, inject mice intraperitoneally with 0.5 mL pristane/mouse using a 26-gauge needle and an all-plastic syringe.
2. Grow hybridoma cells in 100-mm culture dishes in 15% FBS DMEM until almost confluent to obtain at least 5×10^6 cells/mouse.
3. Transfer the cultures to 50-mL centrifuge tubes. Centrifuge at 200×*g* for 5 min and resuspend the cells in 3 mL of DMEM.
4. Count the cells and determine viability by the trypan blue exclusion method (*see* **Section 3.3.2**). Cells should be nearly 100% viable.
5. Centrifuge the cells at 200×*g* for 5 min. Aspirate the supernatant to remove serum completely. Adjust the cell concentration to 1×10^7 cells/mL with DMEM.
6. Inject mice intraperitoneally with 0.5 mL of cell suspension using a 26-gauge needle. The number of mice depends on the amount of the monoclonal antibody needed. Typically, five mice are injected at one time. Wait 1–2 weeks for ascites to form (5–10 mL/mouse).
7. Euthanize mice by cervical dislocation. Lift the abdominal skin using a forceps wiped with an alcohol swab and cut

the skin away with a sterile scissors. Make a small cut in the peritoneum and insert a Pasteur pipette to harvest the ascites fluid.

8. Inject a volume of ice-cold normal saline equal in volume to the ascites fluid removed from the mouse into the mouse's peritoneal cavity washing the inside of the mouse. Combine the wash with the ascites fluid.
9. Centrifuge 3,000×*g* for 20 min at 4°C. Harvest the ascites fluid layer in the middle layer between the pristane and cell pellet (*see* **Note 17**).
10. Filter the ascites fluid using No.2 filter paper to remove residual pristane and large cell debris.
11. Filter-sterilize the ascites fluid using a 0.45-μm filter. Freeze at −80°C. Keep a part of the ascites fluid (<50 μl) before freezing at 4°C to determine the antibody subclass (*see* **Section 3.8**, Step 1).

3.8. Purification of mAb Using Affinity Chromatography

1. Determine the monoclonal antibody subclass using the IsoStrip Mouse Monoclonal Antibody Isotyping Kit (Roche).
2. Equilibrate the HiTrap Protein A column (1 mL) with 10 mL of binding buffer appropriate to the IgG subclass. The column should be used for ascites fluid from a single mouse (5–15 mg of immunoglobulin/single mouse ascites fluid).
3. Dilute the ascites fluid with an equal volume of binding buffer and apply to the column at a flow rate of 0.5 mL/min. Keep the flowthrough fraction to avoid losing the unbound antibody.
4. Wash the column with 10 mL of binding buffer at a flow rate of 1 mL/min.
5. Elute the bound protein with 2–5 column volumes of elution buffer appropriate to the IgG subclass and collect 0.5 mL fractions using a fraction collector. Measure A_{280} to locate the eluted protein. Combine the protein fractions and neutralize with 1 M Tris–HCl buffer, pH 9.0 (check with pH test paper).
6. Dialyze eluted protein overnight at 4°C against 2 L of PBS.
7. Freeze 1 mL aliquots at −80°C

4. Notes

1. Other mouse strains can be used as fusion partners with the P3X63-Ag8.653 murine myeloma cell line (derived from

Balb/c). If other mice are used there is a risk that ascites fluid may not accumulate following injection of mAb producing hybridoma

2. Freunds complete adjuvant (FCA) elicits a strong response to most antigens. It, however, may cause inflammation, induration, pain, and necrosis at the injection site, and its use is therefore ethically controversial. It has also the risk to workers, as it can cause localized soft tissue damage following accidental needlestick injuries. Due to the harmful responses produced by this adjuvant, the ability of other adjuvants such as Titermax Gold (Titermax USA, Norcross, GA) to elicit the desired immune response should be considered.
3. Syringes should be all-plastic and Luar-lock type to prepare adjuvant and pristane. Plastic syringes with rubber pistons contain a lubricant that fails in the presence of them and causes syringes to stick.
4. All FBS used in this chapter are heat inactivated (56°C for 30 min) and aliquots are prepared in a sterile hood and stored at −80°C.
5. The myeloma cells used for cell fusion should lack hypoxanthine–guanine phosphoribosyltransferase (HPRT) activity due to mutation. Myeloma cells are cultured with 8-azaguanine to prevent growth of HPRT-positive revertants when the cells are subcultured. Use of HPRT-negative cells ensures that only splenocyte–myeloma fusion cells will grow during HAT selection. In detail, aminopterin blocks the synthesis of DNA by inhibiting dihydrofolate reductase. Cells that lack the ability to utilize the salvage pathway for nucleotide synthesis are eliminated in the HAT selection. Only cells that possess both HPRT and thymidine kinase can utilize the salvage pathway if supplied with hypoxanthine and thymidine. One week before cell fusion experiment, myeloma cells are grown in medium without 8-azaguanine (*see* **Section 3.3.1**).
6. BriClone is an IL-6-enriched hybridoma cloning medium. It eliminates the need for feeder cells for the generation of hybridomas.
7. The PEG solution needs to be freshly prepared.
8. Freund's incomplete adjuvant can be used instead of pristane. Inject 0.5 mL Freund's incomplete adjuvant/mouse once 1–2 weeks before injecting mice with hybridoma.
9. The ELISA assay is a rapid, sensitive, and quantitative assay making it the most popular immunoassay for hybridoma selection. However, ELISA-positive mAb's may not be suitable for immunoelectron microscopy. Monoclonal

antibodies to be used for immunoelectron microscopy should be identified using both ELISA and a method such as Western blotting which identifies antibodies that have a high affinity for denatured proteins. Western blotting could also detect if the nonspecific binding for non-desired protein exists or not.

10. The antibody dilution should be according to the manufacturer's instructions.
11. A total of 4×10^7 myeloma cells with viability greater than 95% are used for cell fusion. Healthy myeloma cells have a doubling time of about 20 h.
12. If you need to obtain antisera from the mouse, anesthetize the mouse and collect blood prior to killing the mouse and removing the spleen.
13. 8×10^7–4×10^8 immune cells will be obtained from one mouse spleen. Ratios for mixing myeloma and splenocytes should be around 1:5.
14. The timing and the rate of liquid addition are critical and should be measured with a stopwatch. They ensure higher cell viability and higher fusion efficiency.
15. To ensure that selection of hybridomas will be successful, confirm that the myeloma cells that have not undergone the fusion process cannot grow in the HAT medium. In detail, prepare 2 mL of HAT medium in a sterile 35-mm culture dish in the sterile hood and add several drops of myeloma culture. After 1 week, myeloma cells should die and float on the culture medium.
16. Cloning the hybridoma is a critical step required to determine whether a hybridoma producing the desired mAb has been obtained. A positive signal in the indirect ELISA prior to cloning the hybridoma may be a false positive due to the presence of antibody-producing immune cells that have not fused with myeloma cells.
17. The pristane was injected at the start of ascites production and it stays in the peritoneal cavity during the ascites production.

Chapter 7

Production of Antipeptide Antibodies

Bao-Shiang Lee, Jin-Sheng Huang, G.D. Lasanthi P. Jayathilaka, Syed S. Lateef, and Shalini Gupta

Abstract

Peptides (8–20 residues) are as effective as proteins in raising antibodies, both polyclonal and monoclonal with a titer above 20,000 easily achievable. A successful antipeptide antibody production depends on several factors such as peptide sequence selection, peptide synthesis, peptide–carrier protein conjugation, the choice of the host animal, and antibody purification. Peptide sequence selection is likely the most difficult and critical step in the development of antipeptide antibodies. Although the format for designing peptide antigens is not precise, several guidelines can help maximize the likelihood of producing high-quality antipeptide antibodies. Typically, 5–20 mg of peptide is enough for raising an antibody, for preparing a peptide affinity column, and for antibody titer determination using an enzyme-linked immunosorbent assay (ELISA). Usually, it takes 3 months to raise a polyclonal antipeptide antibody from a rabbit that yields ~90 mL of serum which translates into approximately 8–10 mg of the specific antibody after peptide affinity purification.

Key words: Antipeptide antibody, peptide–carrier protein conjugation, keyhole limpet hemocyanin (KLH), polyclonal and monoclonal antibodies, phosphospecific antibodies, antibody titer.

1. Introduction

With their high specificity and binding ability (dissociation constant of 10^{-6}–10^{-12}), antibodies are the most widely used reagents for protein recognition in many biochemical applications (1–4). Native or recombinant proteins are used traditionally to produce antibodies. Generating polyclonal antibodies against a protein yields antibodies against multiple epitopes, which maximize the chance of recognizing the protein. However, this pool of antibodies does increase the cross-reactivity with other proteins

S.D. Schwartzbach, T. Osafune (eds.), *Immunoelectron Microscopy*, Methods in Molecular Biology 657, DOI 10.1007/978-1-60761-783-9_7, © Springer Science+Business Media, LLC 2010

(*see* **Note 1**). Generating polyclonal antibodies against a synthetic peptide (1, 2, 5, 6) (*see* **Note 2**), by contrast, will produce antibodies that are specific to the target protein. There are instances in which a peptide serves as a better choice than a protein, e.g., raising antibodies for a specific protein isoform or for a phosphorylated protein and in cases where the protein is not available.

1.1. Overview of Producing Antipeptide Antibodies

Peptides at low cost with high purity can be obtained from numerous companies [a list of companies can be found on the Peptide Resource Page (www.peptideresource.com)]. Also, there are many companies that produce antipeptide antibodies for a fee [a list of companies can be found on the Antibody Resource Page (www.antibodyresource.com)]. Nevertheless, not all antipeptide antibodies recognize native proteins and have a high titer. The potential drawback of choosing a peptide sequence that does not possess a high antigenicity and will not correspond to an exposed region of the native protein can be reduced substantially by carefully analyzing the protein sequence and structure using the large number of protein structure and antigenicity prediction software. Additionally, co-immunization of multiple peptide antigens from a protein will statistically increase the chances of obtaining antibodies that will recognize the target protein (*see* **Note 3**). Peptides containing phosphorylated amino acids can also be used to produce phosphospecific antibodies (*see* **Note 4**). However, a peptide by itself is typically too small to induce an immune response producing high titer antibodies. The minimum molecular weight needed to induce sufficient immune response is 5–15 kDa. A carrier protein is therefore conjugated to the peptide to induce an immune response in the animal. The most commonly used carrier protein is keyhole limpet hemocyanin (KLH, MW 4.5×10^5–1.3×10^7) that has been shown to aid in the production of high-titer antipeptide antibodies (*see* **Note 5**). KLH contains numerous exposed lysine residues, which allow for the covalent attachment of large numbers of peptide molecules. With the advancement in peptide synthesis, peptide selection, and peptide–carrier protein conjugation, peptides rather than proteins are becoming the method of choice for antibody production.

1.2. Analysis of Protein Sequence and Structure

Before producing a new antipeptide antibody, it is essential to know some basic features of the protein. Ensuring that the correct species and protein sequence have been identified is the first step. Information of the protein's structure can aid in choosing epitopes that are readily accessible to the antibodies. Any potential cross-reactivity with other closely related proteins, e.g., domain structures, should be avoided. Searching the protein NCBI database (www.ncbi.nlm.nih.gov), Uni-Prot (www.uniprot.org), PIR (pir.georgetown.edu), ExPASy (us.expasy.org/tools), etc. might be helpful. Searching a list of commercially available antibodies at www.antibodyresource.com/findantibody.html may help you find the antibody that you need.

1.3. Selection of Peptide for Antibody Production

The choice of peptide sequence for antipeptide antibody production is the single most important step in the process. Despite there being no infallible method for predicting antigenic peptides, there are several guidelines that can be followed to determine the peptide epitopes in a protein that are likely to be antigenic. These guidelines also increase the odds of an antibody recognizing the native protein. Other crucial parameters are the ease of peptide synthesis and conjugation, peptide stability and solubility in water, immunogenicity in the host animals, and specificity for the target protein.

1.3.1. Peptide Antigenicity/Accessibility Factors

It is unpredictable whether a designed peptide antibody will recognize the native protein due to conformation and other structural differences between synthetic peptides and peptide epitopes in native proteins. In general, ideal antigenic epitopes are hydrophilic, surface orientated, and flexible. This is because in most natural environments, hydrophilic regions tend to reside on the surface, whereas hydrophobic regions are likely found hidden in the protein conformation. Antibodies can therefore only bind to epitopes found on the surface of proteins and tend to bind with higher affinity when those epitopes are flexible enough to move into accessible positions. The following are some guidelines to increase the likelihood of successful peptide selection.

1. All recommended peptides are compared for sequence homology with other proteins by Basic Local Alignment Search Tool (BLAST) database searches.
2. BLAST protein search with Protein Lounge (database containing numerous antigenic peptide targets, http://www.proteinlounge.com).
3. Choosing peptides that are in the N- or C-terminal region of the protein is effective because these regions of proteins are usually solvent accessible and unstructured and antibodies developed against these peptides are also more likely to recognize the native protein. The N-terminal capped with an acetyl group and the C-terminal with an amide group will make the peptide appear more like a native protein and will reduce the degradation of the peptide in the animal.
4. Peptides lying in long loops connecting secondary structure motifs are preferable, avoiding those that are located in helical regions. This will increase the odds of the antibody recognizing the native protein. For proteins with known 3D coordinates, secondary structures can be obtained from the sequence link of the relevant entry at the Brookhaven data bank (www.rcsb.org/pdb/home/home.do). Solvent

accessibility can be determined using a variety of programs such as DSSP (swift.cmbi.ru.nl/gv/dssp), NACESS (www.bioinf.manchester.ac.uk/naccess), or WHATIF (swift.cmbi.kun.nl/whatif; swift.cmbi.ru.nl/servers/html/index.html).

5. When no structure information is available, secondary structure and accessibility predictions can be obtained from the following servers with 80% accuracy in predicting α-helix, β-strand, and loop (7): PHD (www.predictprotein.org), JPRED (www.compbio.dundee.ac.uk/~www-jpred), PSIPRED (bioinf.cs.ucl.ac.uk/psipred), PredAcc (bioserv.rpbs.jussieu.fr/RPBS/html/fr/T0_Home.html), SSPRED (coot.embl.de/~fmilpetz/SSPRED/sspred.html), PREDATOR (www-db.embl-heidelberg.de/jss/servlet/de.embl.bk.wwwTools.GroupLeftEMBL/argos/predator/down_predator.html), etc.
6. Web sites for antigenicity prediction such as www.innovagen.se, immunax.dfci.harvard.edu/Tools/antigenic.pl (8),www.immuneepitope.org/home.do, www.innovagen.se/custom-peptide-synthesis/peptide-property-calculator/peptide-property-calculator.asp, and ca.expasy.org/tools/protscale.html are useful as they check the peptide sequence against a number of criteria to estimate its suitability for antipeptide antibody production. Several software programs such as MacVector, DNAStar, and PC-Gene that incorporate one or more of several accepted algorithms for predicting peptide antigenicity are also very useful.
7. Not all areas of the protein such as the transmembrane regions are accessible to antibodies and should be avoided. Accessible (hydrophilic, surface-oriented, and flexible) regions of the native protein are exposed on the surface of the protein and in contact with the aqueous environment. Antigenic peptides are generally located in solvent accessible regions and may contain both hydrophobic and hydrophilic residues.
8. Generally, 8–20 amino acid long peptides should be used. Longer peptides increase the risk of losing specificity and shorter peptides may elicit antibodies that would not recognize the native protein with sufficient affinity.
9. Incorporation of proline and tyrosine residues confers some structural motif to the immunogen which is likely to be found in the native protein.

1.3.2. Peptide Synthesis and Solubility Factors

Investigating protein sequence to identify the best candidate peptides that optimize synthesis success and solubility is an essential step in antipeptide antibody production. Several observations are listed as follows:

1. Peptide solubility is strongly influenced by amino acid composition. Peptides with a high content of hydrophobic residues, such as Leu, Val, Ile, Met, Phe, and Trp, have either limited solubility in aqueous solution or are completely insoluble. It is advisable to keep the hydrophobic amino acid content below 50% and to ensure that there is at least one charged residue for every five amino acids. A single conservative replacement, such as replacing Ala with Gly, or adding a set of polar residues to the N- or C-terminal, may also improve solubility.
2. During synthesis, β-sheet formation causes incomplete solvation of the growing peptide and results in a high degree of deletion sequences in the final product. This problem can be avoided by choosing sequences that do not contain multiple and adjacent residues comprising Val, Ile, Tyr, Phe, Trp, Leu, Gln, and Thr. If sequences cannot be chosen to avoid stretches of these residues, it often helps to break the pattern by making conservative replacements, for example, inserting a Gly or Pro at every third residue, replacing Gln with Asn, or replacing Thr with Ser.
3. Peptides containing multiple Cys, Met, or Trp residues are also difficult to obtain in high purity partly because these residues are susceptible to oxidation and/or side reactions.
4. The following amino acids or sequences are best avoided:
   a. A sequence starting or ending with Pro.
   b. Ala, Val, Thr, Pro, or Ser doublets for synthetic purposes and ending sequences in Val, Ile, Trp, Tyr, and Phe.
   c. Extremely long repeats of the same amino acid (e.g., ArgArgArgArgArg) and Gln or Asn at the N-terminal.
5. A peptide having an overall charge close to neutral is desirable.
6. Ending sequences with hydrophilic residues are favorable as side groups will promote increased solubility along with free alpha reactive groups. Terminal ends are highly exposed to their environment.
7. Limiting the number of contiguous charged or hydrophobic residues is helpful as they can isolate a portion of the peptide leading to increased solubility problems
8. Due to the nature of glycine and its lack of a side group, it does not behave as a hydrophobic residue unless contiguous stretches exist.

1.4. Peptide Synthesis

Usually peptides used for antipeptide antibody production contain 10–25 amino acid residues and are obtained either from a core facility or a commercial vendor (9–12). A state-of -the-art peptide synthesizer would have little problem producing 50 mg of peptide with >90% purity even with several phosphorylated amino acids in the peptide. Peptides are routinely synthesized using stepwise Fmoc solid-phase synthesis chemistry starting from the C terminus. The procedure for synthesizing a peptide is as follows (*see* **Note 6**): (I) the Fmoc group of the amino acid-preloaded resin is removed by 20% piperidine, (II) Fmoc-amino acid (with or without modification) is coupled 60 min to the resin-bound peptide using 0.1 M 2-(1H-benzotriazole-1-yl)-1,1,3,3,-tetramethyluronium hexafluorophosphate (HBTU) in dimethylformamide (DMF) containing 0.4 M 4-methylmorpholine, (III) Steps 1 and 2 are repeated until the last amino acid is added, (IV) the Fmoc group of the resin-bound peptide is removed by 20% piperidine, (V) the peptide is then deprotected and cleaved from the resin using trifluoroacetic acid (TFA), (VI) ethyl ether is added to precipitate the peptide from the TFA solution and the precipitated peptide is lyophilized, (VII) the crude peptide is purified on a reversed-phase C18 column using a preparative HPLC system. A flow rate of 20 mL/min with solvent A (0.1% TFA in deionized water) and solvent B (0.1% TFA in acetonitrile) is used. The column is equilibrated with 5% solvent B. After sample loading, the column is eluted with a linear gradient from 5% solvent B to 100% solvent B in 60 min, and (VIII) the pure peptide fraction is identified by matrix-assisted laser desorption/ionization time-of-flight mass spectrometry (MALDI-TOF MS).

Dissolving peptides in aqueous solutions is not trivial. In practice, the peptide sequence should contain at least 20% charged residues to facilitate solubilization. Hydrophilic peptides containing >25% charged residues (Glu, Asp, Lys, Arg, and His) and <25% hydrophobic residues also generally dissolve in aqueous media provided that the charged residues are fairly distributed throughout the sequence. Both acidic peptides (Glu+Asp residues> Lys+Arg+His residues) and basic peptides (Lys+Arg+His residues> Glu+Asp residues) are more soluble at neutral pH than at acidic pH. Hydrophobic peptides containing 50–75% hydrophobic residues may be insoluble or only partially soluble in aqueous solutions even if the sequence contains 25% charged residues. It is best to first dissolve these peptides in a minimal amount of stronger solvents such as DMF, acetonitrile, isopropyl alcohol, ethanol, acetic acid, 4–8 M guanidine HCl or urea, dimethyl sulfoxide (DMSO), and then slowly add (drop wise) the solution to a stirred aqueous buffer solution. Very hydrophobic peptides containing >75% hydrophobic residues will generally not dissolve in aqueous solutions. These peptides generally require initial solubilization in very strong solvents such as TFA and formic acid and may precipitate when added into an aqueous buffered solution. Peptide sequences containing a very high (>75%) proportion of Ser, Thr, Glu, Asp, Lys, Arg, His, Asn, Gln, or Tyr are capable of forming extensive intermolecular hydrogen bond networks and have a tendency to form gels in concentrated aqueous solutions. Sonication may increase solubility to an extent. Acetic acid (10%) in the solvent will help dissolve basic peptides, whereas 10% ammonium bicarbonate will help dissolve acidic peptides. For peptides with extremely low solubility in aqueous solutions, organic solvents (such as DMSO, isopropanol, methanol, and acetonitrile) should be used first. Once the peptides are completely dissolved, deionized water may be gradually added until the desired concentration is obtained. If peptide samples need to be frequently or periodically taken from a stock at −20°C, it is recommended to make a series of aliquots from the stock. Peptide sequences containing Cys, Met, or Trp are prone to air oxidation. It is recommended to purge the air out of the vial and replace it with a blanket of nitrogen or argon. Lyophilized peptides can be stored with drierite long term at −20°C.

1.5. Animal Immunization

A number of animals are suitable hosts for antipeptide antibody production, including mice, guinea pigs, rats, hamsters, rabbits, chickens, pigs, goats, sheep, bovines, donkeys, and horses (*see* **Note 7**). Selection of the appropriate animal species is dependent on several factors: the presence of a homologous protein in the species being immunized, the amount of antibody required, the amount of antigen available for immunization, the time required to obtain an antibody response, and the cost. Information on emulsion of adjuvant mixed with antigen, immunization routes, bleed, and antiserum preparation are widely available (1, 2). The most common host animal for polyclonal antipeptide antibody production is the New Zealand White rabbit (female, 8 weeks of age, used 95% of the time). It has the ability to respond to broad classes of antigens and can produce good antibodies in 77 days. For rabbit projects, the titers remain relatively level after the second booster and additional immunizations are used to maintain antibody titers rather than to increase them. Freund's adjuvant (*see* **Note 8**) is used in these rabbit projects. An adjuvant is a substance that serves to enhance the immune response against the antigen. Freund's adjuvant should be emulsified aseptically using syringes or sonication. An insoluble antigen is as good as a soluble one. In the case of rabbits, 20 mL of serum is collected prior to the initial immunization by injection at four separate subcutaneous (s.c.) sites (two inguinal, two axillary) of 0.25 mg peptide-KLH emulsified with Freund's complete adjuvant (FCA). A booster injection with 0.25 mg peptide-KLH emulsified with Freund's incomplete adjuvant (FIA) is given 14 days after the initial immunization. Subsequent booster injections with 0.25 mg peptide–KLH emulsified with Freund's incomplete adjuvant (FIA) are given every 4 weeks and 20 mL of serum is collected 10 days after each booster injection. At the conclusion of the animal immunization at day 77, a large-volume terminal bleed (90 mL) is collected by the exsanguination of the rabbit.

2. Materials

2.1. Peptide–Carrier Protein Conjugation

1. Phosphate buffer (0.01 M), pH 7.0. Prepare by using 0.01 M Na_2HPO_4 to adjust the pH of a 0.01 M solution of NaH_2PO_4 to pH 7.0.
2. Phosphate buffer (0.05 M), pH 6.0. Prepare by using 0.05 M Na_2HPO_4 to adjust the pH of a 0.05 M solution of NaH_2PO_4 to 6.0.
3. Dimethylformamide (DMF, Sigma, St. Louis, MO).
4. *m*-Maleimidobenzoic acid *N*-hydroxysuccinimide ester (MBS, Sigma, St. Louis, MO).
5. Keyhole limpet hemocyanin (KLH, Pierce, Rockford, IL).
6. Bovine serum albumin (BSA, Sigma, St. Louis, MO).
7. PD-10 column (Pharmacia Bioscience, Piscataway, NJ).

2.2. Antipeptide Antibody Titer Determination by ELISA

1. Carbonate buffer: 15 mM Na_2CO_3, 35 mM $NaHCO_3$, 0.02% NaN_3, adjust pH to 9.6 with 1 N NaOH.
2. Synthetic peptide (0.2–2.5 μM) in carbonate buffer.
3. Phosphate-buffered saline (PBS): 140 mM NaCl, 27 mM KCl, 7.2 mM Na_2HPO_4, 14.7 mM K_2HPO_4, pH 7.2.
4. PBS containing 0.05% Tween-20 (PBST).
5. Blocking solution: 10 mg/mL BSA in PBST.
6. Secondary antibody: Goat anti-rabbit globulin conjugated to alkaline phosphatase (Sigma, St. Louis, MO).
7. Enzyme substrate: 1 mg/mL *p*-nitrophenyl phosphate (Sigma, St. Louis, MO), 0.2 M Tris buffer, 5 mM $MgCl_2$.
8. Stopping solution: 0.01 M ethylenediaminetetraacetic acid or 3N NaOH.
9. Immulon 2 or equivalent 96-well flat bottom microtiter plate (Dynetech Laboratory, Chantilly, VA; BD, Franklin Lakes, NJ).
10. Microtiter plate reader (Dynetech Laboratory, Chantilly, VA): spectrophotometer with 405-nm filter.

2.3. Peptide Affinity Purification

1. CNBr-activated Sepharose 4B (Pharmacia Bioscience, Piscataway, NJ).
2. Eppendorf Centrifuge 5804R (Brinkmann Instrument, Westbury, NY).

3. 1 mM HCl.
4. Synthetic peptide powder.
3. Coupling buffer: 0.1 M $NaHCO_3$, pH 8.3, 0.5 M NaCl.
4. Washing buffer 1: 50 mM Tris–HCl, pH 8.0, 0.1% Triton X-100, 0.5 M NaCl.
5. Washing buffer 2: 50 mM Tris–HCl, pH 9.0, 0.1% Triton X-100, 0.5 M NaCl.
6. Washing buffer 3: 50 mM sodium phosphate, pH 6.3, 0.1% Triton X-100, 0.5 M NaCl. Prepare by using 50 mM Na_2HPO_4 containing 0.1% Triton X-100 and 0.5 M NaCl to adjust the pH of a 50 mM solution of NaH_2PO_4 containing 0.1% Triton X-100 and 0.5 M NaCl to 6.3.
7. Elution buffer: 50 mM glycine–HCl, pH 2.5, 0.1% Triton X-100, 0.15 M NaCl.
8. Phosphate-buffered saline (PBS): 140 mM NaCl, 27 mM KCl, 7.2 mM Na_2HPO_4, 14.7 mM K_2HPO_4, pH 7.2.
9. Chromatography column (1 cm ×10 cm).
10. Syringe filter (0.2 μm) (Whatman, Piscateway, NJ).

3. Methods

3.1. Peptide–Carrier Protein Conjugation

Molecules smaller than 5–15 kDa may not elicit any significant immune responses. To stimulate antibody responses to smaller peptides, peptides need to be covalently conjugated to a larger immunogenic carrier protein (KLH, BSA, etc.) prior to immunization (1–2, 5–6, 13). Poor conjugation of the peptide to the carrier protein is one of the reasons why peptides fail to induce antibody production. It is critical that the peptide to carrier protein molar ratio be very high and that all epitopes on the peptides be properly oriented in order to induce a high-titer-specific immune response. Several popular coupling methods which couple through sulfhydryl, amino, carboxyl, or hydroxyl groups are available. However, most peptides contain several side chains such as amino, carboxyl, and hydroxyl groups resulting in a multipoint attachment. It is preferable to attach a carrier protein through a sulfhydryl group present in a cysteine residue at either the N or C terminus of a peptide. The procedure for coupling peptides to KLH or BSA through a cysteine is as follows (13):

1. Dissolve 5 mg of KLH or BSA in 0.5 mL of 0.01 M phosphate buffer, pH 7.
2. Dissolve 3 mg of MBS in 200 μL DMF.

3. Add 70 μL of MBS solution to 0.5 mL of KLH or BSA solution and stir or rotate for 30 min at room temperature. Add 2 mL of 0.05 M phosphate buffer, pH 6.
4. Equilibrate a PD-10 column using approximately 25 mL of 0.05 M phosphate buffer, pH 6. Add 2.5 mL of the MBS/KLH or BSA/MBS solution to the column and elute with 3.5 mL of 0.05 M phosphate buffer, pH 6. Add 0.5 mL of deionized water to the 3.5 mL of purified KLH/MBS or BSA/MBS.
5. Dissolve 5 mg of peptide in 100 μL of DMF. Rapidly add 1 mL of purified KLH/MBS or BSA/MBS. Shake rapidly and immediately add 11 μL of 2N NaOH.
6. Check the pH with pH paper. It should be 7.0–7.2. Too high a pH or too a low pH will stop the reaction between KLH/MBS or BSA/MBS and peptide. If needed, immediately add an appropriate amount of 0.5N HCl or 2N NaOH to change the pH.
7. Stir or rotate the solution for 3 h or overnight at 4°C. Finally, add 3 mL of ammonium bicarbonate (0.1 M) before lyophilizing the coupled peptide.

3.2. Antipeptide Antibody Titer Determination by ELISA

An antibody titer is defined as the highest antibody dilution that still yields a positive reactivity of a particular epitope in an assay system such as ELISA. This value gives an indication of the concentration of an antibody preparation. This section describes an assay used to determine the titer of antipeptide antibodies in serum raised against a peptide or a protein containing the peptide sequence.

1. Coat the wells of a mictrotiter plate with 300 μL of 0.2–2.5 μM synthetic peptide, leaving wells at the end as blanks. Incubate overnight at 4°C (*see* **Note 9**).
2. Discard the unbound synthetic peptide.
3. Wash the wells three times with PBST.
4. Block the unoccupied sites with 300 μL/well of blocking solution.
3. Wash the wells three times with PBST.
4. Prepare serial dilutions of antiserum with PBST ranging from 1:300 to 1:300,000.
5. Add the serial dilutions of the antiserum to the wells and incubate for 2 h at 37°C.
6. Wash the wells three times with PBST.
7. Dilute the secondary antibody 1:7,000 with PBST.
8. Add the secondary antibody to wells and incubate at 37°C for 2 h.

9. Wash the wells three times with PBST.
10. Add 50 μL enzyme substrate. Incubate for 10–30 min at 37°C.
11. Terminate color development by addition of 100 μL of stopping solution.
12. Measure absorbance at 405 nm with a microtiter plate reader. The titer corresponds to the highest dilution that still yields a positive reading.

3.3. Peptide Affinity Purification

Among many antibody purification methods (1–6), peptide affinity purification is the most effective technique to purify the antipeptide antibody. The peptide affinity purification is used for isolating those antibodies that recognize a specific epitope with about the same specificity as that of monoclonal antibodies.

1. Resuspend 1 g dried CNBr-activated Sepharose 4B in 50 mL of 1 mM HCl for 30 min.
2. Centrifuge for 5 min at 1,000×*g* and discard the supernatant.
3. Wash the CNBr-activated Sepharose 4B by resuspending the resin in 50 mL of 1 mM HCl and after 15 min spin at 1,000×*g* discarding the supernatant. Repeat this process twice.
4. Dissolve 10 mg of synthetic peptide in 5 mL of coupling buffer.
5. Mix the synthetic peptide solution with the swollen gel. Stir gently for 1 h.
6. Centrifuge for 5 min at 1,000×*g* and discard the supernatant.
7. Wash excess synthetic peptide with 20 mL coupling buffer. Centrifuge for 5 min at 1,000×*g* and discard the supernatant.
8. Block remaining active groups by transferring the resin to 0.1 M Tris–HCl, pH 8.0, stand for 2 h.
9. Wash the resin with 0.1 M acetate buffer containing 0.5 M NaCl, pH 4.0. Centrifuge for 5 min at 1,000×*g* and discard the supernatant.
10. Wash the resin with 0.1 M Tris–HCl pH 8.0 containing 0.5 M NaCl. Centrifuge for 5 min at 1,000×*g* and discard the supernatant.
11. Transfer the resin into PBS.
12. Pack the peptide affinity column by pouring the resin into a vertically held column.
13. Wash the column with 100 bed volumes of PBS.

14. Filter 15 mL rabbit serum through a 0.2-μm filter.
15. Dilute the serum with PBS to 50 mL.
16. Load the filtered serum onto the peptide affinity column.
17. Wash the column with 20 mL PBS.
18. Wash the column with 20 mL washing buffer 1.
19. Wash the column with 20 mL washing buffer 2.
20. Wash the column with 20 mL washing buffer 3.
21. Elute the antibodies from the column with 20 mL elution buffer and collect in a tube containing 4 mL of 1 M Tris–HCl, pH 9.0.
22. Wash the column with 20 mL PBS (*see* **Note 10**).
23. Use multiple PD-10 columns to desalt the antibodies by loading 2.5 mL antibody per column and eluting with 3.5 mL PBS as the desalting buffer (*see* **Note 11**).
24. Calculate the antibody concentration (mg/mL) by A_{280} (the absorbance reading at 280 nm for a 1 mL solution) × 0.7.
25. Measure the titer of the purified antibody with the synthetic peptide using ELISA (*see* **Note 12**).

4. Notes

1. Although monoclonal antibodies with their high specificity can be the antibody type of choice, they generally have a lower affinity, narrower utility, longer response time, and are more expensive than polyclonal antibodies.
2. Although pure peptides are always better, it is not a requirement to have >90% pure peptides for antibody production. Even ~70% pure peptides can be used to generate antibodies successfully. However, if affinity purification is required, the more pure the antigen that is coupled to the peptide affinity column, the higher the specificity of the purified antibodies.
3. In order to improve the chances of producing high-titer antibodies, using several peptide sequences from a protein as antigens in a co-immunization protocol is beneficial. Individual KLH conjugates are mixed in equimolar ratios and used as immunogen. Having several peptides derived from a single protein will mathematically increase the chances of obtaining antibodies that will recognize the target protein. Peptide affinity columns can be used to isolate an antibody against each peptide.

4. The generation of a phophospecific antibody includes the synthesis of phosphorylated and non-phosphorylated peptides, immunization with the phosphorylated peptide, and fabrication of phosphorylated and non-phosphorylated peptide affinity columns for successive purification. The serum undergoes two rounds of purification. The first is the purification of antibodies that recognize the phosphopeptide, followed by a depletion step whereby the antibodies that bind to the non-phosphorylated peptide are removed. The flow-through of the non-phosphopeptide column is repurified by the phosphopeptide column. To get the anti-non-phosphopeptide antibody, the serum should be applied to the non-phosphopeptide column first, and then remove the non-specific antibodies by the phosphopeptide column. Co-immunization with either different length of peptides in the case of a known phosphorylation site or peptides containing predicted phosphorylation sites in cases of unknown phosphorylation sites may be used to increase the chance of success.
5. Proteins such as thyroglobulin, rabbit serum albumin, bovine serum albumin, ovalbumin, and human gamma globulin are all very effective as carrier proteins. The resultant peptide–carrier protein complex is able to stimulate the immune system to produce antibodies against both the peptide and the carrier proteins. Thus it is important that ELISA analysis be performed using peptide or peptide conjugated to a different carrier protein. Finally, there is some evidence that antibodies can also be produced that react to the cross-linker used to couple the peptide to the carrier proteins. There is an alternative multiple antigenic peptide (MAP peptide, juxtaposition of four or eight peptide molecules on a cross-linked lysine core) method (14–15) for preparing peptide antigens, its advantage being that the conjugation step is not necessary. However, MAP is not a better method than KLH and it can bypass the immune response system in some hosts.
6. All reagents are of highest purity to ensure superb quality in synthesis. Water content is minimized in solvents and all containers are purged with nitrogen and sealed to ensure the longest possible shelf life.
7. There is little need to prescreen animals prior to immunization in most instances, but prescreening can be extremely important in studies of certain organisms such as yeast. All animals used for antibody production are certified specific pathogen-free, which provides a cleaner basis for antibody production. However, animals may already have generated closely related antibodies to something in their

environment or their feed which may be similar to the antigen of choice. It is therefore a good idea to immunize more than one animal using a standard protocol. At the conclusion, the animal(s) with the best antibody response are continued on extended protocols.

8. Freund's complete adjuvant (FCA) should be used for priming (first immunization) only. Freund's incomplete adjuvant (FIA) should be used for subsequent immunizations to prevent lesions at the sites of injection. Data shows positive and negative aspects about using FCA which was developed in the 1930s. It contains killed mycobacteria tuberculosis, paraffin oil, and mannide monoosleate and elicits a delayed hypersensitivity reaction. The water-in-oil emulsion using FCA is stable and provides a slow release of antigen and protects the antigen from degradation. The drawback with FCA is that it may cause granulomas and inflammation at the injection site with an intradermal injection. Note that one should avoid using FCA for studies of *Mycobacterium tuberculosis.* Though FCA has been a mainstay and shown consistently superior results than do alternative adjuvants in antibody production, some animal care and use committees reject the use of FCA due to its toxicity to the host animal. Therefore, other adjuvants such as Ribi, TiterMax, and Adjuvax (1) should be use in this case.
9. Most peptides can be coated on ELISA plates using carbonate buffer at pH 9.6. If the peptide does not adsorb completely, try other buffers in the pH range of 4–8. The coupling of peptides to BSA to facilitate coating is usually not necessary.
10. A typical peptide affinity purification of 20 mL of serum yields approximately 2 mg of specific antibody. Although the capacity does decrease slightly with each use, the column can often be used for many additional purifications if stored at 4°C with sodium azide.
11. For short-term storage of antibodies, 4°C is recommended. For longer term storage, −20°C or −80°C is recommended. It is important, however, to avoid repeated cycles of freezing and thawing, as this will lead to partial denaturation of the antibodies. Sodium azide is an antimicrobial agent that prevents the growth of bacteria in the serum or purified antibody. Antibody should be stored in ready-to-use aliquots, thawing them only when they are needed. The addition of 50% glycerol to the PBS is preferred, but not necessary.
12. In cases where an antibody does not work in a particular assay, likely explanations are the following: (I) the peptide

sequence corresponds to a nonexposed region of the native protein, (II) the protein's conformation in the peptide region differs enough that the antibody has trouble recognizing the native protein, and (III) the target protein is not present in the sample. It is not uncommon to see multiple bands in protein Western blots even when affinity-purified antibody is used. This is not indicative of a problem with the antibody's specificity. Rather, this typically occurs for one of the following reasons: (I) the antipeptide antibody recognizes a homologous protein in the sample that shares one or more epitopes with the peptide sequence, (II) the protein has a different molecular weight than previously predicted, (III) the antibody recognizes either cleaved fragments of the protein at lower molecular weights or aggregated multimers of the native protein at higher molecular weights. The advantage of using a Western blot (native or sodium dodecyl sulfate polyacrylamide gel electrophoresis; 1D or 2D) (2, 3, 16, 17) to evaluate the antibody over the ELISA method is that proteins are separated and probed by antibody individually. The downside is that Western blotting is a labor-intensive technique.

Acknowledgments

We thank the support of the Research Resources Center at the University of Illinois at Chicago.

References

1. Howard, G. C. and Kaser, M. R., eds. (2007) *Making and Using Antibodies: A Practical Handbook.* CRC Press, Taylor & Francis, Boca Raton, FL.
2. Harlow, D. L. (1988) *Antibodies: A Laboratory Manual.* Cold Spring Harbor Laboratory, New York.
3. Rosenberg, I. M. (1996) *Protein Analysis and Purification Benchtop Techniques.* Birkhauser, Boston, MA.
4. Walker, J. M., ed. (1996) *The Protein Protocols Handbook.* Humana Press, Totowa, NJ.
5. Coligan, J. E., Kruisbeek, A. M., Margulies, D. H., Shevach, E. M., and Strober, W. (1996) *Current Protocols in Immunology.* John Wiley & Sons, New York, vol. 2, 9.0.1–9.8.15.
6. Delves, P. J. (1997) *Antibody Production Essential Techniques.* John Wiley & Sons, New York.
7. Tsigelny, I. F., ed. (2002) *Protein Structure Prediction: Bioinformatic Approach.* International University Line, San Diego, CA.
8. Kolaskar, A. S. and Tongaonkar, P. C. (1990). A semi-empirical method for prediction of antigenic determinants on protein antigens. *FEBS Lett.* **276**, 172–174.
9. Benoiton, N. L., ed. (2006) *Chemistry of Peptide Synthesis.* Taylor & Francis, New York.
10. Howl, J., ed. (2005) *Peptide Synthesis and Applications.* Humana Press, Totowa, NJ.
11. Kates, S. A. and Albericio, F. (2000) *Solid Phase Synthesis: A Practical Guide.* Marcel Dekker, New York.

12. Fields, G. B. (1997) *Solid-Phase Peptide Synthesis*. Academic Press, New York.
13. Lateef, S., Gupta, S., Jayathilaka, G., Krishnanchettiar, S., and Lee, B. S. (2007) An improved protocol of coupling synthetic peptides to KLH for antibody production using MBS as bifunctional linker. *J. Biomol. Tech.* **18**, 173–176.
14. Posnett, D., McGrath, H., and Tam, J. P. (1988). A novel method for producing anti-peptide antibodies. *J. Biol. Chem.* **263**, 285–288.
15. Tam. J. P. (1988) Synthetic peptide vaccine design: synthesis and properties of a high-density multiple antigenic peptide system. *PNAS USA* **85**, 5409–5413.
16. Westermeier, R. (2001) *Electrophoresis in Practice*. Wiley-VCH, Weinheim, Germany.
17. Hames, B. D. (1998) *Electrophoresis of Protein*. Oxford University Press, Oxford.

Chapter 8

Preparation of Colloidal Gold Particles and Conjugation to Protein A, IgG, F(ab')$_2$, and Streptavidin

Sadaki Yokota

Abstract

Colloidal gold probes, including protein A-, IgG-F(ab')$_2$-, and streptavidin-labeled gold particles, are useful tools for localization of antigens in cells and tissues by immunoelectron microscopy (IEM). This chapter describes different methods for the preparation of colloidal gold and conjugation of colloidal gold to protein A, IgG, and streptavidin.

Key words: Preparation of colloidal gold, sizing of gold particles, protein A-gold, IgG-gold, streptavidin-gold.

1. Introduction

Colloidal gold is an electron-dense spherical particle of 2–40 nm in diameter that can be conjugated to proteins and is clearly recognizable in the electron microscope. These properties can be exploited to localize target molecules in thin sections of cells and tissues by immunoelectron microscopy (IEM). IEM uses colloidal gold probes conjugated with antibodies or protein A/G to detect antigens in thin plastic (1) or frozen sections (2). In addition, colloidal gold particles of various sizes can be separately produced to allow labeling of two or more different target molecules in the same section. Finally, gold particles are easily counted in electron microscopic images, enabling accurate and useful quantitative analysis of the labeling (1).

Preparing gold probes for IEM essentially involves two steps: preparation of colloidal gold particles of specific sizes and

S.D. Schwartzbach, T. Osafune (eds.), *Immunoelectron Microscopy*, Methods in Molecular Biology 657,
DOI 10.1007/978-1-60761-783-9_8, © Springer Science+Business Media, LLC 2010

conjugation of the gold particles to proteins such as protein A/G, IgG, and streptavidin. In the first step, monodispersed colloidal gold is prepared by reduction of tetrachloroauric acid. The size of the colloidal gold particles produced is determined by the reducing reagent used and its concentration (3). In the second step, colloidal gold particles of uniform size are conjugated to primary antibodies with a specific binding affinity for the target molecule or to protein A/G, streptavidin, or a secondary antibody that has a specific affinity for the primary antibody recognizing the target. Although a range of pre-prepared immunogold probes are commercially available, they can be easily produced at low cost in the ordinary cell biology laboratory. This chapter first details the preparation of colloidal gold particles of various sizes, including handling of the glass reaction vessels used, the general properties of colloidal gold particles, and how to check the size of gold particles prepared. Methods for conjugating the colloidal gold particles to proteins, including protein A, IgG, $F(ab')_2$, and streptavidin, are then described with a brief review of the properties of these proteins. Finally, the chapter will note methods for titrating the minimum amount of protein needed to stabilize colloidal gold preparations and for removing non-specific proteins from the colloidal gold probes before use.

2. Materials

2.1. Colloidal Gold Preparation

1. Double-distilled water (DDW) is used for all solutions and procedures.
2. 4% gold(III) chloride trihydrate (Sigma-Aldrich, St. Louis, MO): Prepare a 4% stock solution and store at 4°C (*see* **Note 1**). Dilute to final concentration just before use.
3. 0.2 M potassium carbonate: Store at room temperature (RT).
4. 1 M sodium thiocyanate. Harmful irritant. Store at RT.
5. 1 M sodium borohydride. Highly flammable, toxic. Store at RT.
6. 2.2 mM and 38.8 mM trisodium citrate: Store at RT.
7. Siliconized flasks (*see* **Note 2**). In a hood with gloves and safety glasses, soak the inner surface of the glassware with 5% dimethyldichlorosilane in heptane until the heptane evaporates. Wash the glassware with DDW and use only for colloidal gold preparation.
8. Magnetic stirring plate and stir bars.

2.2. Titration to Determine Minimum Amount of the Protein to be Conjugated Needed to Stabilize the Colloidal Gold

1. 10% NaCl. Store at RT.
2. 0.1 M acetic acid. Store at RT.

2.3. Preparation of Protein A-Colloidal Gold Probes

1. Protein A (GE Healthcare Bio-Sciences Corp. Piscataway, NJ). Store at −20°C.
2. BSA, bovine serum albumin (Sigma-Aldrich, St. Louis, MO). Store at −20°C.
3. 50% glycerol. Store at RT.
4. Phosphate-buffered saline (PBS): Prepare a 10× stock solution by mixing 80 g NaCl, 2 g KCl, 11.5 g $Na_2HPO_4{\cdot}7H_2O$, 2 g KH_2PO_4, and DDW for a final concentration of 1 L. Autoclave and store at 4°C. Dilute 10-fold to use.

2.4. Conjugation of IgG or F(ab')$_2$ with Colloidal Gold

1. 2 mM borate–HCl buffer (pH 9.0). Prepare a 12.5 mM stock solution by mixing 4.6 mL of 0.1 M HCl, 50 mL of 0.025 M sodium borate, and add DDW for a final volume of 100 mL. Autoclave and store at RT. Dilute 6.25-fold to use.
2. Millipore filters with 0.45-μm pore size (Millipore Corporation, Billerica, MA).
3. 20 mM Tris–HCl, pH 8.2. Prepare a 100 mL 10× stock solution of 0.2 M Tris–HCl, pH 8.2, autoclave and store at 4°C. Dilute 10-fold to use.
4. Triton X-100 (Sigma-Aldrich, St. Louis, MO). Harmful irritant. Store at RT.
5. Polyethylene glycol 20,000 (Sigma-Aldrich, St. Louis, MO). Store at RT.
6. Polyvinylpyrrolidone (M.W. 10,000) (Sigma-Aldrich, St. Louis, MO). Store at RT.
7. 5 mL ultraclear centrifuge tubes (Beckman Coulter, Fullerton, CA).
8. SW50.1 ultracentrifuge rotor (Beckman Coulter, Fullerton, CA).
9. Ultracentrifuge (Beckman Coulter, Fullerton, CA).
10. Affinity-purified IgG (MP Biochemicals, Solon, OH).
11. Affinity-purified F(ab')$_2$(Abcam, Cambridge, MA).

2.5. Conjugation of Streptavidin with Colloidal Gold

1. Streptavidin (Sigma-Aldrich, St. Louis, MO). Store at −20°C.
2. 1 M sodium bicarbonate. Store at RT.
3. Polyethylene glycol 6,000 (Sigma-Aldrich, St. Louis, MO). Store at RT.
4. Sodium azide. Highly toxic and corrosive. Hazardous material for the environment. Store at RT.

3. Methods

3.1. Colloidal Gold Preparation

3.1.1. Preparation of Colloidal Gold with a Particle Size of 2–3 nm in Diameter

1. Put 50 mL of 1% gold(III) chloride trihydrate into a clean 200 mL siliconized flask (4).
2. Add 0.75 mL of 0.2 M K_2CO_3 to this flask.
3. Stir the mixture on a magnetic stirrer while quickly adding 0.3 mL of 1 M sodium thiocyanate.
4. Allow the mixture to stand in the dark without stirring for 12–15 h. The final color of the colloidal gold solution should be yellow (*see* **Note 2**).
5. Check the size of the colloidal gold particles with an electron microscope (*see* **Section 3.1.7**)

3.1.2. Preparation of Colloidal Gold with a Particle Size of 4–5 nm in Diameter

1. Add 0.5 mL of 1% gold(III) chloride trihydrate solution to 50 mL of DDW in a 100 mL siliconized flask (5).
2. Stir this solution with a magnetic stirrer and quickly add 0.3 mL of 1 M sodium borohydride.
3. The color of the mixture should change immediately to reddish brown then to a dark red wine color within 20 min.
4. Keep the mixture overnight at 4°C.
5. Check the size of the colloidal gold particles with an electron microscope (*see* **Section 3.1.7**).

3.1.3. Preparation of Colloidal Gold with a Particle Size of 8.5 nm in Diameter

1. Heat 106 mL of 2.2 mM trisodium citrate in a 200 mL flask until boiling (6).
2. Immediately add 1 mL of 24.3 mM gold(III) chloride trihydrate and stir.
3. Continue to boil the mixture with stirring for 15 min.
4. The color should change to yellowish red during heating
5. Maintain the starting volume of the mixture by addition of DDW as necessary during boiling.
6. Cool the mixture in an ice bath.

7. Check the size of the colloidal gold particles with an electron microscope (*see* **Section 3.1.7**).

3.1.4. Preparation of Colloidal Gold with a Particle Size of 12 nm in Diameter

1. Heat 243 mL of 0.3 mM gold(III) chloride trihydrate in a clean 400 mL siliconized flask until boiling (6).
2. With vigorous stirring, quickly add 8.5 mL of 38.8 mM trisodium citrate to the boiling solution.
3. Continue to boil for 15 min.
4. Maintain the volume of the mixture with DDW as necessary.
5. The color should change from dark blue to wine red during boiling. Cool the mixture in an ice bath and check the size of the colloidal gold particles with an electron microscope (*see* **Section 3.1.7**).

3.1.5. Preparation of Colloidal Gold with a Particle Size of 15 nm in Diameter

1. Heat 243 mL of 0.3 mM gold(III) chloride trihydrate in a clean 400 mL siliconized flask until boiling (6).
2. While stirring vigorously, quickly add 7.5 mL of 38.8 mM trisodium citrate.
3. Continue to boil for 15 min.
4. Maintain the volume of the mixture with DDW as necessary.
5. The color should change from blue to wine red during boiling.
6. Cool the mixture in an ice bath and check the size of the colloidal gold particles with an electron microscope (*see* **Section 3.1.7**).

*3.1.6. Preparation of Colloidal Gold with a Particle Size of 15–150 nm in Diameter (see **Note 3**)*

1. Heat 50 mL of 0.01% gold(III) chloride trihydrate solution in a 100 mL flask until boiling (3)
2. While boiling, quickly add the volume (0.16–1 mL) of 1% trisodium citrate solution (**Table 8.1**) required to produce the desired size colloidal gold particles. The color of the boiling mixture should turn faint blue within 25 s and then 70 s later suddenly change to brilliant red.
3. Continue to boil for a further 5 min until reduction of the gold chloride is complete as indicated by the solution becoming the color shown in **Table 8.1**.
4. Cool the mixture in an ice bath and check the size of the colloidal gold particles with an electron microscope (*see* **Section 3.1.7**)

3.1.7. Determining the Size of Colloidal Gold Particles

1. Float a Formvar-coated nickel grid on a drop of 10% BSA–PBS solution for 10 min to coat the membrane with BSA.

Table 8.1
The amount of trisodium citrate to add to 50 mL of 0.01% gold(III) chloride trihydrate to obtain colloidal gold particles of the specified size[a]

Diameter of colloidal gold particles (nm)	Amount (mL) of 1% trisodium citrate	Color of colloidal gold solution
16.0	1.00	Orange
24.5	0.75	Red
41.0	0.50	Red
71.5	0.30	Dark red
97.5	0.21	Violet
147.0	0.16	Violet

[a]Modified from Frens (7)

2. Wash the grid with a jet of DDW for several seconds.
3. Float the grid on a drop of the colloidal gold solution for 5 min. The gold particles will bind the BSA coating the membrane.
4. Wash the grid with a jet of DDW for several seconds and air dry.
5. Take electron micrographs at ×80,000 magnification and enlarge them photographically or digitally to ×400,000.
6. Measure the diameter of the gold particles and calculate the actual size of the colloidal gold particles. Electron microscopic images of colloidal gold particles of different sizes are shown in **Fig. 8.1**.

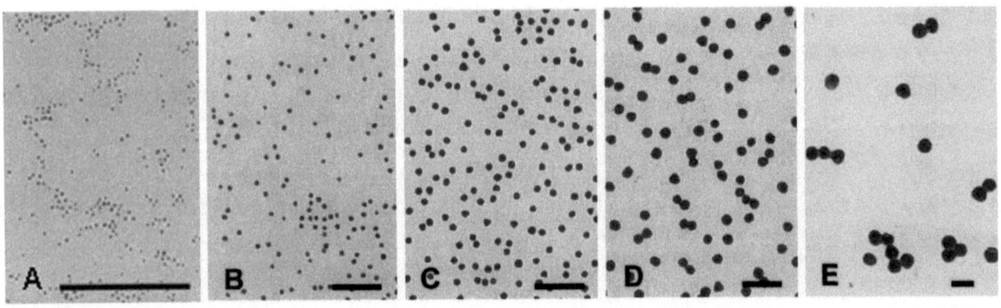

Fig. 8.1. Electron microscopic images of gold particles. (**A**) 2–3-nm gold particles. (**B**) 4–5-nm gold particles. (**C**) 8.5-nm gold particles. (**D**) 15-nm gold particles. (**E**) 28-nm gold particles. Bar = 50 nm.

3.2. Titration to Determine Minimum Amount of the Protein to be Conjugated Needed to Stabilize the Colloidal Gold

1. Pipette 100 μL of DDW containing 0, 0.5, 1, 2, 3, or 4 μg of the protein to be conjugated into six test tubes, respectively.
2. Adjust the pH of the colloidal gold solution to pH 5.2 using 0.1 M acetic acid. Measure the pH by pipetting a drop of the solution (2–3 μL) onto a piece of litmus paper (*see* **Note 4**).
3. Into each tube, pipette 0.5 mL of colloidal gold solution. After 10 min at RT, pipette 0.5 mL of 10% NaCl into each tube with agitation.
4. The tube containing the minimum amount of protein to be conjugated required to stabilize 0.5 mL colloidal gold solution is the tube containing the least protein and no bluish black precipitate after the addition of NaCl.

3.3. Preparation of Protein A-Colloidal Gold Probes

3.3.1. Conjugation of Protein A to Colloidal Gold (see Notes 5 and 6)

1. Place 20 mL of the colloidal gold solution you prepared into a beaker and adjust the pH to 5.2 using 0.1 M acetic acid. Measure the pH by pipetting a drop of the solution (2–3 μL) onto a piece of litmus paper.
2. Prepare a solution of 1 mg/mL protein A in distilled water.
3. Based on the least amount of protein A needed to stabilize 0.5 mL colloidal gold (determined as described in **Section 3.2**), pipette 10% more than this amount of protein A into the beaker containing 20 mL colloidal gold solution you prepared (*see* **Note 7**).
4. Allow to stand for 10 min.
5. Add 1 mL of 10% BSA to the gold solution to further stabilize the gold conjugate.
6. Using a fixed angle rotor, centrifuge the gold-protein A conjugate at the following settings for 30 min: 15,000×*g* for 15-nm gold conjugate and larger, 35,000×*g* for 8-nm gold conjugate, and 100,000×*g* for 2-3-nm gold conjugate. Aggregated protein A-gold conjugate forms a tight pellet on the side of the tube while the non-aggregated conjugate pools loosely on the bottom of the tube.
7. Slowly remove the supernatant from the centrifuge tube using a capillary tube connected to an aspirator. Remove as much of the non-bound protein A containing supernatant as you can being careful to avoid disturbing the conjugated protein A pooled at the bottom of the tube.
8. Collect the dark red mass of gold-protein A conjugate pooled at the bottom of tube by resuspending it in 50% glycerin–PBS (*see* **Note 8**).

3.3.2. Removal of Non-conjugated Protein A (see ***Note 9****)*

1. In a 5 mL ultraclear centrifuge tube, overlayer a 1 mL 50% glycerol–PBS cushion with 3.5 mL of 10% glycerol–PBS.
2. Layer 0.4 mL of the non-aggregated protein A colloidal gold conjugate onto the top of the tube.
3. Centrifuge in a Beckman SW50.1 rotor for 30 min at 15,000×*g* for 15-nm gold conjugate and larger, 35,000×*g* for 8-nm gold conjugate, and 100,000×*g* for 2–3-nm gold conjugate.
4. Using an aspirator, remove the supernatant leaving the 50% glycerol cushion containing the colloidal gold-conjugated protein A. Collect the colloidal gold-conjugated protein A and mix it with 50% glycerol-PBS.
5. Add 10% BSA to the protein A-gold conjugate to a final concentration of 0.5%.
6. Store the protein A-gold conjugate at −20°C until use (*see* **Note 9**).

3.4. Conjugation of Affinity-Purified IgG and F(ab')$_2$ with Colloidal Gold

1. Dialyze IgG or F(ab')$_2$ solution (1 mg/mL) against 2 mM borate–HCl buffer (pH 9.0). Remove precipitates by centrifugation at 10,000×*g* for 1 h (7, 8).
2. Place 1 mL of the colloidal gold solution you prepared into a beaker. Adjust the pH to 9.0 with 0.2 M K_2CO_3 just before use.
3. Based on the least amount of IgG or F(ab')$_2$ needed to stabilize 0.5 mL colloidal gold (determined as described in **Section 3.2**), pipette 10% more than this amount of IgG or F(ab')$_2$ into the beaker containing 1 mL colloidal gold solution you prepared (*see* **Note 7**). Allow to stand for 10 min.
4. Add BSA to the mixture to a concentration of 1% BSA.
5. Filter the mixture through a 0.45-μm Millipore filter.
6. Using a fixed angle rotor, centrifuge the gold–IgG or gold–F(ab')$_2$ conjugate at the following settings for 30 min: 15,000×*g* for 15-nm gold conjugate and larger, 35,000×*g* for 8-nm gold conjugate, and 100,000×*g* for 2–3-nm gold conjugate. Aggregated IgG and F(ab')$_2$–gold conjugates form a tight pellet on the side of the tube while the non-aggregated conjugates pools loosely on the bottom of the tube.
7. Slowly remove the supernatant from the centrifuge tube using a capillary tube connected to an aspirator. Remove as much of the non-bound IgG or F(ab')$_2$ containing supernatant as you can, being careful to avoid disturbing the conjugated IgG or F(ab')$_2$ pooled at the bottom of the tube (*see* **Note 11**).

8. Collect non-aggregated conjugate at the bottom of the tube, mix with 0.5 mL 20 mM Tris–HCl buffer (pH 8.2) containing 1% BSA, and re-centrifuge as before. Repeat twice.
9. Mix conjugate with a small amount of 20 mM Tris–HCl (pH 8.2), 40% glycerol, 1% BSA, 0.01% Triton X-100. Store the conjugate at −20°C (*see* **Note 12**).

3.5. Conjugation of Streptavidin with Colloidal Gold

1. Dissolve 500 μg streptavidin in 0.5 mL of 1 mM phosphate buffer (pH 7.4) (9).
2. Add 200 μL of 1 M $NaHCO_3$ and the streptavidin solution to 20 mL of the colloidal gold solution you prepared.
3. After 10 min of stirring, add 200 μL of 2% polyethylene glycol 6,000.
4. Using a fixed angle rotor, centrifuge the gold–streptavidin conjugate at the following settings for 30 min: 15,000×*g* for 15-nm gold conjugate and larger, 35,000×*g* for 8-nm gold conjugate, and 100,000×*g* for 2–3-nm gold conjugate. Aggregated streptavidin-gold conjugate forms a tight pellet on the side of the tube while the non-aggregated conjugate pools loosely on the bottom of the tube.
5. Slowly remove the supernatant from the centrifuge tube using a capillary tube connected to an aspirator. Remove as much of the non-bound streptavidin containing supernatant as you can being careful to avoid disturbing the conjugated streptavidin pooled at the bottom of the tube.
6. Suspend the non-aggregated conjugate at the bottom of the tube in 0.1 M phosphate buffer (pH 7.4), 0.02% polyethylene glycol 6,000, 0.05% NaN_3. Store at 4°C.

4. Notes

1. The concentration of gold(III) chloride trihydrate solutions can be measured by absorption at 290 nm. The absorption coefficient for a 0.3 mM gold(III) chloride trihydrate solution is 0.878 ± 0.009. Using different reducing agents on gold(III) chloride trihydrate produces varying sizes of gold colloid.
2. Aggregation and precipitation of the colloidal gold during preparation can be avoided by siliconizing the glassware used. The color change of the colloidal gold solution must be carefully watched during the reaction. When the reduction is completed, the color of the solution is the color indicated in the text.

3. Gold particles with a diameter >30 nm are unstable even after protein conjugation and therefore should not be stored for long periods. In addition, large-size gold probes are dissociated relatively easily from sections during the IEM procedure (10).
4. Standard electrodes may be damaged by colloidal gold; therefore litmus paper is used for measuring the pH of colloidal gold solutions. Colloidal gold of smaller particle size requires more protein to stabilize it than those preparations containing larger colloids.
5. Colloidal gold is a negatively charged suspension and its stability is maintained by static repelling forces. Therefore, the suspension is unstable and precipitates immediately in the presence of electrolyte. However, when macromolecules such as proteins are absorbed onto the surface of colloidal gold particles by electrostatic van-der-Waals forces, a stable complex is formed that is no longer precipitated in solutions containing electrolytes. Proteins absorbed by gold particles do not lose their enzymatic activity or ligand affinity making them useful for detecting enzyme substrate reactions and receptor ligand interactions.
6. Protein G and protein L are conjugated to colloidal gold by the same method.
7. For example, if the minimum amount of protein A needed is 2 μg/0.5 mL colloidal gold, 80 μg of protein A is needed to saturate 20 mL of colloidal gold. The 10% extra thus requires a total of 88 μg.
8. The color of the colloidal gold conjugate prepared using thiocyanate is dark yellow.
9. When the protein A-gold probes are contaminated with even small amounts of free protein A, the binding of protein A-gold probes to IgG is greatly decreased due to the binding of free protein A to IgG. The most effective way to remove free protein A is the discontinuous glycerol gradient centrifugation method. Repeating the differential centrifugation twice can also be used to eliminate free protein A.
10. The protein A-gold conjugates retain its biological activity for 5 years under these storage conditions.
11. After centrifugation, excess medium attached to the tube wall is wiped off with tissue paper. This almost completely avoids contamination with free IgG or $F(ab')_2$.

12. Under these storage conditions, the conjugates retain binding activity for 1 year. During storage, the conjugates slowly precipitate and should be centrifuged to remove the aggregated conjugates immediately before use.

References

1. Roth, J. (1982) The protein A-gold (pAg) technique – A qualitative and quantitative approach for antigen localization on thin sections. In *Techniques in Immunocytochemistry*, Bullock, G. R. and Petrusz, P., eds. Academic Press, London, UK, vol. 1, 108–133.
2. Tokuyasu, K. T. (1980) Immunochemistry on ultrathin frozen sections. *Histochem. J.* **12**, 381–403.
3. Frens, G. (1973) Controlled nucleation for the regulation of particle size in monodisperse gold suspensions. *Nat. Phys. Sci.* **241**, 20–22.
4. Baschong, W., Lucocq, J. M., and Roth, J. (1985) "Thiocyanate gold": small (2–3 nm) colloidal gold for affinity cytochemical labeling in electron microscopy. *Histochemistry* **83**, 409–411.
5. Tschopp, J., Podack, E. R., and Müller-Eberhard, H. J. (1982) Ultrastructure of the membrane attack complex of complement: detection of the tetramolecular C9-polymerizing complex C5b–8. *Proc. Natl. Acad. Sci. USA* **79**, 7474–7478.
6. De Roe, C., Courtoy, P. J., and Baudhuin, P. (1987) A model of protein-colloidal gold interaction. *J. Histochem. Cytochem.* **35**, 1191–1198.
7. De Mey, J., Moeremans, M., Geuens, G., Nuydens, R., and De Brabander, M. (1981) High resolution light and electron microscopic localization of tubulin with the IGS (immuno gold staining) method. *Cell Biol. Int. Rep.* **5**, 889–899.
8. Ackerman, G. A., Wolken, K. W., and Gelder, F. B. (1980) Surface distribution of monosialoganglioside Gm1 on human blood cells and the effect of exogenous GM1 and neuraminidase on cholera toxin surface labeling. A quantitative immunocytochemical study. *J. Histochem. Cytochem.***28**, 1100–1112.
9. Bonnard, C., Papermaster, D. S., and Kraehenbuhl, J.-P. (1984) The streptavidin-biotin bridge technique: application in light and electron microscope immunocytochemistry. In *Immunolabeling for Electron Microscopy*, Polak, J. M. and Varndell, I. M., eds. Elsevier, Amsterdam, The Netherlands, 95–111.
10. Yokota, S. (1988) Effect of particle size on labeling density for catalase in protein A-gold immunocytochemistry. *J. Histochem. Cytochem.* **36**, 107–109.

Part II

Microscopy Toolbox

Chapter 9

Immunoelectron Microscopy of Chemically Fixed Developing Plant Embryos

Tetsuaki Osafune and Steven D. Schwartzbach

Abstract

The hydrophobic plant cell wall, large acidic central vacuole, diverse secondary compounds, intercellular airspaces, and rigid starch granules present obstacles to ultrastructure preservation and specimen sectioning. We describe modifications of fixation and embedding procedures successfully used with microbes, protists, and mammalian tissues that have overcome these obstacles. Vacuum infiltration is used to remove intercellular air rapidly replacing it with fixative and buffer preserving cellular ultrastructure while neutralizing the acidic vacuole. Vacuum infiltration of embedding resin ensures uniform embedding resin permeation allowing production of intact ultrathin sections that are stable under the electron beam and suitable for immunolabeling. The methodology described has been used for immunolocalization of non-specific lipid transfer proteins in the diverse cell types found in developing castor bean fruits but is suitable for all plant tissues.

Key words: Castor bean embryo, immunogold, immunolocalization, lipid transfer protein.

1. Introduction

The overall strategy of sample preparation for immunoelectron microscopy, fixation, dehydration, embedding in plastic resin, sectioning, and immunolabeling are independent of the biological material under investigation. The biological properties and ultrastructure of plants present technical challenges requiring plant-specific modifications of the sample preparation protocols utilized for microbes, protists, and mammalian tissue. Plants contain a hydrophobic cuticle, a rigid cell wall, and starch granules that present barriers to penetration of fixatives and embedding resins. Cells within some plant tissues are not packed together but are separated by intercellular air spaces which present

S.D. Schwartzbach, T. Osafune (eds.), *Immunoelectron Microscopy*, Methods in Molecular Biology 657,
DOI 10.1007/978-1-60761-783-9_9, © Springer Science+Business Media, LLC 2010

further barriers to fixative and embedding resin penetration. In contrast to other organisms, plant cells have an acidic central vacuolar compartment occupying a majority of the cell's volume. Fixation buffers must have sufficient buffering capacity to neutralize this acidic vacuolar compartment. The vacuole also contains numerous secondary products that can react with fixatives producing precipitates. Lysis of the vacuole during fixation not only releases precipitate forming compounds but also results in a collapse of the cell membrane altering cellular ultrastructure. The vacuole, cytoplasm, and cellular organelles are soft structures while the cell wall and starch grains are hard rigid structures. These rigidity differences complicate sectioning. Ultrathin sections from poorly embedded samples often have holes and they may be unstable in the electron beam.

Although formidable, the challenges of fixing and sectioning plant material have been overcome. Cutting tissue into small pieces facilitates rapid permeation of buffer, fixative, and embedding resin. Appropriate buffers neutralize the acidic vacuolar compartment. Vacuum infiltration of fixatives rapidly replaces intercellular air with the fixative preserving cellular ultrastructure. Vacuum infiltration of embedding resin ensures that the embedding resin permeates the sample uniformly allowing production of intact ultrathin sections that are stable in the electron beam and suitable for protein immunogold localization. This chapter describes the method we have successfully utilized for immunolocalization of non-specific lipid transfer proteins in developing castor bean fruits to the dense granules and vacuoles of the endosperm and the cell wall of pericarp sclerenchyma cells (1).

2. Materials

2.1. Sample Preparation

1. 50% glutaraldehyde (EM grade, Electron Microscopy Sciences, Hatfield PA) stored at 0–4°C (*see* **Note 1**).
2. 0.1 M potassium phosphate buffer, pH 7.2: Prepare by using 0.1 M KH_2PO_4 to adjust the pH of a 0.1 M solution of K_2HPO_4 to pH 7.2.
3. 0.1 M potassium phosphate buffer, pH 7.2, containing 2% glutaraldehyde: Prepare just prior to use by mixing 16 mL 0.1 M potassium phosphate buffer, pH 7.2, with 4 mL 50% glutaraldehyde.
4. 50, 70, and 90% ethanol.
5. Acetone.
6. Vacuum desiccator (Fisher Scientific, Pittsburgh, PA).
7. 25 mL borosilicate glass shell vial (Electron Microscopy Sciences, Hatfield PA, *see* **Note 2**).
8. Razor.

2.2. Embedding Samples in Epon

1. Plastic disposable Tri-Corn beakers (Electron Microscopy Sciences, Hatfield, PA).
2. Wooden tongue depressors (Electron Microscopy Sciences, Hatfield, PA).
3. Epon (Embed-812, Electron Microscopy Sciences, Hatfield, PA).
4. NMA (Nadic Methyl Anhydride, Electron Microscopy Sciences, Hatfield, PA).
5. DDSA (Dodecenyl Succinic Anhydride, Electron Microscopy Sciences, Hatfield, PA).
6. DMP-(2,4,6-Tri(dimethylaminomethyl) phenol), Electron Microscopy Sciences, Hatfield, PA).
7. Embedding mix A: Prepare by separately warming the Embed-812 and DDSA to 60°C in disposable beakers to reduce viscosity. Combine in a disposable beaker 44 mL prewarmed Embed-812 with 67 mL prewarmed DDSA. Use a wooden tongue depressor to mix thoroughly (*see* **Note 3**).
8. Embedding mix B: Prepare by separately warming the Embed-812 and NMA to 60°C in disposable beakers to reduce viscosity. Combine in a disposable beaker 67 mL prewarmed Embed-812 with 56 mL prewarmed NMA. Use a wooden tongue depressor to mix thoroughly (*see* **Note 3**).
9. Embedding resin: Prepare in a disposable beaker by mixing equal volumes of room temperature embedding mix A and embedding mix B and add DMP-30 for a final concentration of 2%. Use a wooden tongue depressor to mix thoroughly. Degas for 15 min in a vacuum desiccator, store tightly sealed at room temperature and use that day (*see* **Note 4**).
10. Vacuum desiccator (Fisher Scientific, Pittsburgh, PA).
11. Tweezers with flat gripping surface (Style 2A PTFE coating, Electron Microscopy Sciences, Hatfield, PA).
12. Gelatin capsules size 00 (Electron Microscopy Sciences, Hatfield, PA).

2.3. Preparation of Ultrathin Sections

1. 200 mesh Formvar/carbon-coated nickel grids (Electron Microscopy Sciences, Hatfield, PA).
2. Diamond knife (Electron Microscopy Sciences, Hatfield, PA).
3. Block Trimmer (Electron Microscopy Sciences, Hatfield, PA).
4. Eyelash with handle (Ted Pella, Inc., Redding, CA).

5. Whatman #1 filter paper (Fisher Scientific, Pittsburgh, PA).
6. Anti-capillary self-closing tweezers (Ted Pella Inc., Redding, CA).

2.4. Immunogold Labeling of Sections

1. 0.01 M phosphate buffer pH 7.4: Prepare by using 0.01 M KH_2PO_4 to adjust the pH of a 0.01 M solution of Na_2HPO_4 to pH 7.4.
2. PBS (0.01 M phosphate-buffered saline, pH 7.4): Prepare by mixing 0.85 g NaCl with 100 mL 0.01 M phosphate buffer, pH 7.4.
3. PBS–BSA: Prepare by adding BSA (Sigma, St. Louis, MO) for a final concentration of 1% to PBS.
4. PBS–Tween: Prepare by adding Tween 20 (Sigma, St. Louis, MO) to PBS for a final concentration of 0.05%.
5. Antibody to protein to be localized diluted to appropriate concentration with PBS–BSA immediately before use.
6. Protein-A gold: Protein-A conjugated to 15 nm gold particles (EY laboratories, San Mateo, CA) (*see* **Note 5**).
7. 3% uranyl acetate (Electron Microscopy Sciences, Hatfield, PA): Prepare by mixing 1.5 g uranyl acetate with 50 mL deionized water. Stir the solution overnight in a foil-covered container, add 10 drops of glacial acetic acid, and store for no more than 3 months at 4°C.
8. Parafilm (Fisher Scientific, Pittsburgh, PA).

3. Methods

3.1. Sample Preparation

1. Cut approximately 1–2 g fresh weight plant tissue with a razor into 1–2 mm^3 pieces (*see* **Note 6**) and fully immerse them in vials containing 0.1 M potassium phosphate buffer, pH 7.2, 2% glutaraldehyde (*see* **Note 7**).
2. Place the samples in a vacuum desiccator and incubate under vacuum for 120 min at 4°C (*see* **Note 8**).
3. Remove the glutaraldehyde solution and immediately replace it with 0.1 M potassium phosphate buffer, pH 7.2. Incubate for 20 min and repeat twice (*see* **Note 9**).
4. Dehydrate the samples by removing the phosphate buffer and immediately replacing it with 50% ethanol. Incubate for 20 min and repeat once.
5. Dehydrate the samples by removing the 50% ethanol and immediately replacing it with 70% ethanol. Incubate for 20 min and repeat once.

6. Dehydrate the samples by removing the 70% ethanol and immediately replacing it with 90% ethanol. Incubate for 20 min and repeat once.
7. Remove the 90% ethanol and immediately replace it with acetone. Repeat three times (*see* **Note 10**).

3.2. Embedding Samples

1. Remove the acetone and immediately replace it with a 1:2 embedding resin:acetone mixture. Incubate 4 h to overnight (*see* **Note 11**).
2. Remove the 1:2 embedding resin:acetone mixture with a plastic pipette (*see* **Note 12**) and immediately replace it with a 2:1 embedding resin:acetone mixture. Incubate 4 h to overnight.
3. Remove the 2:1 embedding resin:acetone mixture with a plastic pipette and immediately replace it with 100% embedding resin. Incubate in a vacuum desiccator under vacuum for 1–2 h.
4. Remove the 100% embedding resin and resuspend the sample in a small volume of 100% embedding resin (*see* **Note 13**). Using a tweezers with a flat surface, orient one to three tissue pieces at the bottom of a half full gelatin capsule (*see* **Note 14**). Top the capsule off with embedding resin, insert a sample identification label into the embedding resin at the top of the capsule, close the capsule, and polymerize by incubation in a 60°C oven for 24 h (*see* **Note 15**).
5. Allow the polymerized block to cool for 24 h. Remove the block from the gelatin capsule by placing in water at 37°C until the capsule dissolves.

3.3. Preparation of Ultrathin Sections

1. Mount the block in a block trimmer and shape the end into a four-sided pyramid with walls at a 45° angle and a 0.5–0.75 mm square top surface.
2. Mount the trimmed block on a microtome making sure that the block face is parallel to the knife edge. Fill the diamond knife trough with distilled water so that it is level with the cutting edge.
3. Cut a ribbon of silver ultrathin sections approximately 80–90 nm thick (*see* **Note 16**).
4. Use an eyebrow tool to separate the ribbon into five or six sections and align them in the trough.
5. Place a grid held with a pair of tweezers under the sections and raise it positioning the sections in the middle of the grid (*see* **Note 17**).

6. Blot the jaws of the tweezers and the bottom of the grid with Whatman #1 filter paper to absorb all of the liquid. Place the grid sample side up on a dry piece of filter paper in a Petri dish and allow to dry overnight.

3.4. Immunogold Labeling of Sections

1. Wash ultrathin sections on grids four times with PBS (*see* **Note 18**) by floating grids section side down on a 300 μL drop of PBS for 30 min (*see* **Note 19**).
2. Block ultrathin sections on grids by floating grids section side down on a 300 μL drop of PBS–BSA for 30 min at room temperature.
3. Incubate ultrathin sections on grids with primary antibody by floating grids section side down on a 300 μL drop of primary antibody diluted in PBS–BSA and incubating at 37°C for 20 min (*see* **Note 20**).
4. Wash ultrathin sections on grids twice with PBS–Tween by floating grids section side down on a 300 μL drop of PBS–Tween for 30 min.
5. Incubate ultrathin sections on grids with protein-A gold by floating grids section side down on a 300 μL drop of protein-A gold diluted 1:10 or 1:20 in PBS and incubating for 20 min at room temperature.
6. Wash ultrathin sections on grids twice with PBS–Tween by floating grids section side down on a 300 μL drop of PBS–Tween for 10 min
7. Wash ultrathin sections on grids twice with deionized water by floating grids section side down on a 300 μL drop of deionized water for 10 min.
8. Stain ultrathin sections on grids with uranyl acetate by floating grids section side down on a 300 μL drop of 3% uranyl acetate for 10 min (*see* **Note 21**).
9. Wash ultrathin sections on grids twice with deionized water by floating grids section side down on a 300 μL drop of deionized water for 10 min.
10. Blot the bottom of the grid with Whatman #1 filter paper and place sample side up on a dry piece of filter paper in a Petri dish. Allow to dry overnight.
11. Examine sections in the electron microscope. **Figure 9.1** shows that non-specific lipid transfer proteins are localized to cytoplasmic dense vesicles (DV) some of which are at the vacuole (V) surface in fruit endosperm while **Fig. 9.2** shows that non-specific lipid transfer proteins localize to the cell wall of pericarp tissue. Dense vesicles containing non-specific lipid transfer proteins were not found in pericarp tissue.

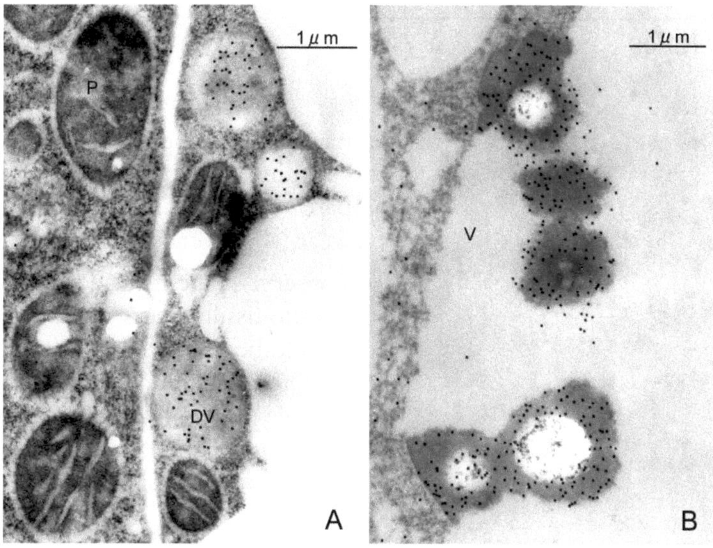

Fig. 9.1. Immunolocalization of non-specific lipid transfer proteins in endosperm cells from developing castor bean fruit at 14 days (**A**) and 28 days (**B**) after flowering. Sections were prepared and immunolabeled with antibody to castor bean non-specific lipid transfer proteins. At 14 days after flowering (**A**), immunogold label localized the non-specific lipid transfer proteins to cytoplasmic dense vesicles (DV), some of which are at the vacuole (V) surface. Immunogold label is absent from the plastid (P). By 28 days after flowering (**B**), immunogold label showed that the non-specific lipid transfer proteins formed aggregates in the central vacuole.

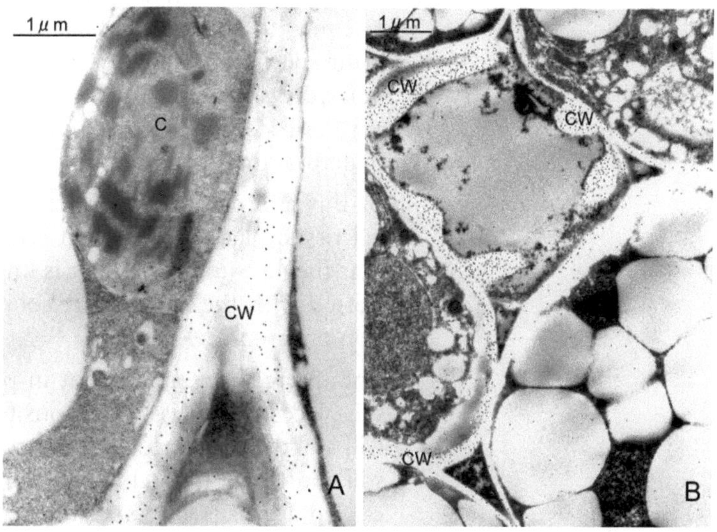

Fig. 9.2. Immunolocalization of non-specific lipid transfer proteins in the pericarp trichome (**A**) and the pericarp sclerenchymatous tissue (**B**) from a developing castor bean fruit at 28 days after flowering. Sections were prepared and immunolabeled with antibody to castor bean non-specific lipid transfer proteins. In the pericarp trichome (**A**), immunogold label localized the non-specific lipid transfer proteins to the cell wall (CW) of fruit trichomes. Staked thylakoids are visible in the chloroplast (**C**) which is devoid of immunolabel. In the pericarp sclerenchymatous tissue (**B**), immunogold label showed that the non-specific lipid transfer proteins were also localized to the thickened cell wall (CW) of the sclerenchyma cell. Note that the non-specific lipid transfer proteins are not found in the cytoplasm.

4. Notes

1. Glutaraldehyde and the other chemicals used for fixation and staining are extremely hazardous. Solutions should be prepared and used in a hood. Wear gloves and protective eyewear whenever handling chemicals.
2. To conserve reagents, the smallest diameter container that allows the tissue pieces to be well separated without overlap should be used. Glass containers must be used as plastic containers are incompatible with acetone which is used during sample preparation and embedding. Petri dishes can be used instead of shell vials and some investigators feel that fluid exchange is easier with Petri dishes due to their low sidewall.
3. Unused embedding mix can be stored for up to 6 months at −20°C but it is preferable to use freshly prepared mix.
4. Resin waste should be collected and allowed to polymerize before disposal.
5. By using two different-sized gold particles for sequential immunolabeling, two antigens can be co-localized on a single section.
6. It is critical that tissue be cut into pieces no larger than 1–2 mm^3. Penetration of fixative and embedding resin into plant specimens is relatively slow. Cutting the plant tissue into very small pieces allows rapid penetration of the fixative and embedding resin ensuring maintenance of ultrastructural integrity.
7. In this and all subsequent steps, ensure that the individual tissue pieces are fully in contact with the fixative and that they do not overlap. It is important to keep the tissue wetted with the appropriate solution throughout the fixation protocol.
8. The intercellular air present in plant tissues can make tissue pieces float. Tissue sections fully infiltrated with buffer will sink to the bottom of the container. Floating sections still contain air in the intercellular spaces and should be discarded at this point as fixative has not fully permeated the tissue piece.
9. The simplest way to change solutions is to slowly tilt the vial and use a disposable glass or plastic pipette with a pipetting bulb to remove the solution. The new solution is then slowly added to the vial, the vial is swirled to resuspend the tissue pieces adhering to the bottom of the vial and, if necessary, a tweezers is used to distribute the tissue pieces throughout the vial so that there is no overlap. Tissue can

be stored in 0.1 M potassium phosphate buffer, pH 7.2, for up to 1 week at 0–4°C.

10. Do not use plastic pipettes to remove the acetone solution as they will dissolve.
11. It is important to add enough acetone embedding resin mixture to the dish so that the plant tissue is not exposed to air when the acetone evaporates from the mixture.
12. Plastic pipettes can be used to remove but not to add acetone embedding resin mixtures because the acetone evaporates during the incubation. The use of the plastic pipette provides a way to monitor that all the acetone has evaporated.
13. Add sufficient embedding resin so that the tissue remains fully immersed. Swirl the sample and then tilt the vial so that the tissue collects at the edge of the vial. Ensure that the tissue pieces remain completely covered in embedding resin.
14. Many investigators prefer Beem capsules over gelatin capsules. The advantage of the Beem capsule is that they have a preformed capsule tip in the shape of a truncated pyramid with a flat top which facilitates block trimming. The disadvantage is that the block must be cut from the capsule with a razor while the gelatin capsule simply dissolves. If using a Beem capsule, simply place the tissue pieces at the tip.
15. Epoxy resins are easy to use and provide excellent preservation of microstructure after sectioning. The necessity to heat the embedding resin for polymerization does however lead to a loss of antigenicity making their use most suitable for localization of high-abundance proteins. Hydrophilic embedding resins such as Lowicryl K4M and LR White which can be polymerized by ultraviolet light at low temperature are better at preserving antigenicity. They are more widely used for immunoelectron microscopy, although they exhibit a lower level of ultrastructure preservation and the sections are more easily damaged by the electron beam.
16. Thick sections can initially be cut and examined in the light microscope to identify the region of the block containing the sample.
17. The most convenient way to pickup and transfer grids is to use self-closing anti-capillary tweezers. Extra care must be taken to remove all moisture when other types of instruments are used for transfer.
18. To increase antigenicity, ultrathin sections on grids can be floated section side down on a 300 μL drop of freshly prepared 0.3–3% (w/v) hydrogen peroxide in distilled water

for 10 min prior to washing the sections with PBS. Incubation in hydrogen peroxide increases immunoreactivity by etching the resin surface making the embedded antigen more accessible to the antibody. Incubation in hydrogen peroxide will not increase the antigenicity of samples embedded in hydrophilic resins such as Lowicryl K4M and LR White.

19. A piece of parafilm should be placed in a Petri dish containing a piece of buffer saturated Whatman #1 filter paper. Place drops of washing solution side by side on the parafilm sheet and move the grid sequentially from drop to drop incubating with the lid closed. Incubation in a water saturated environment prevents the drops from evaporating.

20. Depending on the scarcity of the reagent and time of incubation, drop size can be varied from 100 to 1,000 μL. The optimal antibody dilution and incubation conditions must be determined empirically for each antibody. Typical dilutions are 50–1,000-fold for a 20 min incubation with antibody at 37°C. Lower antibody concentrations can be used with longer incubation times at lower temperatures. Incubation times as long as 24 h at 4°C can be used. The lower the abundance of the antigen, the higher the antibody concentration and/or incubation time needed. When excessive background labeling is seen, increasing the amount of BSA in the blocking solution, lowering the antibody concentration, and/or shortening the incubation time may decrease non-specific labeling. Control experiments with pre-immune serum should be routinely performed to verify the specificity of the immunoreaction.

21. Care must be taken to withdraw the solution from the top of the bottle to avoid depositing precipitated crystals on the grid.

Acknowledgments

This work was supported by National Science Foundation Grant MCB-0080345 and MCB-0196420 to S.D.S and Grant in Aid for Scientific Research No. 20570062 from the Ministry of Education, Sciences, Sports and Culture Japan to T.O.

Reference

1. Osafune, T., Tsuboi, S., Ehara, T., Satoh, Y., and Yamada, M. (1996) The occurrence of non-specific lipid transfer proteins in developing castor bean fruits. *Plant Sci.* **113**, 125–130.

Chapter 10

Pre-embedding Immunogold Localization of Antigens in Mammalian Brain Slices

Thomas Schikorski

Abstract

The detection of proteins with antibodies that are conjugated to gold particles has been a major asset to cell biology and the neurosciences, and knowledge about the subcellular location of antigens has formed the basis for many hypotheses regarding protein function. Many protocols have been developed since the introduction of colloidal gold to immunocytochemistry. The two most widely used techniques, however, are based on transmission electron microscopy and consist of either immunolabeling before the specimens are embedded in resin (pre-embedding immunogold labeling) or immunolabeling after embedding in resin (post-embedding immunogold labeling). The following protocol describes a pre-embedding procedure that gives reliable results with all antibodies that produce adequate staining as observed with a light microscope. This procedure results in almost perfect preservation of the ultrastructure. The procedure employs thick sectioning using a vibratome, permeabilization of membranes with Triton X-100, and immunolabeling with fluorescently conjugated Nanogold antibodies, followed by gold enhancement and embedding for electron microscopy. We also discuss some limitations inherent to pre-embedding immunogold labeling.

Key words: Immunohistochemistry, electron microscopy, fluoro-nanogold, gold enhancement, synaptic vesicle protein, permeabilization, correlated light and electron microscopy.

1. Introduction

Knowledge of protein location and distribution provides the structural basis and enables testability for many hypotheses and theories in cell biology and the neurosciences. For example, the presence of a protein at synapses is a prerequisite for any hypothesis proposing a synaptic function for that particular protein. Therefore, the localization of proteins at the subcellular level

S.D. Schwartzbach, T. Osafune (eds.), *Immunoelectron Microscopy*, Methods in Molecular Biology 657,
DOI 10.1007/978-1-60761-783-9_10, © Springer Science+Business Media, LLC 2010

has played a pivotal role in cell biology. Immunogold labeling has been widely used in many laboratories and EM facilities. Nonetheless, and despite its wide use, immunogold histochemistry has been one of the most difficult techniques in cell biology. In addition to the general limitations inherent to immunocytochemistry, such as antigen preservation and antibody specificity, the major difficulties for electron microscopic localization of antibodies arise from a simple constraint: antibodies and their conjugated gold particles need to cross membranes to reach their antigen. Thus, membranes must be disrupted, or alternatively, cells must be cut open to allow antibodies to enter the intracellular space. Two strategies have been developed to deal with this constraint. One approach is to permeabilize membranes using a detergent, which dissolves parts of the lipid bilayer so that membranes become porous, or to use freeze–thaw cycles in which quickly forming and thawing ice crystals disrupt the membranes. The permeabilization step is followed by immunolabeling and embedding for EM. Since immunocytochemical labeling precedes embedding in resins, this technique is called pre-embedding immunogold labeling. Unfortunately, both permeabilization methods interfere with the structural integrity of membranes and thus are detrimental to the ultrastructure. Consequently, the art of pre-embedding immunogold histochemistry for EM lies in the balance between achieving the best possible permeabilization (and consequently the most successful immunolabeling) with the minimum disruption of ultrastructure. This can be achieved by controlling the number of freeze-thaw cycles or the concentration of the detergent, with less detergent leading to better structural preservation but weaker immunolabeling. A second approach to facilitate the penetration of membranes by gold-conjugated antibodies is to use the smallest gold particles available. This strategy led to the development of ultrasmall gold particles (1). These Nanogold particles are later enhanced in size by chemical deposition of metal ions (either silver or gold) to become visible by EM. Despite these advances, in practice only the first few micrometers of a tissue section can be permeabilized and used for immunolabeling. Stronger permeabilization allowing for deeper penetration of antibodies would compromise the ultrastructure. The superficial labeling of mildly fixed tissue slices is a constraint on the protocol described here when the distribution of a protein is to be measured in larger structures (e.g., an entire cell or a dendritic tree).

One way to circumvent the problems associated with pre-embedding immunogold histochemistry is by using post-embedding immunogold labeling. Here, specimens are first embedded in resin and sectioned at 60–90 nm. The structures that were "cut open" can be directly exposed to antibodies, and permeabilization that conflicts with structural preservation is not needed. However, post-embedding immunocytochemistry has its own limitations. It not only requires special and costly equipment but also poses major technical difficulties. One major limitation is that antibodies cannot penetrate into the resin, and consequently only antigens that are exposed at the surface of a thin section can be labeled. Another limitation is that the tissue must be dehydrated before embedding into resin, and dehydration compromises antigenicity. As a consequence of these limitations, most binding sites remain undetected, resulting in very weak or often unsuccessful labeling. However, when detecting abundant antigens such as the neurotransmitter gamma-aminobutyric acid (GABA), post-embedding immunogold labeling can be implemented successfully.

Since post-embedding poses more and bigger challenges than pre-embedding, we will describe the pre-embedding technique in the following protocol. The reader who is interested in the post-embedding procedure is directed to publications that deal extensively with the inherent problems of post-embedding and offer a large number of various protocols (2). Using the following pre-embedding protocol, we have detected antigens at the ultrastructural level with any antibody that has resulted in successful staining using light microscopy (LM). Implementation of the protocol begins with the mild fixation of tissue (in our example, the brain) via transcardial perfusion. The fixed tissue is then thick-sectioned with a vibratome (100–150 μm) and permeabilized with detergent. The protocol exploits indirect immunocytochemistry with fluorescently conjugated Nanogold antibodies that allow correlated LM and EM. After the documentation of the LM data, the Nanogold particles are gold-enhanced and the slices are processed for EM. Our protocol allows the penetration of antibodies into the slice up to 5 μm, while at the same time preserving the fine structure at very high quality.

2. Materials

2.1. Perfusion Fixation of Mice

1. Nembutal for anesthesia.
2. Oxygenated saline: 135 mM NaCl, 5 mM KCl, 10 mM $NaHPO_4$, 25 mM $NaHCO_3$, and 4 mM $MgCl_2$. Bubble with carbogen (95% O_2, 5% CO_2) until a neutral pH is reached, then add 1 mM $CaCl_2$.
3. HEPES buffered saline (HBS): 154 mM (0.9%) NaCl, 0.2 mM $CaCl_2$, and 20 mM HEPES-NaOH pH 7.4.
4. Fixative: freshly prepared 4% paraformaldehyde, 0.1% glutaraldehyde (Electron Microscopic Sciences (EMS),

Hatfield, PA, *see* **Note 1**), HBS, pH 7.4. For the perfusion of an adult mouse, ~50 mL of the fixative is needed.

5. Peristaltic pump (Fisher Scientific, Pittsburgh, PA).
6. Tubing ~3 mm in inner diameter (Fisher Scientific, Pittsburgh, PA).
7. Needle (20 gauge) (Fisher Scientific, Pittsburgh, PA).
8. A pair of small scissors (EMS, Hatfield, PA).
9. A pair of very fine scissors (EMS, Hatfield, PA).
10. Tweezers (EMS, Hatfield, PA).

2.2. Vibratome Sectioning

1. Vibratome (EMS, Hatfield, PA)
2. Double-edge stainless steel razor blades (EMS, Hatfield, PA)
3. HBS
4. Crazy glue (EMS, Hatfield, PA)
5. 24-well cell culture plate to collect individual sections in series
6. Brush, size 0–1.

2.3. Immunocytochemistry

1. Shaker (LabScientific, Livingston, NJ)
2. HBS
3. Bovine serum albumin (BSA, Sigma-Aldrich, St. Louis, MO)
4. Triton X-100 (Roche, Indianapolis, IN, *see* **Note 2**)
5. Permeabilization solution: HBS, 10% BSA, and 0.025–0.05% Triton X-100
6. Primary antibody solution: antibodies against the antigen being detected (in our example, the monoclonal anti-synaptophysin antibody from Sigma-Aldrich, St. Louis) in HBS plus 1% BSA and 0.0025–0.005% Triton X-100. An antibody concentration of ~1 μg/mL is a good starting point; however, the optimal concentration (that results in a strong label with no background) should be determined experimentally using a dilution series
7. Refrigerator (The Lab Depot, Dawsonville, GA)
8. HBS–0.05%BSA: HBS plus, 0.05% BSA
9. Parafilm
10. Secondary antibody solution: anti-mouse FluoroNanogold Fab fragment antibody (Nanoprobes, Yaphank, NY, *see* **Note 3**) in HBS plus 1% BSA and 0.0025–0.005% Triton X-100. We use the secondary antibody at a dilution between 1:50 and 1:100. The optimal concentration should be determined experimentally using a dilution series

11. Karnovsky's fixative (3): 4% paraformaldehyde, 5% glutaraldehyde, 0.2 mM calcium chloride, and 90 mM cacodylate–HCl pH 7.4

2.4. Gold Enhancement of Nanogold Particles

1. Goldenhance EM® (Nanoprobes, Yaphank, NY)
2. Distilled water
3. Parafilm
4. Shaker (LabScientific, Livingston, NJ)
5. Brush to transfer slices
6. Fluorescence microscope with imaging system (Nikon, Melville, NY)

2.5. Processing Slices for EM

1. Cacodylate buffer: 90 mM sodium cacodylate–HCl, pH 7.4, 0.2 mM $CaCl_2$.
2. 1% osmium tetroxide (OsO_4, EMS, Hatfield, PA, *see* **Note 4**), 1.5% potassium ferrocyanide, 90 mM cacodylate buffer.
3. 1% OsO_4, 90 mM cacodylate buffer.
4. 2% aqueous uranyl acetate (EMS, Hatfield, PA).
5. Ethanol 100% and a series of 50%, 70%, 90% ethanol for dehydration.
6. Propylene oxide (EMS, Hatfield, PA).
7. 8-mL snap-cap glass vials (EMS, Hatfield, PA).
8. Resin mixture for embedding. Prepare separately (1) Epon resin: 20 g EMbed-812, 11 g DDSA, 9 g NMA, and 0.8 mL DMP-30 and (2) Spurr resin: 10 g ERL, 6 g DER, 26 g NSA, and 0.4 mL DMAE (all reagents are from EMS, Hatfield, PA). When both resins are completely mixed, combine Epon and Spurr resin batches and mix thoroughly.
9. Paper and pencil to create labels.
10. Small brush or toothpick.
11. Rotator (Ted Pella, Redding, CA).
12. Flat embedding mold (EMS, Hatfield, PA).

2.6. Sectioning and On-Section Staining

1. Diamond knife (EMS, Hatfield, PA).
2. Ultramicrotome (Leica, Bannockburn, IL).
3. Double-edge stainless steel razor blades (EMS, Hatfield, PA).
4. Formvar-coated copper slot grids (2×1 mm) (EMS, Hatfield, PA).
5. Forceps for grid handling (EMS, Hatfield, PA).
6. Grid storage box (EMS, Hatfield, PA).

7. Oven (temperature range up to 300°C, The Lab Depot, Dawsonville, GA).
8. Volumetric flask (BLD Science, Garner, NC).
9. Amber glass vial with screw cap (BLD Science, Garner, NC).
10. Hanaichi lead solution (*see* **Note 5**): Calcined lead citrate 0.20 g, lead nitrate 0.15 g, lead acetate 0.15 g, sodium citrate 1.00 g, distilled water 41.00 mL.
    Prepare calcined lead citrate by heating crystal lead citrate for several hours in a melting pot (200–300°C) until the color changes to a light brownish yellow. Overheated lead citrate with a dark brownish or black color cannot be used.
    Place the cooled-down calcined lead citrate and the other reagents in a 50 mL volumetric flask and mix well to produce a yellowish milky solution. Add 9.0 mL of 1 N NaOH to the solution and mix well until the solution becomes clear with a light yellowish color. Transfer the solution to an amber glass bottle with a screw cap bottle for storage.
11. Parafilm.

3. Methods

3.1. Perfusion Fixation of Mice

1. Fill the tubing of the peristaltic pump with oxygenated saline. Stop the pump and place the free end of the tubing into the fixative. Avoid air bubbles in the tubing!
2. Anesthetize the mouse by injecting an overdose of Nembutal (30 mg/kg) into the abdominal cavity.
3. Working quickly, open the chest and expose the heart. Once the chest cavity has been opened, the following steps should be done as quickly as possible, since the animal can no longer breathe, and anoxia quickly causes artifacts of the fine structure.
4. Cut the right atrium using a pair of fine scissors.
5. Insert a blunt needle connected to the end of the tubing into the left ventricle. Turn on the pump and perfuse the animal with oxygenated saline and fixative (*see* **Note 6**).
6. Continue perfusion for ~3–5 min. Longer perfusion will result in stronger fixation and better structural preservation, but may compromise antigenicity.
7. Dissect the brain and place it into fixative for immersion fixation.

3.2. Vibratome Sectioning

1. Wash the brain twice in HBS.
2. Separate the hemispheres with a razor blade along the mid-sagittal plane.
3. Cut off the cerebellum and mount one hemisphere on the vibratome's specimen holder using the crazy glue. For frontal sections, the anterior cortex should be facing up, and the posterior end of the hemisphere should be mounted onto the specimen holder.
4. Fill the vibratome reservoir with HBS and cut 100–150-μm thick sections (*see* **Note 7**).
5. Collect the sections in series with a brush, and place each section into a well of the 24-well plate containing HBS.

3.3. Immunohistochemistry

1. Wash the slices in the wells of the 24-well plate thoroughly with HBS three times for 5 min or longer.
2. Permeabilize and block nonspecific binding sites by adding 350–500 μL of permeabilization solution to each well of the 24-well plate. Place the 24-well plate on a shaker for 30 min.
3. Replace the permeabilization solution with ice-cold primary antibody solution. Do not wash the slices between steps. Place the 24-well plate on a shaker and incubate at 4°C overnight. Compared to incubation at room temperature (RT), prolonged incubation at low temperature with gentle agitation produces stronger specific immunolabeling while keeping the background low (*see* **Note 8**).
4. Wash three times for 5 min with ice-cold HBS–0.05%BSA.
5. Replace the buffer with anti-mouse FluoroNanogold antibody solution (*see* **Note 8**). Return the 24-well plate to the shaker and incubate at 4°C overnight, or alternatively, at RT for 2–3 h.
6. Wash three times in HBS–0.05%BSA for 5 min.
7. Place sections in HBS–0.05%BSA buffer under an epifluorescence microscope and observe the immunofluorescence. Photograph the staining (*see* **Note 9**).
8. Return the slices to the 24-well plate and replace the HBS with Karnovsky's fixative. Fix at RT for 20 min or longer.

3.4. Gold Enhancement of Nanogold Particles

1. Wash the slices in the 24-well plate with four changes of distilled water for a total of at least 2 h.
2. Mix Goldenhance EM® according to the manufacturer's instructions.
3. Place a slice onto a piece of parafilm mounted onto a shaker (*see* **Note 8**).

4. Add 100 µL of Goldenhance EM® to the slice. Quickly turn on the shaker and agitate vigorously (the slice should swirl around in the solution). We stop the gold enhancement when the first hint of the typical purple-gold color becomes visible. Depending on the batch of Goldenhance EM®, this will take between 1 and 4 min. The resulting gold particle size will average around 10–20 nm (*see* **Note 10**).
5. Wash thoroughly with agitation in ice-cold water to stop the gold enhancement.
6. Return the slice to the 24-well plate in HBS and wash twice with HBS for 5 min (*see* **Note 11**).
7. Replace HBS with Karnovsky's fixative and store until processing for EM.

3.5. Processing Slices for EM

1. Wash the brain slices in the 24-well plate three times in cacodylate buffer.
2. Post-fix the brain slices in 500 µL of 1% OsO_4 plus 1.5% potassium ferrocyanide in cacodylate buffer. Place the 24-well plate on a shaker and incubate at room temperature with shaking at 150 rpm for 20–60 min. Longer postfixation times result in better fixation and higher contrast of the final EM graphs.
3. Wash three times in cacodylate buffer.
4. Post-fix the brain slices in 500 µL of 1% OsO_4 in cacodylate buffer. Place the 24-well plate on a shaker and incubate at room temperature with shaking at 150 rpm for 20–60 min.
5. Wash once in cacodylate buffer, and then wash thoroughly three times for 5 min with distilled water by placing the 24-well plate on a shaker.
6. Stain neurons in 500 µL of 2% aqueous uranyl acetate on a shaker at RT for 1 h.
7. Wash with distilled water three times for 5 min by placing the 24-well plate on a shaker.
8. Dehydrate the brain slices in an ascending ethanol series (50%, 70%, 90%, and 100%) for 5–10 min in each solution. Repeat the 100% ethanol step three times for 10 min to ensure complete dehydration.
9. In the meantime, prepare 8-mL snap cap glass vials with ~2 mL of propylene oxide, and place a pencil-written paper label in each vial. The paper label should be small enough to fit later into the flat-embedding mold.
10. Transfer the slices from the 24-well plate to the glass vials containing propylene oxide. Either a brush or a toothpick can be used. Caution! The slices are very brittle after

dehydration and break easily during handling. Place the glass vials on a rotator for 10 min.

11. Replace the propylene oxide with a mixture of propylene oxide and resin (1:1). Place on a rotator for 20 min.
12. For resin infiltration, replace the propylene oxide/resin mixture with pure resin. Place on a rotator for 1 h. Exchange pure resin and infiltrate on a rotator for 2 h or overnight. Replace the resin a third time and allow the pure resin to infiltrate on a rotator for another 2 h or longer.
13. Using a dry toothpick, place each slice into a flat embedding mold together with its label. Orient the slices so that the plane of thin-sectioning is perpendicular to the slice's surface.
14. Polymerize at 60°C for 18–48 h.

3.6. Sectioning and On-Section Staining

1. Cut into the slice along the longitudinal axis of the block until the area of interest is reached. Trim the block as close as possible to the top and bottom surface of the slice without cutting into the surface. The surface of the slice is the area that has been labeled and is later imaged in the EM.

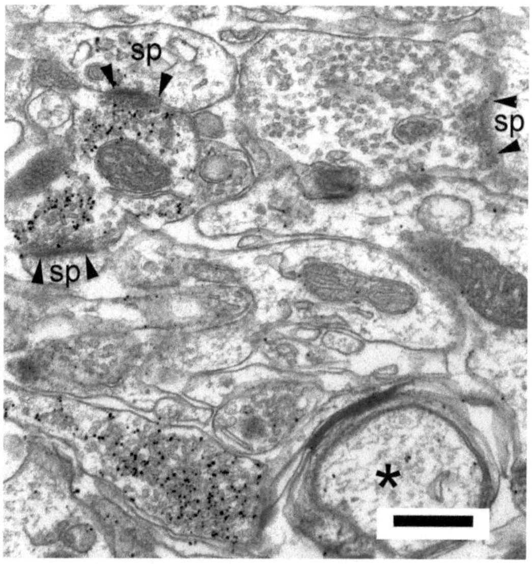

Fig. 10.1. Brain tissue labeled with antibody to the synaptic vesicle protein, synaptophysin. The tissue slice was taken from the hippocampal area CA1 of a mouse. Six presynaptic boutons are shown. The majority of the gold particles are associated with synaptic vesicles but can also be found within axons (*asterisk*) and in post-synaptic profiles. The labeling intensity is high in the bouton at the lower left corner of the micrograph, whereas the labeling is very sparse within the bouton at the upper right corner. This type of high variability among boutons is typical for hippocampal synapses. Arrowheads point to the post-synaptic density of a synapse located on spines (*sp*). Bar = 500 nm.

2. Fasten the block into the ultramicrotome and align the block face with the diamond knife.
3. Cut 500-nm sections until full sections are obtained.
4. Section 50–60 nm thin sections (silver).
5. Collect short ribbons of sections on Formvar-coated slot grids and let dry.
6. Place a 40-μL droplet of the lead solution on a flat piece of parafilm. Place a slot grid with the sections facing down on top of the droplet and stain for 2 min (*see* **Note 12**).
7. Wash thoroughly with distilled water and let dry.
8. Examine sections in the electron microscope. **Figure 10.1** shows an example of synapses that were labeled with antibody to synaptophysin, a synaptic vesicle protein that abundantly co-localizes with synaptic vesicles.

4. Notes

1. EM grade 16% paraformaldehyde solution and 25% glutaraldehyde solution can be purchased in 10-mL ampoules.
2. Membrane proteins are highly sensitive to peroxides and carbonyl compounds. Depending on the production process, detergents of the polyoxyethylene type (e.g., Triton X-100) may be contaminated with peroxides and carbonyl compounds. The extent of peroxide formation is increased by light and long storage times. Triton X-100 containing these compounds is widely used for LM and is very deleterious to the fine structure. Roche's Triton X-100, however, has been purified to reduce the amount of peroxides and carbonyl compounds and is packaged in airtight injection vials. These vials should not be opened, because exposure to air facilitates the formation of peroxides and carbonyl compounds. Instead, a syringe should be used to remove the solution through the cap. The purified Triton X-100 is a superior permeabilization reagent compared to other detergents used in EM and has almost no effect on the ultrastructure.
3. Several fluorescent tags are available. We have been using the Alexa Fluor® 488 tag because the green emission can easily be observed both by eye and with any imaging system.
4. Osmium can conveniently be bought in 2-mL ampoules that ensure optimal quality and safety in the laboratory.

5. A stable lead solution (4) is based on calcined lead citrate. The solution is stable during storage and prevents the formation of lead precipitates during staining. It also provides high contrast on-section staining. This lead solution is stable and free from precipitates when kept at room temperature or in the refrigerator for 1 year.
6. The saline in the tubing will expel the blood out of the cardiovascular system and enrich the tissue with oxygen before fixation. The amount of saline can be varied by changing the length and diameter of the tubing.
   The pressure of the solution entering the cardiovascular system affects fixation. If the pressure is too low, fixation is slow and structural preservation is compromised. If the pressure is too high, blood vessels may burst and leaking will lower the pressure within the systemic system, resulting in slow and insufficient fixation. To adjust the pressure, vary the speed of the peristaltic pump so that with a 20-gauge needle, a water beam of about 6 in. is created. With the right pressure, fixation should be evident within 1 min.
7. The section thickness is not critical for the immunostaining itself. Because only the surface will be labeled, thicker sections do not increase labeling intensity. However, vibratome sections thinner than 50 μm pose an increased risk that the tissue will tear and separate during handling. On the other hand, thick vibratome sections (>200 μm) turn very dark during osmification and dehydration, which prohibits the observation of the slice's gross structure in the light microscope. In our experience, a section of 100–150 μm is an ideal compromise that provides easy handling while allowing recognition of the various brain regions with the LM after embedding.
8. For the cost-effective use of reagents, slices can be placed in a 100-μL droplet on top of a piece of parafilm that has been taped to a culture plate. The culture plate, together with the slices, should be placed in a moist chamber on top of a shaker for overnight incubation. Shake at a moderate speed of about 150 rpm.
9. Alexa Fluor 488 is very stable in the fluorescence microscope when moderate excitation light intensities are used. Therefore, no precautions are needed, and anti-fading reagents can be omitted. The staining, however, is weaker compared to immunocytochemistry for LM because slices have been permeabilized to a smaller extent.
10. Enhancement times also depend on the concentration of the enhancement solution and will be longer for lower

concentrations. Because only a small amount of enhancement solution is used, carrying over water to the parafilm together with the slice will dilute the enhancement solution and thus extend enhancement time. For example, when 100 μL of enhancement solution is used, 50 μL of water will reduce its concentration by 33%.

11. During washing and storing, gold-enhanced slices will darken even though enhancement has stopped. The color intensity also depends on the intensity of the immunolabel, with controls being less strongly colored.
12. Moderate on-section staining with lead will result in EM micrographs with dark black gold particles, while the ultrastructure falls in the middle range of gray values. Separating the density of the gold particles and the ultrastructure by placing them into different ranges of gray values allows for the unambiguous identification of gold particles and also has the advantage that EM micrographs can more easily be thresholded for automated particle analysis.

Acknowledgments

This work was supported by the NIH grant U54 NS039408 and the NIH grant R21 NS0263208.

References

1. Hainfeld, J. F. and Furuya, F. R. (1992) A 1.4-nm gold cluster covalently attached to antibodies improves immunolabeling. *J. Histochem. Cytochem.* **40**, 177–184.
2. Newman, G. R. and Hobot, J. R. (2001) *Resin Microscopy and On-Section Immunocytochemistry*. Springer Verlag, Berlin.
3. Karnovsky, M. J. (1965) A formaldehyde-glutaraldehyde fixative of high osmolarity for use in electron microscopy. *J. Cell. Biol.* **27**, 137A.
4. Hanaichi, T., Sato, T., Iwamato, T., Malavasi-Yamashiro, J., Hoshino, M., and Mizuno, M. (1986) A stable lead by modification of Sato's method. *J. Electron Microsc.* **35**, 304–306.

Chapter 11

Pre-embedding Immunoelectron Microscopy of Chemically Fixed Mammalian Tissue Culture Cells

Haruo Hagiwara, Takeo Aoki, Takeshi Suzuki, and Kuniaki Takata

Abstract

Immunoelectron microscopy is one of the best methods for detecting and localizing protein molecules in cells and tissues. Gold particles of 1.4 nm in diameter (Nanogold) conjugated with Fab' fragments easily penetrate into the cell interior and are used for pre-embedding immunoelectron microscopy. To obtain a contrast for the gold label, silver enhancement of the gold particles is essential. By changing the intensity of the silver enhancement, the size of the granules can be controlled. In this chapter, we described the use of Nanogold for pre-embedding immunoelectron microscopy of paraformaldehyde-fixed cultured cells.

Key words: Pre-embedding method, cultured cells, Nanogold, silver enhancement.

1. Introduction

Immunoelectron microscopy using antibody probes conjugated with gold particles permits high-resolution detection and localization of antigens on and within cells. Successful detection and localization depends on the antigen-recognition specificity of the primary antibodies, preservation of the antigenicity of the antigens, and the ability of the antibodies to penetrate the cell in order to bind to the antigens. The diameters of the colloidal gold particles that are usually used in immunoelectron microscopy range from 5 to 25 nm. Antibody probes conjugated with these gold particles are too large to freely penetrate into fixed cells, even after permeabilization of the cellular membrane. They are useful in cases in which the antigen is exposed on the surface of the

S.D. Schwartzbach, T. Osafune (eds.), *Immunoelectron Microscopy*, Methods in Molecular Biology 657,
DOI 10.1007/978-1-60761-783-9_11, © Springer Science+Business Media, LLC 2010

cell or on the surface of ultrathin sections (1), but are of little use for detecting intracellular molecules with the pre-embedding method (2). Recently, the problem of the lack of penetration of immunogold particles due to their size has been overcome by the development of 1.4 nm-gold (Nanogold)-conjugated immunoprobes (3). As their small size improves penetration, the pre-embedding technique using Nanogold immunoprobes has become commonly used in immunoelectron microscopy for frozen tissue sections (4) and cultured cells (5). Although the 1.4-nm gold particles are discernible in unstained samples at high magnification by electron microscopy, their visibility can be increased with silver enhancement (6). Silver enhancement kits are now commercially available. A pre-embedding method for immunoelectron microscopy using Nanogold probes, which we have used to show the localization of inversin in the primary cilia of paraformaldehyde-fixed tissue culture cells (5), is described in this chapter.

2. Materials

2.1. Pre-embedding Immunolabeling of Cultured Cells

1. Cultured cells, adherent type.
2. Culture medium appropriate to the cell type.
3. 12-well tissue culture plate (1–4×A–C) (Becton Dickinson, Franklin Lakes, NJ) coated with type IV collagen (Sigma-Aldrich, St Louis, MO): To coat the well with type IV collagen, incubate each well with 0.01% type IV collagen in 0.05 N HCl for 15 min. After incubation, remove the collagen solution into a bottle and store it at 4°C. The recovered collagen solution can be used repeatedly. Dry the culture plate on a clean bench at room temperature and store it in the laboratory.
4. Phosphate-buffered saline (PBS): To make 1,000 mL PBS, dissolve 8.0 g of NaCl, 0.2 g of KCl, 1.44 g of Na_2HPO_4, and 0.24 g of KH_2PO_4 in 800 mL of distilled water (7). Adjust the pH to 7.4 with HCl and adjust the volume to 1,000 mL. Store at 4°C or prepare a 10x stock solution and store it at room temperature.
5. 0.2 M sodium phosphate buffer (PB), pH7.4: Prepare by using 0.2 M NaH_2PO_4 to adjust the pH of a 0.2 M Na_2HPO_4 solution to 7.4.
6. 6% paraformaldehyde (PFA) stock solution: To make 100 mL of solution, mix 6 g of PFA (EM grade, Nacalai Tesque, Kyoto, Japan) powder with 80 mL of distilled

water in a Pyrex flask containing a stir bar. Heat to 60°C, and while slowly stirring, add 1 N NaOH dropwise until the solution clears. Cool down to room temperature, add distilled water to make a final volume of 100 mL, and filter with filter paper to remove undissolved PFA (*see* **Note 1**).

7. 3% PFA in 0.1 M PB: To make 100 mL solution, mix 50 mL of 6% PFA stock solution, and 50 mL of 0.2 M PB (*see* **Note 2**).
8. 25% glutaraldehyde (GA, EM grade, TAAB, Aldermaston, England).
9. 1% GA in 0.1 M PB: To make 100 mL solution, mix 4 mL of 25% GA, 50 mL of 0.2 M PB, and 46 mL of distilled water.
10. 0.1% Triton X-100 in PBS (*see* **Note 3**).
11. 0.02 M glycine (Sigma-Aldrich, St. Louis, MO) in PBS.
12. 10% sodium azide: Dissolve 5 g of sodium azide in 50 mL of distilled water. Store at room temperature.
13. 3% bovine serum albumin (BSA) fraction V (Roche, Basel, Swiss) in PBS: Add 3 g of BSA to 100 mL of PBS and allow it to dissolve. Then, add 0.2 mL of 10% sodium azide. Store at 4°C.
14. 1% BSA in PBS: Add 1 g of BSA to 100 mL of PBS and allow it to dissolve. Then, add 0.2 mL of 10% sodium azide. Store at 4°C.
15. 0.1% BSA in PBS: Add 1 g of BSA to 1,000 mL of PBS and allow it to dissolve. Then, add 2 mL of 10% sodium azide. Store at 4°C.
16. Primary IgG antibody of known specificity.
17. Normal IgG solution: Dilute IgG from the same species as the primary antibody to 20 μg/mL in 1% BSA in PBS.
18. Nanogold solution: 1.4-nm gold particles attached to affinity-purified Fab' fragments (Nanoprobes, Yaphank, NY). Select the Nanogold antibody that specifically reacts with the primary antibody.

2.2. Silver Enhancement of Nanogold Particles

1. HQ silver enhancement kit (Nanoprobes, Yaphank, NY): Bottle A (Initiator): Pale yellow-brown solution; Bottle B (Moderator): viscous pale brown solution; Bottle C (Activator): Viscous pale brown solution (*see* **Note 4**).
2. Silver enhancement solution: Prepare the solution immediately before use in a darkroom under a safety lamp by mixing equal amounts of each component. To prepare about 1 mL of the solution, dispense 10 drops of the initiator (A) into a clean tube, add 10 drops of moderator (B), mix

thoroughly, add 10 drops of activator (C), and mix thoroughly again (*see* **Note 5**). The room temperature should be kept between 20°C and 25°C.

3. 50 mM HEPES-NaOH pH 5.8: To make 100 mL of 50 mM HEPES-NaOH (pH 5.8), add 1.192 g of HEPES to 80 mL of distilled water. Adjust the pH to 5.8 with 1 N NaOH and add distilled water to make a final volume of 100 mL. At need, prepare 1 M HEPES-NaOH (pH 5.8) stock solution, store 1 mL aliquots at −20°C, and dilute it before use.
4. 2% osmium tetroxide (OsO_4) solution (TAAB, Aldermaston, England).
5. 0.5% OsO_4 in 0.1 M PB: To make 10 mL solution, mix 2.5 mL of 2% OsO_4, 5 mL of 0.2 M PB, and 2.5 mL of distilled water.

2.3. Embedding in Plastic Resin, Ultrathin Sectioning, and Staining

1. 50, 70, 90, and 95% ethanol in distilled water.
2. 2% uranyl acetate in 50% ethanol: dissolve 1 g of uranyl acetate (TAAB, Aldermaston, England) in 50 mL of 50% ethanol.
3. Epoxy resin kit (Nisshin EM, Tokyo, Japan) containing epoxy resin (Quetol 812), dodecenyl succinic anhydride (DDSA), methyl nadic anhydride (MNA), and 2, 4, 6-tri(dimethylaminomethyl)phenol (DMP-30): To make 100 mL of resin, mix 47 mL of Quetol 812, 24.7 mL of DDSA, and 28.3 mL of MNA and stir it for 2 h. Add 2 mL DMP-30, stir the solution for 30 min, and then degas it with a vacuum pump (*see* **Note 6**).
4. 2% aqueous uranyl acetate: Dissolve 1 g of uranyl acetate (TAAB, Aldermaston, England) in 50 mL of distilled water.
5. Reynolds' lead citrate solution (8): Dissolve 1.33 g of $Pb(NO_3)_2$ and 1.76 g of $Na_3(C_6H_5O_7)2H_2O$ in 30 mL of distilled water. Slowly add 8 mL of 1 N NaOH and add distilled water to make a final volume of 50 mL.

3. Methods

In immunoelectron microscopy of adherent cultured cells, cells can be processed while still attached to the base of the plastic culture dish or as small blocks after being scraped from the plastic culture dishes. We describe a pre-embedding immunoelectron microscopy technique for chemically fixed tissue culture cells attached to the base of a culture dish.

3.1. Pre-embedding Immunolabeling of Cultured Cells (See Note 7)

1. Culture the cells in three wells (1A, 1B, 1C) of a 12-well tissue culture plate until they are confluent. Two wells (1A, 1B) are for immunolabeling, and one (1C) is for the control.
2. Gently decant the tissue culture medium and rinse the wells once with PBS.
3. Aspirate the PBS, add 3% PFA in 0.1 M PB, and gently swirl. Fix for 15 min at room temperature.
4. Remove the fixative by aspiration and rinse the cells twice with PBS.
5. Aspirate the wash buffer and permeabilize the cells by adding 0.1% Triton X-100 in PBS. Incubate for 25 min at room temperature.
6. Aspirate the 0.1% Triton X-100 in PBS.
7. Wash the cells three times for 5 min with 0.02 M glycine in PBS.
8. Aspirate the wash buffer, and add 3% BSA in PBS. Incubate for 30 min at room temperature to block nonspecific antibody binding.
9. Aspirate the blocking buffer and wash the cells with 0.1% BSA in PBS.
10. Dilute the primary IgG antibody with 1% BSA in PBS and add the antibody solution to wells 1A and 1B. Add the normal IgG solution to well 1C (*see* **Note 8**).
11. Incubate for 3 h at room temperature.
12. Aspirate the antibody solution and wash the cells six times for 5 min with 0.1% BSA in PBS.
13. Aspirate the last washing buffer, but do not allow the cells to dry.
14. Dilute the Nanogold solution 1:50 with 1% BSA in PBS and add the solution to the wells (*see* **Note 8**).
15. Incubate overnight at 4°C in a refrigerator.
16. Aspirate the Nanogold solution and wash the specimens six times for 5 min with 0.1% BSA in PBS.
17. Aspirate the last washing buffer. Wash the cells six times for 5 min with PBS.
18. Aspirate the last PBS wash.
19. Add 1% GA in 0.1 M PB and fix the cells for 10 min at room temperature.
20. Remove the fixative by aspiration and wash the cells for 5 min in PBS.

21. Aspirate the PBS. Wash the cells four times for 15 min with 50 mM HEPES-NaOH pH 5.8.
22. Aspirate the last HEPES solution. Wash the cells twice for 5 min with distilled water.

3.2. Silver Enhancement of Nanogold Particles

1. Remove the distilled water from the tissue culture plate in a dark room.
2. Add the silver enhancement solution.
3. Incubate well 1A for 6 min at 20°C and incubate wells 1B and 1C for 8 min (*see* **Note 9**).
4. Aspirate the enhancement solution.
5. Wash the cells five times for 1 min in distilled water to terminate the silver enhancement reaction.
6. Aspirate the last distilled water wash and wash the cells once for 5 min in PBS.
7. Aspirate the PBS, add 0.5% OsO_4 in 0.1 M PB (*see* **Note 10**), and postfix the cells for 1 h at 4°C. Roll the plate occasionally to ensure uniform penetration of the fixative.
8. Aspirate the fixative.
9. Rinse the cells twice for 5 min with 0.1 M PB.

3.3. Embedding in Plastic Resin, Ultrathin Sectioning, and Staining

The following procedure embeds the culture cells with good cutting properties.

1. Aspirate the 0.1 M PB.
2. Dehydrate the cells by incubation for 10 min in 50% ethanol (*see* **Note 11**).
3. Aspirate the ethanol and dehydrate the cells by incubating them for 10 min in 70% ethanol.
4. Aspirate the ethanol and dehydrate the cells by incubating them for 10 min in 90% ethanol.
5. Aspirate the ethanol and dehydrate the cells by incubating them for 10 min in 95% ethanol.
6. Aspirate the ethanol and dehydrate the cells by incubating them three times in absolute ethanol for 15 min. After the third incubation, remove the ethanol, but do not allow the cells to dry.
7. Fill each well about half full with epoxy resin (*see* **Note 12**).
8. Incubate the specimens for 1 h at room temperature.
9. Remove the epoxy resin from the well. Fill the well about half full with new epoxy resin and incubate it overnight at room temperature.
10. Remove the epoxy resin from the well.

11. Fill the well about half full with new epoxy resin.
12. Polymerize the resin at 60°C for 3–4 days.
13. Allow the plate containing the polymerized resin to cool to room temperature.
14. Remove as much of the plastic of the tissue culture plate as possible with a pair of pliers. Separate the wells from each other (**Fig. 11.1**).

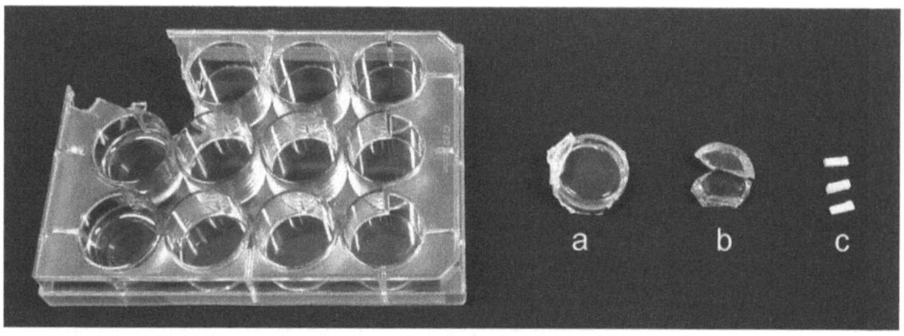

Fig. 11.1. Preparation of resin blocks for ultrathin sections. In order to cut ultrathin sections, polymerized resin specimens must be detached from the tissue culture plate. Using a pliers and a hammer, one of the wells was separated from the plastic plate (**A**). After removal of the well around the polymerized specimens, the resin block was placed on the floor with its bottom facing up. By strongly and repeatedly striking the plastic bottom with a hammer, the polymerized resin containing the resin-embedded cells was detached from the tissue culture plate (**B**). During this procedure, the resin block is usually broken into several pieces of various sizes. Finally, small pieces of specimens, 2×2×5 mm were cut from the resin blocks with a scroll saw (**C**).

15. Put the removed well on the floor with its bottom facing upward. Strongly strike the bottom of the wells with a hammer to detach the epoxy resin together with the resin-embedded cells from the plastic plate. Cut the resin blocks into small pieces (about 2×2×5 mm) with a scroll saw (**Fig. 11.1**) Set the specimen in the specimen holder and trim the top of the specimen to form a pyramid.
16. Adjust the surface of the specimen so that it is parallel with the plane of the diamond knife and start cutting ultrathin sections.
17. Place the ultrathin sections on Formvar-coated grids (150 mesh).
18. Stain the sections with 2% aqueous uranyl acetate solution for 10 min.
18. Wash the grids with distilled water.
19. Stain the sections with Reynolds' lead citrate solution for 5 min.
20. Wash the grids with distilled water and then dry them.

21. Observe the sections with a TEM and take micrographs. **Figure 11.2** shows the localization of inversin to the primary cilia of renal cells by pre-embedding immunoelectron microscopy using Nanogold and silver enhancement. Small silver granules are localized to the primary cilium.

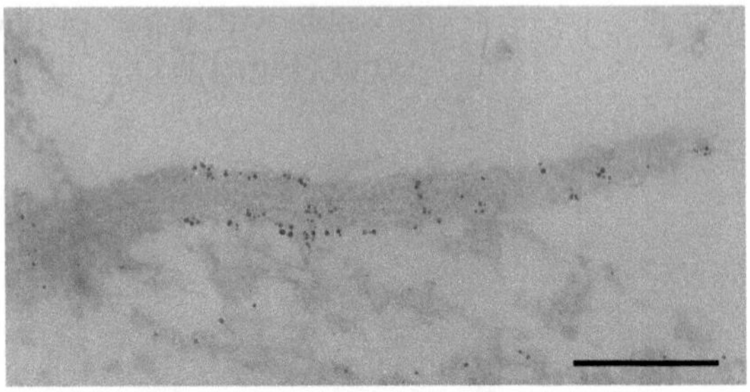

Fig. 11.2. Localization of inversin in the primary cilia of renal cells. Cultured renal cells were fixed with 3% PFA for 15 min at room temperature, permeabilized with 0.1% Triton X-100 for 25 min, and labeled with rabbit anti-inversin antibody. After washing, the specimens were incubated with 1.4 nm Nanogold-labeled Fab' fragments against rabbit IgG at 4°C overnight, washed in PBS, and fixed with 1% GA. The cells were incubated with the silver enhancement solution for 6 min at 20°C, washed, fixed with 0.5% OsO_4 for 1 h at 4°C, and then dehydrated through an ascending series of ethanols. After the embedding in epoxy resin, ultrathin sections were cut, stained with uranyl acetate and lead citrate, and examined with an electron microscope. Gold labeling is seen in a primary cilium. Silver-enhanced granules, 15–20 nm in diameter, are localized in the region between the ciliary membrane and the axoneme of the primary cilium. Bar: 1 μm.

4. Notes

1. Use of a freshly made PFA solution is recommended. A solution stored for a few weeks in the refrigerator can also be used.
2. Chemical fixation is necessary to preserve the structure of the cells and fix the antigens at their proper sites. However, the fixative that is used for regular electron microscopic observation of ultrastructures decreases the antigenicity of antigens. 2–4% PFA without GA or with reduced 0.1–0.2% levels of GA is used for immunoelectron microscopy. After immunoreaction, the specimens are refixed with a higher concentration of GA to preserve the structure and to immobilize the immunoprobes.
3. 0.1% Triton X-100 in PBS is used to permeabilize the cells fixed with PFA.
4. Store the HQ silver enhancement kit at −20°C. Before use, keep the kit at room temperature to melt each solution.

5. The solutions are viscid. It is difficult to push out the solutions from the bottle using your fingers; however, pliers are useful for this purpose.
6. The complete epoxy mixture slowly starts to polymerize at room temperature. The epoxy resin can be stored in a freezer (e.g., −80°C) for months in a closed plastic bottle. The bottle must attain room temperature before the resin is used for embedding.
7. The optimal fixative, fixation time, antibody titer, and antibody incubation conditions must be determined experimentally. The simplest way to do this is to stain cells under different conditions and examine them with a light microscope to determine the best conditions for immunolabeling.
8. About 0.3 mL of the solution is needed for a single well of a 12-well tissue culture plate to immerse the cells. If only a small quantity of antibody is available, use a 24-well tissue culture plate.
9. Silver enhancement is time-dependent. The enhancement time is the time required to obtain adequate amplification of the signal with minimum background staining. When Nanogold particles are enhanced for 4, 6, and 8 min at 20°C, the diameter of the silver-enhanced particles will be 10, 15, and 20 nm, respectively, for a single layer of cultured cells. You should try a series of different development times to determine the optimum time for your experiment.
10. 0.5% OsO_4 in 0.1 M PB is used for post-fixation after silver enhancement of the 1.4-nm gold particles. Since high concentrations of OsO_4 cause the silver granules to disintegrate, adhere strictly to the recommended concentration of 0.5% OsO_4 after silver enhancement.
11. The incubation of specimens with uranyl acetate before dehydration, so-called en bloc staining, improves the contrast in subsequent section staining. For en bloc staining, aspirate the 0.1 M PB, rinse the specimens twice for 5 min in distilled water, and then incubate them with 2% uranyl acetate in 50% ethanol for 30 min at room temperature. During staining, shield the specimens from light to prevent the precipitation of uranyl acetate.
12. Infiltrate the specimens with pure epoxy resin immediately after dehydration. As epoxy resins are more readily soluble in propylene oxide, propylene oxide is frequently used immediately prior to infiltration with the resin. However, it melts the plastic tissue culture plate so you should not use propylene oxide prior to the infiltration with the resin in this method.

Acknowledgments

This study was supported, in part, by Grants-in-Aid for Scientific Research from the Ministry of Education, Culture, Sports, Science, & Technology of Japan.

References

1. Hagiwara, H., Aoki, T., Kano, T., and Ohwada, N. (2000). Immunocytochemistry of the striated rootlets associated with solitary cilia in human oviductal secretory cells. *Histochem. Cell Biol.* **114**, 205–212.
2. Ferguson, D. J., Hughes, D. A., and Beesley, J. E. (1998) Immunogold probes in electron microscopy. *Methods Mol. Biol.* **80**, 297–311.
3. Hainfeld, J. F. and Furuya, F. R. (1992) A 1.4-nm gold cluster covalently attached to antibodies improves immunolabeling. *J. Histochem. Cytochem.* **40**, 177–184.
4. Mizoguchi, A., Nakanishi, H., Kimura, K., Matsubara, K., Ozaki-Kuroda, K., Katata, T., Honda, T., Kiyohara, Y., Heo, K., Higashi, M., Tsutsumi, T., Sonoda, S., Ide, C., and Takai, Y. (2002) Nectin: an adhesion molecule involved in formation of synapses. *J. Cell Biol.* **156**, 555–565.
5. Shiba, D., Yamaoka, Y., Hagiwara, H., Takamatsu, T., Hamada, H., and Yokoyama, T. (2009) Localization of Inv in a distinctive intraciliary compartment requires the C-terminal ninein-homolog-containing region. *J. Cell Sci.* **122**, 44–54.
6. Humbel, B. M., Sibon, O. C., Stierhof, Y. D., and Schwarz, H. (1995) Ultra-small gold particles and silver enhancement as a detection system in immunolabeling and in situ hybridization experiments. *J. Histochem. Cytochem.* **43**, 735–737.
7. Harlow, E. and Lane, D. (1999) *Using Antibodies: A Laboratory Manual.* Cold Spring Harbor Laboratory Press, Cold Spring Harbor, New York, USA.
8. Reynolds, E. S. (1963) The use of lead citrate at high pH as an electron-opaque stain in electron microscopy. *J. Cell Biol.* **17**, 208–212.

Chapter 12

Immunoelectron Microscopy of Cryofixed and Freeze-Substituted Plant Tissues

Miyuki Takeuchi, Keiji Takabe, and Yoshinobu Mineyuki

Abstract

Cryofixation and freeze-substitution techniques preserve plant ultrastructure much better than conventional chemical fixation techniques. The advantage of cryofixation is not only in structural preservation, as seen in the smooth plasma membrane, but also in the speed in arresting cell activity. Immunoelectron microscopy reveals the subcellular localization of molecules within cells. Immunolabeling in combination with cryofixation and freeze-substitution techniques provides more detailed information on the immunoelectron-microscopic localization of molecules in the plant cell than can be obtained from chemically fixed tissues. Here, we introduce methods for immunoelectron microscopy of post-embedded, cryofixed plant tissues by applying an antibody to a thin plastic resin-embedded section prepared by cryofixation followed by freeze-substitution.

Key words: Immunoelectron microscopy, plant tissue, cryofixation, freeze-substitution, high-pressure freezing.

1. Introduction

Immunoelectron microscopy enables us to study the in situ localization of specific molecules within cells. While conventional electron microscopy provides no information about specific molecules, immunoelectron microscopy can help to connect a visible structure with a specific molecule. Colloidal gold particles are often used for labeling antibodies. Among several techniques for immunolabeling are labeling methods using plastic resin-embedded samples, i.e., post-embedding and pre-embedding immunolabeling. Post-embedding labeling (antibody

S.D. Schwartzbach, T. Osafune (eds.), *Immunoelectron Microscopy*, Methods in Molecular Biology 657,
DOI 10.1007/978-1-60761-783-9_12, © Springer Science+Business Media, LLC 2010

labeling on a section of a resin-embedded specimen) is the most widely employed technique. In this method, the most important aspect for success is the balance between preservation of cell structure and retention of the antigenicity against the antibodies applied. Ultrastructural preservation in immunoelectron microscopy is usually inferior compared to that in conventional transmission electron microscopy because fixatives used for structural preservation may prevent the antibody–epitope reaction. Hence, high concentrations of glutaraldehyde and/or osmium tetroxide are avoided in immunoelectron microscopy. In addition, one of the drawbacks of chemical fixation is the speed of diffusion of the fixative into the specimen, and dramatic structural changes occur inside the cell during the penetration and fixation process with chemicals (*see* Mineyuki and Gunning (1)).

Cryofixation is an alternative to chemical fixation. This technique immobilizes molecules in the cell and immediately stops all cell activities. It has been applied in immunoelectron microscopy for plant tissues and good results have been reported, with excellent structural preservation with both the minimum use of fixatives and without fixatives (2–5). Cryofixation is usually followed by freeze-substitution, and during this procedure, water within the frozen samples is gradually substituted with pure acetone or acetone containing a low concentration of fixatives at −80°C. After freeze-substitution, specimens are gradually warmed, embedded in resin, and then an immunoreaction using gold particles is performed on ultrathin sections.

Several cryofixation techniques have been used for biological specimens, such as slam freezing (or cold metal block freezing), plunge freezing, propane jet freezing, and high-pressure freezing. Details of these techniques are summarized in the review by Gilkey and Staehelin (6). The freezing must achieve very rapid cooling rates so as to minimize damage to the sample caused by ice crystal formation. As the cell wall in plant tissues prevents rapid cooling of the sample, the well-preserved region of the specimen is relatively less than that in animal tissues. To preserve plant tissues in good condition, high-pressure freezing is often used (4, 7–9). In high-pressure freezing, specimens are frozen in a liquid nitrogen jet at 2,100 bar to prevent ice crystal nucleation and growth in the specimens. The advantages of high-pressure freezing include high efficiency in achieving good preservation and a well-preserved area that can reach 0.2 mm below the sample surface.

Preparation of ultrathin sections requires that samples be embedded in plastic resins. LR-White and Lowicryl resins are often used for immunolabeling. A good antibody–epitope reaction is expected with LR-White. This is a hydrophilic acryl resin having a low viscosity that is an advantage in infiltrating into plant tissue. Lowicryl resins such as HM20 are also widely used. HM20 is a low-temperature cured resin that can be polymerized at −60°C under UV irradiation minimizing antigenicity loss. Epoxy resins, such as Epon or Spurr, are not often used for immunomicroscopy because heating specimens are required for resin polymerization and heating can compromise the epitope-reducing immunolabeling. However, some antibodies do remain active in these resins presumably because the epitope is not heat sensitive.

In this chapter, we present a procedure for high-pressure freezing. Propane plunge freezing, a conventional method, is also described because it is easy to apply and works well, although it is applicable only to very thin specimens. Procedures for embedding in both LR-White and Lowicryl HM20 are also presented.

2. Materials

2.1. Cryofixation

1. High-pressure freezer, BAL-TEC HPM 010 (BAL-TEC AG, Balzers, Liechtenstein).
2. Specimen carrier: Freezer hats for high-pressure freezing (Brass Planchet "A" and "B"; Ted Pella, Inc., Redding, CA, USA).
3. Plunge freezer, VFZ-1 (Vacuum Devices Inc., Tokyo, Japan).
4. Propane.
5. Liquid nitrogen.
6. Onion seeds (*Allium cepa* L. cv. Highgold Nigou, Sakata Seed Co., Yokohama, Japan).

2.2. Freeze-Substitution

1. 0.5% glutaraldehyde in acetone: dilute a 70% aqueous solution of glutaraldehyde (Electron Microscopy Science, Hatfield, PA, USA) with acetone for a final concentration of 0.5% and chill in liquid nitrogen prior to use.
2. Cryotube vials (Nunc A/S, Roskilde, Denmark).

2.3. Embedding

1. LR-White (Hard) (London Resin, Berkshire, UK).
2. Lowicryl HM20 (Polysciences, Warrington, PA, USA): Mix 2.98 g cross-linker D and 17.02 g monomer E by gently stirring with a glass rod (*see* **Note 1**).
3. Lowicryl HM20 with initiator C: Mix 2.98 g cross-linker D and 17.02 g monomer E by gently stirring with a glass rod. Add 0.1 g initiator C until it is completely dissolved in the resin.
4. Rotator TYPE N (TAAB, Berkshire, UK).

5. Gelatin capsule, 8 mm in diameter (No. 00, Lilly Co., Indianapolis, Indiana, USA).
6. Ultraviolet polymerizer, TUV-200 (Dosaka EM Co., Kyoto, Japan).

2.4. On-Grid Section Immunolabeling

1. Formvar-coated nickel grid (SPI supplies, West Chester, PA, USA).
2. Ultra microtome (Leica, Wetzlar, Germany).
3. Phosphate-buffered saline (PBS): 8 mM disodium hydrogen phosphate, 1.5 mM potassium dihydrogen phosphate, 137 mM sodium chloride, 2.7 mM potassium chloride.
4. PBS-T: PBS containing 0.1% Tween 20.
5. Blocking buffer: PBS containing 0.8% bovine serum albumin (BSA), 0.1% gelatin from cold water fish skin (Sigma-Aldrich), 5% normal goat serum (same species as secondary antibody), and 2 mM NaN_3.
6. Washing buffer: PBS containing 0.8% BSA, 0.1% gelatin from cold water fish skin, and 2 mM NaN_3.
7. Primary antibody to molecule of interest.
8. Colloidal gold (5–15 nm)-conjugated secondary antibody that is appropriate for the primary antibody such as Goat anti-mouse IgG antibody, 10-nm gold conjugate (British Biocell International, Cardiff, UK).
9. 2% glutaraldehyde in PBS.
10. 2% (w/v) uranyl acetate in 60% ethanol. Store at 4°C in the dark.
11. Reynolds' lead citrate (10): dissolve 1.33 g lead nitrate and 1.76 g sodium citrate in 30 mL distilled water. The distilled water is boiled to remove carbon dioxide before use. Clarify the solution by adding 8 mL 1 N NaOH solution. Add distilled water for a final volume of 50 mL seal in a syringe and store at 4°C in the dark.

3. Methods

3.1. Cryofixation

3.1.1. High-Pressure Freezing of Onion Cotyledon

1. Germinate onion seeds in 0.05 M sucrose for 2 days and transfer to 0.1 M sucrose 1 day before freezing (9) (*see* **Note 2**).
2. Set up the high-pressure freezing apparatus according to the manufacturer's instructions.

3. Excise a piece of onion cotyledon that will fit in the sample carrier. The sample carrier, Planchet A and B are both 2.0-mm internal diameter, and the cavity depths are 0.1, 0.2, and 0.3 mm. Using various combinations of Planchet A and B, the cavity thickness can be varied from 0.1 to 0.6 mm. A thickness of 0.2 or 0.3 mm is often used. The free space of the carrier cavity is filled with 0.1 M sucrose (*see* **Note 3**).
4. Place the sample in the high-pressure freezing apparatus and freeze.
5. Immediately after freezing, transfer the sample into liquid nitrogen (*see* **Note 4**). Under liquid nitrogen, open the sample carrier sets and remove a carrier (*see* **Note 5**). Store the frozen sample in liquid nitrogen until freeze-substitution.

3.1.2. Plunge Freezing with Liquid Propane of Poplar Xylem

1. A plunge freezer VFZ-1 device is composed of containers of liquid propane and liquid nitrogen, and a sample plunger equipped forceps (*see* **Note 6**). Place liquid nitrogen in the bottom container.
2. Flush the propane gas slowly and place liquid propane in the cryogen holder (*see* **Note 7**).
3. Excise a small piece of plant tissue (about 5 mm × 2 mm × 0.5 mm).
4. Pick up the specimen with the forceps of the sample plunger and plunge the specimen into the liquid propane.
5. Immediately transfer the frozen specimen into liquid nitrogen and release it. Store the frozen sample in liquid nitrogen until freeze-substitution.

3.2. Freeze-Substitution

Transfer the cryofixed samples into a small screw cap vial containing 0.5% glutaraldehyde in acetone chilled in liquid nitrogen (*see* **Note 8**). Keep the sample at –80°C for more than 72 h to complete the freeze-substitution.

Cryofixation provides excellent ultrastructure preservation as exemplified by the remarkable difference in smooth plasma membrane ultrastructure compared to chemically fixed tissue (**Fig. 12.1**). However, organelle-dependent artifacts can arise during cryofixation and freeze-substitution. For example, microtubules, which are 25-nm diameter hollow, rod-like structures, are easily observed by conventional electron microscopy (**Fig. 12.1A arrowheads**) in chemically fixed tissue and in well-preserved cyrofixed samples (**Fig. 12.1B arrowheads**). They are susceptible to cryodamage. If specimens are not preserved well under cryofixation, their cylindrical structure collapses. Sometimes we find only a few microtubules, or damaged microtubules, in cells whose plasma membrane and other organelles seem to be successfully fixed (11).

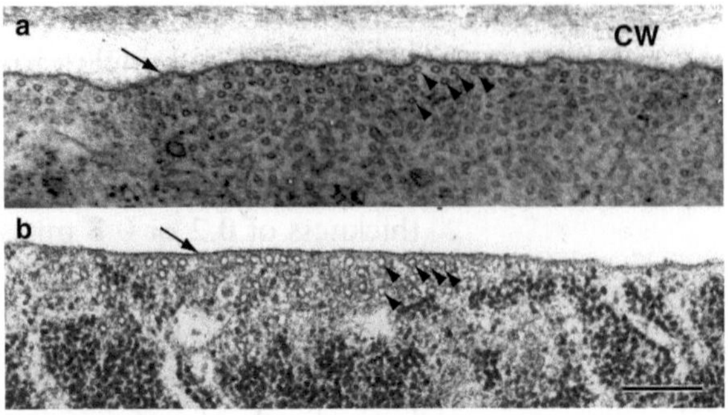

Fig. 12.1. Comparison of cross-sectional images of onion root tip cells fixed by glutaraldehyde/osmium tetroxide (**A**) and by high-pressure freezing and freeze-substitution (**B**) showing the superior ultrastructure preservation of the cryofixed sample. The plasma membrane (*arrows*) is smooth in the cryofixed cell (**B**), although it is wavy in the chemically fixed cell (**A**). These are prophase cells and the preprophase band of microtubules (*arrowheads*) is visible. CW, cell wall. Bar, 200 nm. (Reproduced from (11) with permission from *Plant Morphology*.)

3.3. Embedding in Resin

3.3.1. Embedding in LR-White

1. Gradually transfer the freeze-substituted samples in screw capped vials to room temperature (RT) by incubation at –20°C for 2 h, 4°C for 2 h, then RT for 2 h.
2. When the temperature of the sample reaches RT, remove the 0.5% glutaraldehyde in acetone from the vial and wash the specimens with acetone three times for 15 min.
3. After the third wash, remove the acetone from the vial and wash the specimens with ethanol, three times for 15 min.
4. After the third wash, remove the ethanol and add LR-White: Ethanol = 1:2 to the vial. Incubate for 2 h at RT under gentle agitation with a rotary shaker (*see* **Note 9**).
5. Incubate in LR-White: Ethanol = 1:1 overnight.
6. Incubate in LR-White: Ethanol = 2:1 for 8 h.
7. Incubate in 100% LR-White overnight.
8. Incubate in 100% LR-White for 8 h.
9. Remove the specimen from the screw cap vial and place it at the bottom of a gelatin capsule. Fill the capsule with fresh LR-White resin and close.
10. Seal capsules in a container under nitrogen gas. Polymerize at 50°C for 16 h.
11. Store the polymerized samples in silica gel.

3.3.2. Embedding in Lowicryl HM20

1. Warm freeze-substituted samples in screw capped vials to –60°C (*see* **Note 10**).

2. Remove the 0.5% glutaraldehyde in acetone from the vial and wash specimens with acetone at –60°C, three times for 1 h.
3. After the third wash, remove the acetone and incubate in 12.5% HM20 in acetone for 3 h at –60°C.
4. Incubate in 25% HM20 in acetone overnight at –60°C.
5. Incubate in 50% HM20 in acetone for 4 h at –60°C.
6. Incubate in 75% HM20 in acetone for 4 h at –60°C.
7. Incubate in 100% HM20 in acetone overnight at –60°C.
8. Incubate in 100% HM20 in acetone for 8 h at –60°C.
9. Incubate in 100% HM20 in acetone for 3 days at –60°C.
10. Incubate in 100% HM20 with initiator C in acetone for 2 days at –60°C.
11. Place a specimen in a gelatin capsule. Fill the capsule with fresh HM20 with initiator C and close the cap.
12. Polymerize at –60°C under UV for 2 days using the UV polymerizer.

3.4. On-Grid Section Immunolabeling

1. Prepare the embedded specimen block for sectioning. If it is necessary, fix the specimen block in the desired orientation on a support for the microtome. Trim the specimen surface about 0.5 mm × 0.5 mm and cut ultrathin sections (50–90 nm thick) using an ultra microtome. Pick up the sections on Formvar-coated nickel grids (*see* **Note 11**).
2. Incubate the sections with 50 mM glycine in PBS for 15 min (*see* **Note 12**).
3. Wash the sections for 5 min with PBS.
4. Incubate the sections with blocking buffer for 30 min to block nonspecific binding sites on the sections.
5. Wash the sections for 5 min with PBS.
6. Incubate the sections in the primary antibody diluted with antibody dilution buffer for 1–2 h at 37°C (*see* **Note 13**).
7. Wash the sections three times for 5 min with PBS-T.
8. Incubate the sections in the colloidal gold-conjugated secondary antibody diluted with antibody dilution buffer for 1–2 h at 37°C.
9. Wash the sections three times for 5 min PBS-T.
10. Post-fix the sections with 2% glutaraldehyde in PBS for 5 min.
11. Wash the sections with distilled water with a stream of running water from a bottle.

12. Counterstain the sections with 2% uranyl acetate in 60% ethanol for 10 min followed by Reynolds' lead citrate (10) for 2 min.
13. Examine samples under the transmission electron microscope. An example of a high-pressure frozen HM20-embedded onion cotyledon epidermal cell immunolabeled with a monoclonal anti-α-tubulin antibody is shown in **Fig. 12.2**, while a plunge-frozen LR-White-embedded secondary wall-forming fiber cell of poplar secondary xylem immunolabeled with anti-PRX3 antibody is shown in **Fig. 12.3**. Note that both the cotyledon (**Fig. 12.2**) and hard woody tissue (**Fig. 12.3**) exhibit excellent ultrastructure preservation and immunolabeling.

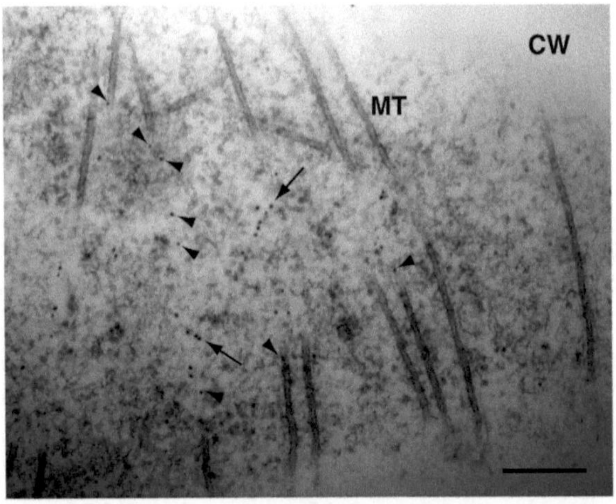

Fig. 12.2. Immunolabeling of an onion cotyledon epidermal cell with the monoclonal anti-α-tubulin antibody. A high-pressure frozen onion cotyledon was freeze-substituted with 0.25% glutaraldehyde/0.1% uranyl acetate in acetone and embedded in HM20 resin. The sample was immunolabeled with a 1/50 dilution of mouse monoclonal anti-α-tubulin primary antibody and a 1/50 dilution of goat anti-mouse IgG secondary antibody conjugated to 10-nm gold particles. Microtubules are clearly visible, but are not labeled because they are within the resin. In contrast, gold particles (10 nm diameter) appear at the end of microtubules (*arrowheads*) and on faintly contrasted linear structures (*arrows*). Microtubules may be exposed on the surface of the section at these positions. MT, microtubule: CW, cell wall. Bar, 200 nm.

4. Notes

1. Use well-ventilated fume hood for mixing resins. Avoid contact with skin and eyes and avoid inhalation of resin vapor.
2. In the case of onion cotyledon, feeding sucrose prior to freezing decreases cryodamage (9). This pretreatment is

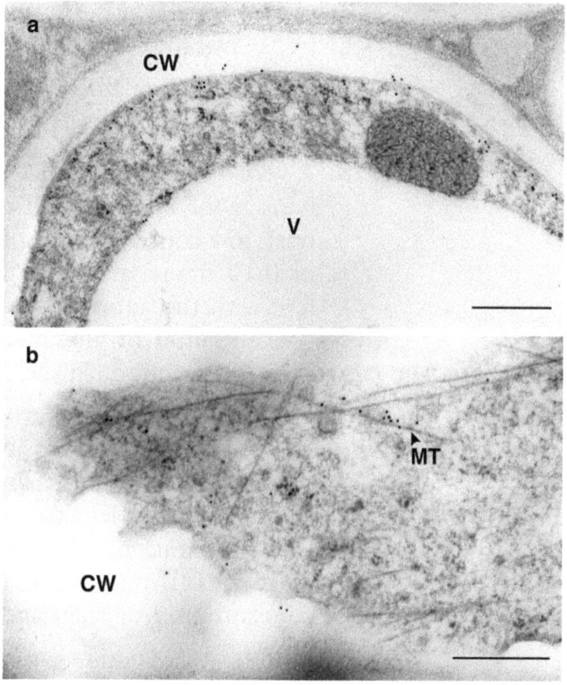

Fig. 12.3. Immunolocalization of a peroxidase with anti-PRX3 antibody in a secondary wall-forming fiber cell of poplar secondary xylem. Plunge-frozen samples were freeze-substituted with 0.5% glutaraldehyde in acetone and embedded in LR-White resin. The sample was immunolabeled with a 1/50 dilution of anti-PRX3 primary antibody and a 1/25 dilution of goat anti-rabbit IgG secondary antibody conjugated to 15-nm gold particles (5). The peroxidase is involved in woody cell wall formation and gold particles (15 nm diameter) are observed in the vicinity of the plasma membrane in the cross section (**A**). In a longitudinal section of a fiber (**B**), the gold particles are also seen in the plasma membrane area, where cortical microtubules are also visible. Note that organelles are well preserved in this method although the woody tissues are relatively hard and not easy to fix. CW, cell wall; V, vacuole; MT, microtubule. Bar, 500 nm. (Reproduced from (5) with permission from the *Journal of Wood Science*.)

not necessary for all plant materials and un-treated materials are generally supplied for freezing.

3. The carrier cavity is filled with a cryoprotectant containing solution. We use 0.1 M sucrose for onion cotyledon. Cryoprotectants, e.g., hexadecane, dextran, polyethylene glycol, and hydroxyethyl starch, have been used for various samples.
4. Work carefully with liquid nitrogen and frozen samples. Forceps and any other tools must be pre-chilled before touching the frozen sample. Wear eyeglasses and gloves to protect yourself.
5. For infiltration of fixatives into the sample, the cover carrier should be removed. The sample can remain on the

carrier for freeze-substitution and the following washing steps until the step where resin is applied.

6. Alternatively, small containers and forceps can be used.
7. Use propane under ventilation.
8. A very weak fixative, e.g., 0.25% (or 0.5%) glutaraldehyde, is used for immunoelectron microscopy. For better contrast, low concentrations of osmium tetroxide (e.g., 0.01%) or 0.1% uranyl acetate can be used but they might interfere with the antibody–antigen reaction. Even a very low concentration of glutaraldehyde can inhibit the antigen–antibody reaction. In such cases, pure acetone can be used for freeze-substitution (2).
9. LR-White is a low-viscosity resin and easily infiltrates into the specimens. The specimens should sink into the resin solution. If samples float on top of the solution at the end of an infiltration step, resin has not penetrated into the specimens. If resin infiltration is problematic, try a lower concentration of resin and/or lengthen the time of each step.
10. For temperature control, an automatic freeze-substitution system is sold by Leica. Alternatively, a −80°C deep freezer can be reset to −60°C.
11. A nickel grid is used to avoid a reaction between grid metal and the solutions.
12. For immunostaining the sections, a grid is incubated in solutions by floating section side down on a drop of solution and transferred sequentially from drop to drop to complete the immunostaining. To cover a grid, 5–50 μL drops of solution are placed on a piece of Parafilm in a Petri dish kept in a moist chamber to avoid drying of the solution. The section must never dry out during the procedure.
13. Optimal conditions for labeling with the primary and the secondary antibodies should be found through preliminary trials. The antibody dilution is often much less than that used for non-embedded samples.

Acknowledgments

This work was supported by JSPS Grant-in-Aid for Scientific Research (A) 17207006 to YM.

References

1. Mineyuki, Y. and Gunning, B. E. S. (1988) Streak time-lapse video microscopy: analysis of protoplasmic motility and cell division in *Tradescantia* stamen hair cells. *J. Microsc.* **150**, 41–55.
2. Lancelle, S. A. and Hepler, P. K. (1989) Immunogold labelling of actin on sections of freeze-substituted plant cells. *Protoplasma* **150**, 72–74.
3. Nakashima, J., Awano, T., Takabe, K., Fujita, M., and Saiki, H. (1997) Immunocytochemical localization of phenylalanine ammonia-lyase and cinnamyl alcohol dehydrogenase in differentiating tracheary elements derived from zinnia mesophyll cells. *Plant Cell Physiol.* **38**, 113–123.
4. Samuels, A. L., Giddings, T. H., Jr., and Staehelin, L. A. (1995) Cytokinesis in tobacco BY-2 and root tip cells: a new model of cell plate formation in higher plants. *J. Cell Biol.* **130**, 1345–1357.
5. Takeuchi, M., Takabe, K., and Fujita, M. (2005) Immunolocalization of an anionic peroxidase in differentiating poplar xylem. *J. Wood Sci.* **51**, 317–322.
6. Gilkey, J. C. and Staehelin, L. A. (1986) Advances in ultrarapid freezing for the preservation of cellular ultrastructure. *J Electron Microsc. Tech.* **3**, 177–210.
7. Craig, S. and Staehelin, L. A. (1988) High pressure freezing of intact plant tissues. Evaluation and characterization of novel features of the endoplasmic reticulum and associated membrane systems. *Eur. J. Cell Biol.* **46**, 81–93.
8. Zhang, G. F. and Staehelin, L. A. (1992) Functional compartmentation of the golgi apparatus of plant cells: immunocytochemical analysis of high-pressure frozen- and freeze-substituted sycamore maple suspension culture cells. *Plant Physiol.* **99**, 1070–1083.
9. Murata, T., Karahara, I., Kozuka, T., Giddings, T. H., Jr., Staehelin, L. A., and Mineyuki, Y. (2002) Improved method for visualizing coated pits, microfilaments, and microtubules in cryofixed and freeze-substituted plant cells. *J. Electron Microsc.* **51**, 133–136.
10. Reynolds, E. S. (1963) The use of lead nitrate at high pH as an electron opaque stain in electron microscopy. *J. Cell Biol.* **17**, 208–212.
11. Mineyuki, Y., Murata, T., Giddings, T. H., Jr., and Staehelin, L. A. (1998) Observation of merismatic cells in seedlings of higher plants using a high pressure freezing method. *Plant Morphol.* **10**, 30–39.

Chapter 13

In Vivo Cryotechniques for Preparation of Animal Tissues for Immunoelectron Microscopy

Shinichi Ohno, Nobuhiko Ohno, Nobuo Terada, Sei Saitoh, Yurika Saitoh, and Yasuhisa Fujii

Abstract

The final goal of immunohistochemical studies is that all findings examined in animal experiments should reflect the physiologically functional background. Therefore, the preservation of original components in cells and tissues is necessary for describing the functional morphology of living animal organs. It is generally accepted that morphological findings of various organs are easily modified during the conventional preparation steps. The quick-freezing method, by which resected tissues are quickly frozen, reduces morphological artifacts resulting in significant findings of native cells and tissues. However, tissues have to first be resected from living animal organs for quick-freezing. We have developed an "in vivo cryotechnique" for immunohistochemistry of some components in living animal organs. All physiological processes are immediately immobilized in the ice crystals by the "in vivo cryotechnique," and every components of the cells and tissues are maintained in situ at the time of freezing. Thus, ischemic or anoxic effects are minimized on immunohistochemical localization of the components. Another new "cryobiopsy" technique will be useful for capturing time-dependent morphological changes in the same animal including humans and for maintaining intracellular components.

Key words: In vivo cryotechnique, cryobiopsy, quick-freezing method, freeze-substitution, living animal organs, immunolocalization, morphofunctional study.

1. Introduction

Conventional chemical fixation and alcohol dehydration have been commonly used as easy preparation procedures for morphological observation but they always produce many technical artifacts of morphology and molecular distributions in dynamically changing cells and tissues. On the other hand, the quick-freezing

S.D. Schwartzbach, T. Osafune (eds.), *Immunoelectron Microscopy*, Methods in Molecular Biology 657, DOI 10.1007/978-1-60761-783-9_13, © Springer Science+Business Media, LLC 2010 methods, by which resected tissues are quickly frozen for physical fixation, are generally accepted to prevent technical artifacts such as shrinkage of tissues during fixation and dehydration (1). Quick-freezing has also allowed us to obtain better immunoreactivity of certain cell and tissue antigens. However, prepared tissues have to first be resected from living animal organs for the quick-freezing and inevitably exposed to the artificial stresses of ischemia and anoxia.

To resolve these technical problems, we have developed a new "in vivo cryotechnique" to strictly clarify morphofunctional significance in living animal organs reporting dynamically changing immunolocalizations of functional proteins in cells and tissues (2, 3). The "in vivo cryotechnique" by which all cells and tissues of living animal organs are immediately cryofixed in vivo prevents the common technical artifacts of resected tissue specimens including ischemia and anoxia described in the previous paragraph. The directly frozen tissue specimens can be processed in the same way as those prepared by conventional quick-freezing methods including slamming or plunging quick-freezing and high-pressure freezing. The "in vivo cryotechnique" has been used for immunohistochemical and morphological examination of cells and tissues by light or electron microscopy providing new insights into organ physiology or pathology (2, 3). The application of the "in vivo cryotechnique" to immunohistochemistry has also revealed more precisely the intracellular or extracellular localizations of dynamically changing molecules (3). The purpose of this chapter is to describe experimental protocols for the "in vivo cryotechnique" including "cryobiopsy" that are suitable for immunohistochemical applications.

2. Materials

2.1. Animal Preparation

1. Animals: Adult C57BL/6 male mice, weighing 20–25 g and Wister male rats weighing 200–250 g.

2.2. Mixed Isopentane–Propane (IP) Cryogen Preparation

1. Isopentane (*see* **Note 1**)
2. Propane (*see* **Note 2**)
3. Liquid nitrogen
4. Styrofoam box with cover
5. 50-mL glass beaker wrapped with wire for suspending in Styrofoam box
6. Stirring bar
7. Magnetic stirrer

2.3. Plunging Quick-Freezing (QF) Method

1. Pentobarbital (*see* **Note 3**)
2. Tweezers and pins
3. Razor blades
4. Filter paper (Type No. 2; TGK Co., Tokyo, Japan)

2.4. Freeze-Substitution (FS) Fixation Method

1. 20% paraformaldehyde (PF) in water: Prepare by adding 20 g EM grade PF (Merk KGaA, Darmstadt, Germany) to approximately 80 mL of water. Heat using a stirring hot-plate in a fume hood to 60°C to dissolve the powdered PF. Cool and add distilled water for a final volume of 100 mL (*see* **Note 4**).
2. 2% PF in acetone (*see* **Note 5**): Dilute the 20% PF stock solution tenfold with pure acetone. Incubate in a refrigerator for a few days with sufficient Molecular Sieves 3A (Nacalai Tesque Inc., Kyoto, Japan) for complete dehydration.
3. 0.25% glutaraldehyde (GL) in acetone (*see* **Note 5**): Prepare by diluting 70% aqueous EM grade GL (TAAB, Berks, UK) with pure acetone and dehydrate in the same way as described above.

2.5. In Vivo Cryotechnique (IVCT)

2.5.1. In Vivo Cryotechnique with IP Cryogen Alone

1. Electric dental drill: BL-F5A (Osada Electric Company, Tokyo, Japan).

2.5.2. In Vivo Cryotechnique with a Cryoknife

1. In vivo cryoapparatus VIO-10 (Eiko Engineering, Hitachinaka, Ibaraki, Japan).
2. Cryoknife (Eiko Engineering) precooled with liquid nitrogen.

2.5.3. Operation of the "In Vivo Cryoapparatus"

1. In vivo cryoapparatus (Eiko Engineering).

2.6. Cryobiopsy for Anesthetized Animal

1. Homemade cryoforceps (**Fig. 13.1**, *see* **Note 6**).

2.7. Embedding of Tissues After FS Step

1. LR-gold resin (*see* **Note 7**) (Polyscience Inc., Warrington, PA, USA).
2. Benzil (*see* **Note 8**) (Nakalai Tesque Inc., Kyoto, Japan).

2.8. Post-embedding Immunostaining

1. 0.1 M phosphate buffer (PB) pH 7.4: Dissolve 11.86 g of $NaH_2PO_4{\cdot}2H_2O$ and 115.99 g of $Na_2HPO_4{\cdot}12H_2O$ in 4 L of distilled water.
2. PBS: PB containing 0.15 M sodium chloride.

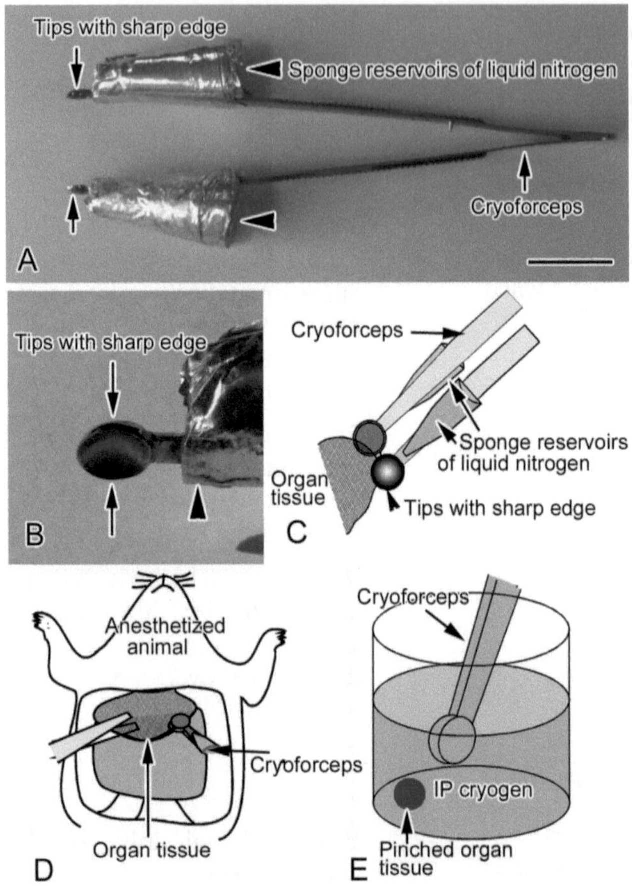

Fig. 13.1. Pictures of the homemade cryoforceps and schematic representation of the cryobiopsy system. (**A**) Homemade cryoforceps showing tips which are round cups whose edges are sharpened (*arrows*) with sponges just above the tips to absorb liquid nitrogen cooling the tips to liquid nitrogen temperature (*arrowheads*). Bar: 2 cm. (**B**) Closeup of rounded tips. (**C**) The cryoforceps contain a tissue-pinching tip and liquid nitrogen reservoirs. (**D**) The precooled cryoforceps are used for pinching off the target organ tissue. (**E**) The cryobiopsied tissue is quickly plunged into the isopentane–propane (IP) cryogen.

3. Rabbit primary antibody (Sigma Chemical, St. Louis, MO, USA) or biotinylated substrate-specific binding protein such as hyaluronic acid binding protein (B-HABP) (Seikagaku Co., Tokyo, Japan).
4. Goat anti-rabbit IgG antibody conjugated with 15-nm gold particles (Amersham, Arlington, IL, USA) or Goat anti-biotin antibody conjugated with 10-nm gold particles (*see* **Note 9**) (British Biocell International, UK).
5. Formvar-coated nickel grid (Electron Microscopy Sciences, Hatfield, PA) (*see* **Note 10**).

3. Methods

3.1. Animal Preparation

1. Adult C57BL/6 mice and Wister rats are housed in an air-conditioned room and freely fed on a commercial diet and supplied with water.

3.2. Mixed Isopentane–Propane (IP) Cryogen Preparation

1. Hang a 50-mL glass beaker containing a magnetic stirring bar in a Styrofoam box (**Fig. 13.2A**).

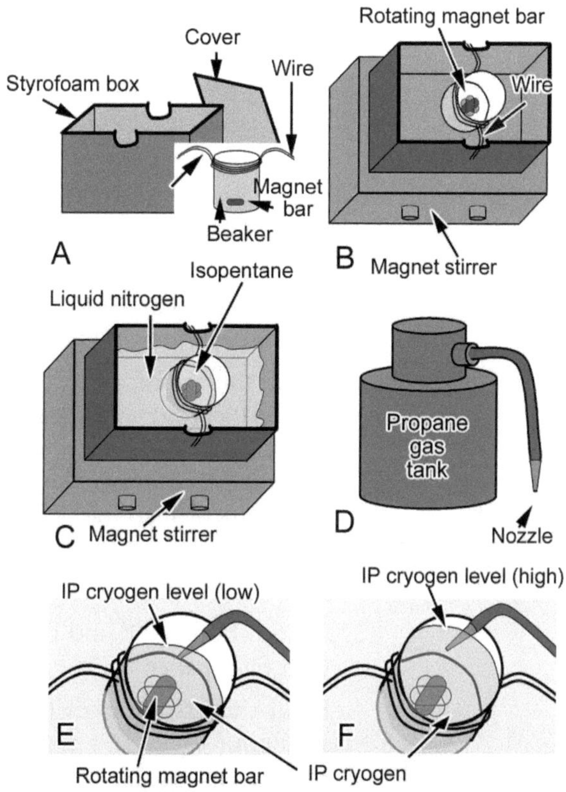

Fig. 13.2. Schematic drawings which show how to prepare the isopentane–propane (IP) cryogen in a homemade plunge-freeze apparatus. (**A**) A Styrofoam box with slits to hold the ends of the wire wrapped around a beaker, the box cover, and a 50-mL beaker containing a magnetic stirring bar wrapped with wire to hang the beaker in the Styrofoam box. (**B**) The beaker wrapped with wire is hung in the box which is on a magnetic stirrer so that the stirring bar can smoothly rotate in the beaker. (**C**) Liquid nitrogen (−196°C) is poured into the box with the hanging beaker which contains about 15 mL of liquid isopentane which is continuously stirred with the stirring bar. When the bottom side of the beaker becomes submerged in the liquid nitrogen, the liquid nitrogen begins to bubble. The magnetic bar is continuously stirring the liquid isopentane alone. (**D**) Propane gas is blown out of a nozzle (*arrow*) to be quickly liquefied. (**E**) The propane gas is jetted out of the nozzle into the cooled isopentane where it liquefies and the mixed IP cryogen level gradually elevates. (**F**) The amount of liquid IP cryogen is finally about 45 mL and the release of propane gas from the nozzle is stopped.

2. Place the box on a magnetic stirrer and check that the stirring bar rotates in the beaker (**Fig. 13.2B**).
3. Remove the beaker from the box and fill the box with an appropriate amount of liquid nitrogen so that when the beaker is placed back in the box, it will be submerged in liquid nitrogen (**Fig. 13.2C**).
4. Pour 15 mL of liquid isopentane into the beaker at room temperature. Hang the beaker in the box containing the liquid nitrogen and stir the isopentane. The initial extensive bubbling of the liquid nitrogen will decrease as the isopentane cools (**Fig. 13.2C**).
5. When the peripheral part of the isopentane cools almost to the temperature of the liquid nitrogen (−196°C) and begins to solidify, immediately place the tip of the nozzle from the propane gas tank in the cooled isopentane (**Fig. 13.2D**) and vigorously blow the propane gas into the cooled isopentane (**Fig. 13.2E**) to liquefy the propane.
6. When the level of IP cryogen in the beaker gradually increases, reaching a volume of about 45 mL (**Fig. 13.2F**), a ratio of isopentane to propane of 1:2, immediately stop the propane gas flow (4) (*see* **Note 11**).
7. Cover the Styrofoam box to minimize exposure of the prepared IP cryogen to humid air (**Fig. 13.2A**) (*see* **Note 12**).

3.3. Plunging Quick-Freezing (QF) Method

1. Anesthetize animals with an intraperitoneal injection of sodium pentobarbital (100 μg/g body weight).
2. Carefully take out target organs and immediately cut them into small tissue pieces with razor blades.
3. Absorb body fluid and blood around the small tissue pieces with filter paper.
4. Hold the tissue pieces lightly with a pair of tweezers or a stick pin.
5. Plunge the tissue pieces quickly into the liquid IP cryogen (about −193°C) cooled with liquid nitrogen (−196°C) (**Fig. 13.2**, *see* **Note 13**).

3.4. Freeze-Substitution (FS) Fixation Method

1. Cool the 2% PF in acetone or the 0.25% GL in acetone to about −80°C in a dry ice–acetone bath just before use.
2. Put the frozen tissues into the cooled 2% PF in acetone or 0.25% GL in acetone and freeze-substitute them at about −80°C for 24 h (*see* **Note 14**).
3. Elevate the temperature of specimens to room temperature simply by sequentially incubating them in a deep-freezer (−25°C) for 2 h, a common refrigerator for 2 h, and finally onto a table at room temperature.

3.5. In Vivo Cryotechnique (IVCT)

The IVCT can be simply performed without any special instruments (*see* **Note 15**).

3.5.1. In Vivo Cryotechnique with IP Cryogen Alone

1. Pour the liquid IP cryogen (−193°C) over a target organ of an anesthetized animal and then pour liquid nitrogen over the organ.
2. Plunge the whole organ into liquid nitrogen contained in a Styrofoam box as soon as pouring of both the IP cryogen and liquid nitrogen is finished.
3. Remove a piece of the frozen tissue with an electrical dental drill under liquid nitrogen.
4. Store the frozen tissue in liquid nitrogen until the next FS step.

3.5.2. In Vivo Cryotechnique with a Cryoknife

1. Cryocut an exposed animal organ with a cryoknife precooled in liquid nitrogen, while simultaneously pouring IP cryogen over the organ (*see* **Note 16**) (2).
2. Crack off the cryocut and frozen organ from the animal body under liquid nitrogen.
3. Remove the well-frozen parts near the tissue surface with the electric dental drill under liquid nitrogen.

3.5.3. Operation of the "In Vivo Cryoapparatus" (see *Note 17*)

1. Pour liquid nitrogen into reservoirs 1 and 2, to cool them down (**Fig. 13.3**).
2. Set the timers on the controller for the desired amount of time (5 s); cryogen will be released and press the foot switch to check that the liquid nitrogen in the reservoirs flows through nozzles 1 and 2.
3. Replace liquid nitrogen in reservoir 1 with the IP cryogen.
4. Surgically expose a target organ of a live animal and put the organ on a small plate (*see* **Note 18**).
5. Set the timers on the controller for the desired amount of time (5 s); cryogen will be released onto the target organ and cool the cryoknife in liquid nitrogen stored in a separate container.
6. Shift the precooled cryoknife onto the target organ and press the foot switch to initiate cryogen release. Immediately after pressing, cut the organ manually with the cryoknife, while the IP cryogen is released onto the organ through nozzle 1 with a little delay.
7. Several seconds after the IP cryogen is poured onto the organ, liquid nitrogen is automatically poured onto the organ through nozzle 2.
8. Put the frozen target organ as a whole in liquid nitrogen.

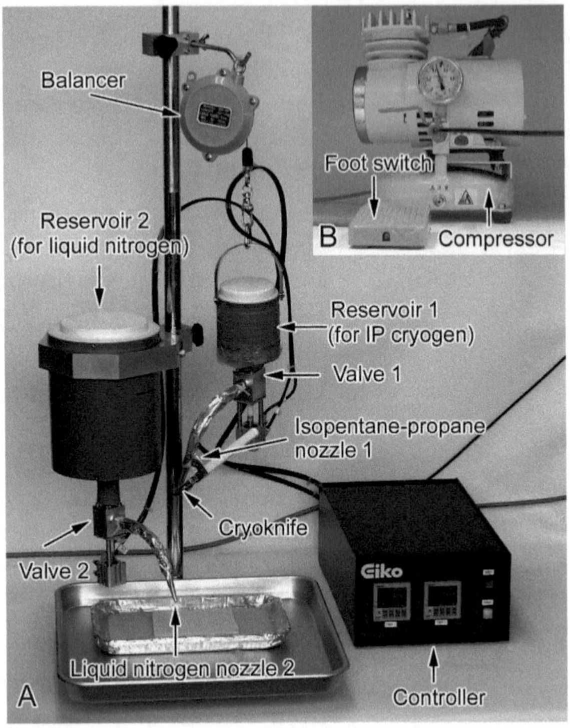

Fig. 13.3. Photographs of the newly developed in vivo cryoapparatus. (**A**) The main part of the apparatus set up on a table. (**B**) The system operating the valves controlling IP cryogen and liquid nitrogen release is placed under the table. As the target organ is cut with the cryoknife, the isopentane–propane (IP) cryogen cooled down in liquid nitrogen in reservoir 1 is immediately poured through nozzle 1 onto it. Then liquid nitrogen is poured out through nozzle 2 from reservoir 2. The opening time of valves 1 and 2 started by pushing the foot switch is automatically regulated by the controller with the support of the air compressor.

9. Get the necessary tissue parts using the electric dental drill under liquid nitrogen.

3.6. Cryobiopsy for Anesthetized Animal

1. Precool metal tips of the cryoforceps in liquid nitrogen (**Fig. 13.1B**).
2. Prepare the liquid IP cryogen beforehand (*see* **Section 3.2**) and expose a target organ of an anesthetized animal (**Fig. 13.1C,D**).
3. As soon as the metal tips are sufficiently cooled with liquid nitrogen, pinch off a piece of the target organ (**Fig. 13.1C,D**) and immediately plung it into the liquid IP cryogen (**Fig. 13.1E**) (*see* **Note 19**).
4. Transfer the frozen specimen into liquid nitrogen.

3.7. Embedding of Tissues After FS Step

1. Wash the freeze-substituted tissues three times for 1 h in pure acetone, incubate them three times for 30 min in ethanol, and finally embed the tissues by incubating them three times for 6 h in LR-gold containing 0.2% w/v benzil.

2. Polymerize the samples in small plastic capsules at −30°C under ultraviolet light for 24 h.

3.8. Post-embedding Immunostaining

1. Cut ultrathin sections at 70 nm thickness on an ultramicrotome and mount them on Formvar-coated nickel grids.
2. Incubate the samples with the B-HABH at a dilution of 1:150 at room temperature for 1 h.
3. After washing in PBS, incubate the samples with the goat anti-biotin antibody conjugated with 10-nm gold particles in a moist chamber at room temperature for 1 h.
4. Wash the samples in PBS five times.
5. Incubate the samples with rabbit anti-fibronectin IgG antibody at a dilution of 1:100 for 1 h.
6. After washing the samples in PBS, incubate them with the goat anti-rabbit IgG antibody conjugated with 15-nm gold particles.
7. Fix the samples with 0.25% GL in PB for 30 min.
8. Counterstain the samples with uranyl acetate for 3 min.

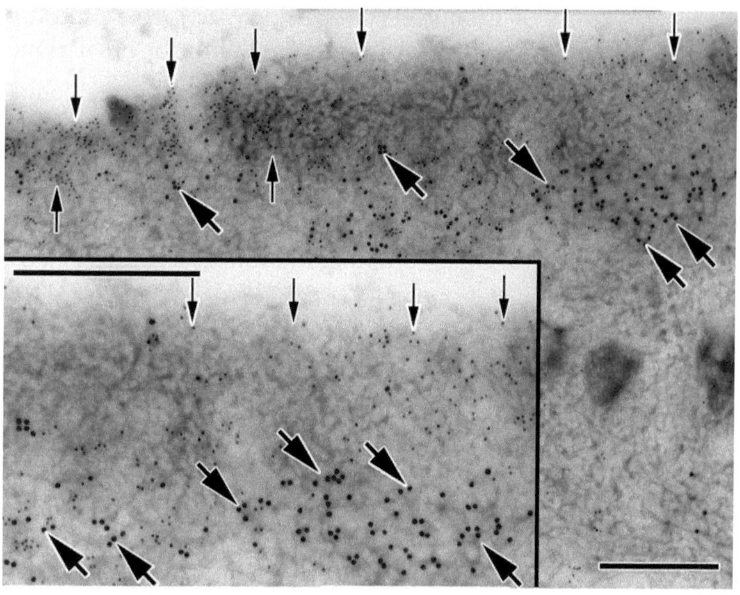

Fig. 13.4. Immunoelectron micrographs of the upper surface layer of rat condylar cartilage prepared by the quick-freeze plunging method, fixed by freeze-substitution with paraformaldehyde, labeled with rabbit anti-fibronectin and biotinylated B-HABP followed by fibronectin detection with goat anti-rabbit IgG conjugated with 15-nm gold particles (*large arrows*) and hyaluronic acid detection with goat anti-biotin antibody conjugated with 10-nm gold particles (*small arrows*). The surface areas appear to be thick and composed of microfibril-like structures. Positive 10-nm gold labeling for hyaluronic acid is observed along the superficial areas (*small arrows*). Immunogold particles of 15 nm in size, corresponding to fibronectin (*large arrows*), are also localized in the upper surface layer. *Inset*: higher magnification. Scale bar: 0.5 μm.

9. Examine the samples in an electron microscope. **Figure 13.4** shows the distribution of fibronectin (15-nm gold particles) and hyaluronic acid (10-nm gold particles) in the upper surface layer of rat condylar cartilage.

4. Notes

1. Isopentane is an extremely volatile and flammable liquid (boiling point, 28°C) at room temperature and pressure. It is commonly used in conjunction with liquid nitrogen to achieve a liquid bath at a temperature of about −160°C.
2. Propane is normally a flammable gas that is compressed to a transportable liquid (boiling point, −42°C). It is used as a cryogen (melting point, −188°C) at liquid nitrogen temperature (−196°C).
3. Pentobarbital is a short-acting barbiturate which is available as both a free acid and a sodium salt. One trade name for this drug is Nembutal which is used as an anesthetic.
4. Paraformaldehyde is toxic and can be absorbed by contact through the skin or by inhalation.
5. A fixative of 2% paraformaldehyde or low concentrations (0.05–0.25%) of glutaraldehyde is often used for immunoelectron microscopy to maintain ultrastructure and antigenicity of various tissues. The optimal concentrations of fixatives must be tailored to individual antigens by determining immunoreactivities in a trial and error manner.
6. The cryoforceps consist of a stainless steel forceps and it has reservoirs for liquid nitrogen (**Fig. 13.1A–C**). It is constructed from a 150-mm stainless steel forceps (model number F-2051; Keisei Medical Industrial Co., Tokyo, Japan) with cupped tips (5 mm in diameter). The handmade reservoirs for liquid nitrogen are made by cutting the bottom off of a plastic microfuge tube (Japan BD Co., Tokyo, Japan), inserting kitchen sponges which absorb the liquid nitrogen into the tubes, and inserting the forceps into the sponge containing tubes. The tubes are wrapped with aluminum foil (Nihonseihaku Co., Tokyo, Japan) (**Fig. 13.1A–C**).
7. LR-gold rapidly penetrates tissues because of its low viscosity. It can be used directly after alcohol dehydration and it is polymerized by ultraviolet light at low temperatures.
8. Benzil is an organic compound, which is employed as a photoinitiator for free-radical curing of polymer networks.

9. The use of primary antibodies raised in two different organisms, for example, goat and rabbit antibodies to two different proteins or a primary antibody and a substrate-specific protein such as a lecithin or B-HABH allows two molecules to be monitored in the same experiment by using different sized gold particles. By using goat anti-rabbit IgG conjugated to 15-nm gold particles the molecule bound by the rabbit primary antibody is detected while by using goat anti-mouse IgG conjugated to 10-nm gold particle the molecule bound by a mouse primary antibody is detected in the same micrograph and distinguished from the molecule bound by the rabbit antibody. Alternatively, a goat anti-biotin antibody conjugated with 10-nm gold particles as shown in **Fig. 13.4** can be used to distinguish the location of a molecule binding a biotinylated probe from the location of the molecule binding a primary antibody.
10. Nickel grids do not interact with the reagents used for immunogold staining.
11. The maximum cooling speed is obtained with a 1:2 isopentane–propane ratio (4). The time necessary for preparation of the liquid IP cryogen is usually less than 5 min. A liquid IP cryogen with high cooling ability plays a crucial role in the successful cryofixation of target organs. Its cooling speed is reported to be higher than that of isopentane or propane alone (4). By plunging resected tissues into the IP cryogen, few visible ice crystals are produced within areas about 10 μm away from the frozen tissue surface at an electron microscopic level.
12. Frost easily forms in the cryogen resulting in a decrease in its cooling speed.
13. Plunge-freezing is a relatively simple and inexpensive procedure. The sample size must be kept small in order to obtain a high surface to volume ratio resulting in an increased freezing speed. A homemade plunge-freeze apparatus has been useful for observation of frozen and hydrated specimens (**Fig. 13.2**).
14. The FS fixation method is used to chemically cross-link structural and soluble molecules in cells and tissues at about −80°C by substituting ice crystals with the organic acetone containing the fixative 2% PF or 0.05–0.25% GL.
15. There are a few technical tips to achieve better freezing by IVCT. The surgically exposed animal organs, naturally wet with body fluid, should not be dried, nor should too much liquid, e.g., blood or poured saline, be allowed to remain on the organ surface. If too much liquid remains on the organ surface, only the liquid will be well-frozen, while

visible ice crystals could be formed in the tissues themselves. Therefore, to obtain well-frozen tissue areas without any visible ice crystals, they should be covered with a little amount of liquid. In such a case, the well-freezing level without any visible ice crystal formation is limited to within less than 10 μm from the frozen tissue surface. To analyze deeper areas of the frozen tissues, anesthetized animal organs must first be cut with a cryoknife precooled in liquid nitrogen (*see* **Section 3.5.2**), instead of simply pouring the liquid IP cryogen over them.

16. When a cryoknife precooled in liquid nitrogen passes through the target organ, the exposed tissue surface in direct contact with the cryoknife is first frozen in the same way as the metal contact method (2). Then, the widely cryocut tissues are additionally frozen by the liquid IP cryogen which is simultaneously poured over them. In this process, the physical speed of the moving cryoknife and the freezing intensity of the cryogen are critical for rapidly freezing the animal organs.
17. It is sometimes difficult for an operator to perform all of the procedures necessary to constantly obtain well-frozen tissue specimens by manual IVCT. To perform IVCT more easily, the "in vivo cryoapparatus" is now commercially available all over the world. By using this cryoapparatus, both outflows of IP cryogen and liquid nitrogen are automatically controlled for the cryofixation of target organs in anesthetized animals.
18. It is recommended to place a small thin plate made of rubber or plastic and wrapped with aluminum foil under the target organ to make it easy to remove. Without such a plate, the cryofixed tissue becomes attached to other adjacent organs in the ice block or pushed deeply into the animal's body making it difficult to handle.
19. Even though the tip of the cryoforceps has sharp edges, some damaged areas are produced in the cryobiopsied specimens. It is essential to find good areas within the specimen without ice crystal damage and mechanical compression. The cooling speed of the metal cryoforceps in liquid nitrogen was measured using a thermocouple and found to be higher than that of liquid nitrogen and lower than that of the IP cryogen. Actually, in the procedure of the cryobiopsy system, the surface areas of pinched tissues are quickly frozen by the metal cryoforceps at the liquid nitrogen temperature and further cooled down by the liquid IP cryogen.

References

1. Menco, B. P. (1986) A survey of ultrarapid cryofixation methods with particular emphasis on applications to freeze-fracturing, freeze-etching, and freeze-substitution. *J. Electron Microsc. Tech.* **4**, 177–240.
2. Ohno, S., Terada, N., Fujii, Y., Ueda, H., and Takayama, I. (1996) Dynamic structure of glomerular capillary loop as revealed by an in vivo cryotechnique. *Virchows Arch.* **427**, 519–527.
3. Ohno, S., Terada, N., Ohno, N., Fujii, Y., and Baba, T. (2006) "In vivo cryotechnique" for examination of living animal organs, further developing to "cryobiopsy" for humans. *Recent Res. Dev. Mol. Cell Biol.* **6**, 65–90.
4. Jehl, B., Bauer, R., Dorge, A., and Rick, R. (1981) The use of propane/isopentane mixtures for rapid freezing of biological specimens. *J. Microsc.* **123**, 307–309.

Chapter 14

Immunoelectron Microscopy of Cryofixed Freeze-Substituted Mammalian Tissue Culture Cells

Akira Sawaguchi

Abstract

Mammalian tissue cultured cells are widely used in cell biology research. Immunoelectron microscopy is a powerful technique to define the subcellular localization of targeted antigens in the cultured cells. Cryofixation is now generally accepted as the best initial fixation step, to preserve the cellular fine structures and antigenicity. This chapter covers the practical procedures for immunoelectron microscopy of cryofixed freeze-substituted mammalian tissue cultured cells processed by high-pressure freezing.

Key words: Immunogold labeling, cryofixation, freeze-substitution, mammalian cultured cells, lowicryl resin, potassium permanganate.

1. Introduction

Mammalian tissue cultured cells are widely used in cell biology research. Recently, most of the immunocytochemical localizations of biochemically defined antigens have been performed with immunofluorescent markers using confocal laser scanning microscopy. Although the resolution of current confocal laser scanning microscopes has remarkably improved, immunoelectron microscopy is still an invaluable technique to define the "subcellular" localization of targeted antigens in the cultured cells.

The goal of specimen preparation for immunoelectron microscopy of mammalian tissue cultured cells is to preserve both subcellular fine structures and antigenicity of targeted antigens as close to their native state as possible. Cryofixation is now generally accepted as the best initial fixation step providing superior

S.D. Schwartzbach, T. Osafune (eds.), *Immunoelectron Microscopy*, Methods in Molecular Biology 657,
DOI 10.1007/978-1-60761-783-9_14, © Springer Science+Business Media, LLC 2010

preservation of fine structure and antigenicity. Currently, high-pressure freezing followed by freeze-substitution is the most reliable method to obtain a high yield of vitreous (ice-crystal damage-free) freezing (1). The application of high-pressure (in the range of 2,100 bars) lowers the freezing point and reduces the rate of ice nucleation and ice-crystal growth. In fact, a variety of methods have been developed for the high-pressure freezing of cultured cell monolayer in the last decade (2–4). In this chapter, practical procedures are described for immunoelectron microscopy of mammalian tissue cultured cells processed by high-pressure freezing/freeze-substitution.

2. Materials

2.1. High-Pressure Freezing of Cultured Cells

1. Specimen carrier: Type B aluminum planchette (M. Wohlwend, Switzerland) for high-pressure freezing machine HPM 010.
2. Tissue culture medium (*see* **Note 1**).
3. Plastic tissue culture dishes or 24-well culture plates.
4. High-pressure freezing machine (HPM 010; BAL-TEC, Liechtenstein, *see* **Note 2**).
5. Liquid nitrogen.

2.2. Freeze-Substitution Followed by Resin Embedding

1. Automatic freeze-substitution machine (AFS; Leica, Vienna, Austria, *see* **Note 3**).
2. Acetone.
3. Ethanol.
4. 0.5% glutaraldehyde in acetone: Prepare by diluting 10% glutaraldehyde in acetone (Electron Microscopy Sciences, Hatfield, PA) with acetone.
5. 1% osmium tetroxide in acetone.
6. Cryogenic vials (1.5 mL; Nalge Nunc Int., Rochester, NY).
7. Flat bottom embedding capsule (TAAB, Berks, England).
8. Lowicryl resin kit (K4M or HM20; Electron Microscopy Sciences, Hatfield, PA) (*see* **Note 4**): Prepare Lowicryl K4M resin by mixing 2.7 g of crosslinker, 17.3 g of monomer, and 0.1 g of initiator in Lowicryl K4M kit or Lowicryl HM20 resin by mixing 2.98 g of crosslinker, 17.02 g of monomer, and 0.1 g of initiator in Lowicryl HM20 kit, in a disposable beaker using a wooden tongue depressor.

9. Ultraviolet light (*see* **Note 5**).
10. Epoxy resin (Epon 812, DDSA, MNA, DMP-30). In a disposable beaker, combine 20.3 g of Epon 812 (TAAB, Berks, England), 5.0 g of DDSA (dodecenyl succinic anhydride; TAAB, Berks, England), and 15.2 g MNA (nadic methyl anhydride; TAAB, Berks, England) and mix thoroughly using a wooden tongue depressor. Then add 0.8 g of DMP-30 (2,4,6-trimethylamino methyl phenol; TAAB, Berks, England) and mix thoroughly using a wooden tongue depressor. After degassing in a vacuum desiccator, store tightly sealed at room temperature, and use that day.
11. Oven (can be set around 60°C).
12. Ultramicrotome.
13. Diamond knife.
14. Formvar-coated 150-mesh gold (or nickel) grids (Electron Microscopy Sciences, Hatfield, PA).

2.3. Immunolabeling Ultrathin Sections

1. Primary monoclonal or polyclonal antibody.
2. Secondary antibody conjugated with colloidal gold (*see* **Note 6**).
3. Bovine serum albumin (BSA, *see* **Note 7**).
4. Normal whole serum, from the species in which the secondary antibody is raised.
5. Phosphate buffer saline (PBS): Prepare 10 times concentrated stock solution by dissolving 23.0 g of disodium hydrogen phosphate (Na_2HPO_4), 5.24 g of sodium dihydrogen phosphate ($NaH_2PO_4{\cdot}H_2O$), 175.32 of sodium chloride separately in distilled water then mix, make up to 2 L. Store at room temperature and dilute 10 times with distilled water for use.
6. 5% sodium meta-periodate in distilled water.

2.4. Contrasting Immunolabeled Embedded Specimens

1. 5% uranyl acetate in water.
2. 2% uranyl acetate in 70% methanol.
3. Reynolds' lead solution: Prepare by adding 1.33 g of lead nitrate and 1.76 g of tri-sodium citrate dihydrate to distilled water and shake vigorously for several min. A milky white suspension will be formed. Agitate the solution over 30 min and then add 8 mL of freshly prepared 1 N NaOH with agitation. The white suspension will turn clear during this step. Make up to 50 mL with distilled water.
4. A stock solution of 1% (w/v) $KMnO_4$ is prepared by dissolving $KMnO_4$ in distilled water. Just before staining, add

100 μL of the $KMnO_4$ stock solution and 100 μL of 1 N H_2SO_4 to 800 μL of distilled water for a final concentration of 0.1% $KMnO_4$ in 0.1 N H_2SO_4.

3. Methods

3.1. High-Pressure Freezing of Cultured Cells (see Note 8)

1. Cultivate mammalian cells on the specimen carriers placed in culture dishes or wells containing culture medium (*see* **Note 9**).
2. Assemble the specimen carriers in the specimen holder of the HPM 010.
3. Cryofix the cells at 2,100 bar in the HPM 010.
4. Immediately transfer the assembly of specimen carriers to liquid nitrogen.

3.2. Freeze-Substitution Followed by Resin Embedding (see Note 10)

3.2.1. Freeze-Substitution with Glutaraldehyde Followed by Lowicryl Resin Embedding

1. Within liquid nitrogen vapor, transfer the specimen carriers with cryofixed cultured cells into 1.5 mL cryogenic vials filled with 1.0 mL of 0.5% glutaraldehyde in acetone.
2. Place the cryogenic vials containing the specimen carriers and 0.5% glutaraldehyde in acetone into the AFS machine at −155°C.
3. Warm from −155°C to −90°C at 5°C/h in the AFS machine.
4. Keep at −90°C for 48 h.
5. Warm to −30°C at 10°C/h in the AFS machine and replace the glutaraldehyde solution with pure ethanol and incubate for 10 min at −30°C. Repeat this ethanol wash twice.
6. Warm to room temperature at 10°C/h and maintain at room temperature for at least 2 h (*see* **Note 11**).
7. Transfer the specimen carriers to flat bottom embedding capsules and infiltrate with pure Lowicryl resin for 30 min at room temperature (*see* **Note 12**).
8. Lower the temperature to −35°C by placing embedding capsule in AFS or a freezer.
9. Polymerize the resin under ultraviolet light for 24 h at −35°C in the AFS or freezer.
10. Continue polymerization for 8 h at room temperature with UV light in order to facilitate sectioning.
11. Remove the specimen carrier from the block (*see* **Note 13**).

12. Use a diamond knife to cut ultrathin sections (60–80 nm in thickness) and transfer them to 150-mesh formvar-coated gold (or nickel) grids (*see* **Note 14**).

3.2.2. Freeze-Substitution with Osmium Tetroxide Followed by Epoxy Resin Embedding

1. Within liquid nitrogen or its vapor, transfer the specimen carriers with cryofixed cultured cells into 1.5 mL cryogenic vials filled with 1.0 mL of 1% osmium tetroxide in acetone.
2. Place the cryogenic vials containing the specimen carriers and 1% osmium tetroxide in acetone into the AFS machine at −155°C.
3. Warm from −155°C to −90°C at 5°C/h in AFS machine.
4. Keep at −90°C for 48 h.
5. Warm to room temperature at 10°C/h.
6. Replace the osmium solution with pure acetone and incubate for 10 min at room temperature. Repeat this acetone wash twice.
7. Transfer the specimen carriers to flat bottom embedding capsules and infiltrate with pure epoxy resin for 60 min at room temperature.
8. Polymerize in an oven at 60°C at least for 18 h.
9. Remove the specimen carrier from the block (*see* **Note 13**).
10. Use a diamond knife to cut ultrathin sections (60–80 nm in thickness) and transfer them to 150-mesh formvar-coated gold (or nickel) grids (*see* **Note 14**).

3.3. Immunolabeling Ultrathin Sections

1. Prior to the start of immunolabeling, pretreat epoxy ultrathin sections with 5% sodium meta-periodate (*see* **Note 15**) in distilled water for 30 min (*see* **Note 16**) and wash with distilled water (*see* **Note 17**).
2. Begin immunolabeling both epoxy and Lowicryl ultrathin sections by incubating for 30 min with 5% normal serum/1% BSA/PBS to block non-specific binding.
3. Incubate sections with monoclonal primary antibody at 4°C overnight or polyclonal primary antibody at room temperature for 2 h (*see* **Note 18**).
4. Wash the sections with PBS.
5. Incubate the sections with 1% BSA/PBS for 10 min to block non-specific binding.
6. Incubate the sections with secondary antibody conjugated with colloidal gold (diluted with 1% BSA/PBS) for 30 min at room temperature.
7. Wash the sections with distilled water.
8. Dry the sections (*see* **Note 19**).

3.4. Contrasting Immunolabeled Embedded Specimens

Non-osmicated specimens embedded in Lowicryl K4M exhibit poor contrast in the electron microscope especially specimens processed by cryofixation and freeze-substitution. To overcome this problem, a simple contrast enhancement method has been developed by Sawaguchi et al. (5) using acidified $KMnO_4$ oxidation prior to uranyl acetate staining to provide greater contrast without a reduction of immunolabeling.

3.4.1. Contrasting Lowicryl Embedded Specimens

1. Incubate the sections in 0.1% $KMnO_4$ in 0.1 N H_2SO_4 for 1 min (*see* **Note 20**).
2. Wash the sections with distilled water.
3. Incubate the sections in 5% aqueous uranyl acetate for 5 min.
4. Wash the sections with distilled water.
5. Incubate the sections in Reynolds' lead citrate for 1 min.
6. Wash the sections with distilled water.
7. Dry the sections.
8. Examine the sections in the electron microscope. **Figure 14.1A** shows an electron micrograph of Lowicryl HM20-embedded cultured rabbit gastric parietal cells processed by high-pressure freezing/freeze-substitution and immunolabeled with monoclonal mouse anti-ezrin antibody

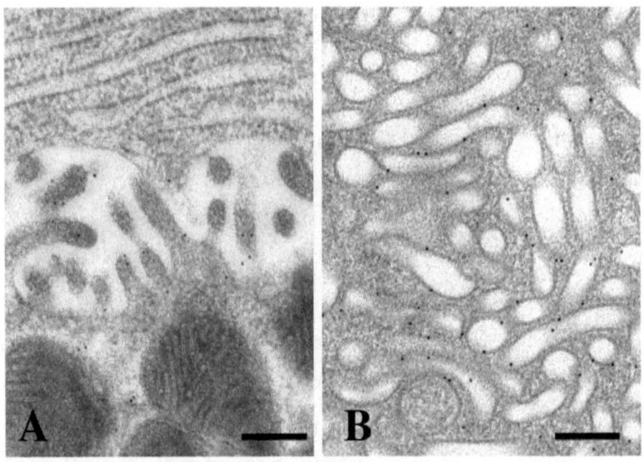

Fig. 14.1. Immunolabeling of cultured rabbit gastric parietal cells processed by high-pressure freezing/freeze-substitution and embedded in Lowicryl HM20 resin (**A**) or epoxy resin (**B**). Bars = 200 nm. (**A**) Cultured cells were embedded in Lowicryl HM20 resin, immunolabeled with monoclonal anti-ezrin followed by 10-nm colloidal gold conjugated with anti-mouse IgG, and treated with $KMnO_4$ to enhance section contrast. The basolateral membrane is intensely labeled with anti-ezrin. Note that the contrast of the sections is comparable to that of the epoxy resin-embedded osmium stained sections in panel B. (**B**) Cultured cells were embedded in epoxy resin, treated with sodium meta-periodate to enhance immunogenicity of osmium-treated samples and immunolabeled with monoclonal anti-H^+/K^+-ATPase antibody followed by 10-nm colloidal gold conjugated with anti-mouse IgG. The tubulovesicular membranes are specifically labeled with the monoclonal antibody to the H^+/K^+-ATPase.

followed by 10-nm colloidal gold conjugated to anti-mouse IgG. Compare the high contrast obtained by the $KMnO_4$ oxidation to the contrast observed in the osmium tetroxide stained epoxy resin embedded cells (**Fig. 14.1B**). Note the intense labeling around the basolateral membrane folds.

3.4.2. Contrasting Epoxy Resin Embedded Specimens

1. Incubate the sections in 2% uranyl acetate in 70% methanol for 4 min.
2. Incubate the sections in a graded series of 70%, 50%, and 30% methanol for 3 min each.
3. Wash the sections with distilled water.
4. Incubate the sections in Reynolds' lead citrate for 3 min.
5. Wash the sections with distilled water.
6. Dry the sections.
7. Examine the sections in the electron microscope. **Figure 14.1B** shows an electron micrograph of epoxy resin-embedded cultured rabbit gastric parietal cells processed by high-pressure freezing/freeze-substitution and immunolabeled with mouse monoclonal anti-H^+/K^+-ATPase antibody followed by 10-nm colloidal gold conjugated to anti-mouse IgG. The tubulovesicular membranes are specifically immunolabeled. Note that after 5% sodium meta-periodate treatment, the labeling efficiency of the epoxy resin-embedded cells (**Fig. 14.1B**) is comparable to that of the Lowicryl-embedded cells (**Fig. 14.1A**).

4. Notes

1. The conventional culture medium can be applied for cryofixation of mammalian tissue culture cells as long as the contents of medium are adjusted for the cell type.
2. A new high-pressure freezing machine, HPM 100, has been released from BAL-TEC. Another high-pressure freezing machine series, EM PACT, is also commercially available from Leica. Both the HPM 100 and the new EM PACT2 are compact mobile workstations allowing cryofixation of cultured mammalian cells in the laboratory.
3. If AFS is not available, freeze-substitution can be carried out in a deep freezer at −80°C or in dry ice-cooled acetone at −78°C. Briefly, place the cryogenic vials in a deep freezer or dry ice-cooled acetone for 24–48 h and then transfer the vials into a conventional freezer at −20°C. After 2 h, transfer the vials into a refrigerator at 4°C. After 2 h,

take the vials and process the freeze-substituted samples for embedding.

4. It is well known that hydrophobic apolar Lowicryl HM20 is easier to section compared to hydrophilic polar Lowicryl K4M. However, in my experience, immunolabeling density on Lowicryl K4M embedded specimens is higher than on Lowicryl HM20. The differences in labeling density could be explained by a cleavage process in Lowicryl K4M sectioning which may increase accessibility of antibody to its specific binding sites as suggested by Kellenberger et al. (6).
5. The Leica AFS is equipped with an ultraviolet light for low temperature embedding. An ultraviolet light TUV-200 (Dosaka EM, Osaka, Japan) is commercially available for use in a freezer for low temperature embedding.
6. It has been reported that commercially prepared colloidal gold complexes with secondary antibodies frequently contain free active antibody (7). Monodisperse colloidal gold can be prepared by the tannic acid/citrate reduction method described by Slot and Geuze (for particle diameter 3–15 nm, (8)) or the citrate reduction method by Frens (for particle diameter over 15 nm, (9)). Then, colloidal gold–antibody complexes can be prepared by the method of De Mey et al. (10). Consult practical procedures in the literature by Griffiths (11).
7. Bovine serum albumin can be replaced with fish gelatin, ovalbumin, or non-fat dried milk to prevent non-specific immunolabeling.
8. Sawaguchi et al. (3, 4) provide a more comprehensive discussion of the conditions affecting high-pressure freezing of mammalian tissue cultured cells on type B aluminum planchette specimen carriers.
9. Several methods have recently been developed for high-pressure freezing of mammalian cultured cells (reviewed in (2)) which are growing on different supports. Choose the best method for your experiment because each of the methods has merits and demerits depending on the cell type and objective of the experiment.
10. The recipe for the freeze-substitution medium varies considerably depending on the nature of the antigens. Freeze-substitution should be carried out in acetone alone for glutaraldehyde-sensitive antigens. On the other hand, freeze-substitution of osmium tetroxide-resistant antigens can be carried out in 1% osmium tetroxide in acetone. Uranyl acetate (0.1% at final concentration) can be added into the freeze-substitution medium to improve membrane contrast as long as the target antigen is resistant to it.

11. It is recommended to keep the freeze-substituted cells at room temperature at least for 2 h to remove the remaining hydration shell of proteins (12).
12. Phased infiltrations from lower concentration of resin are not necessary for thin cultured cell monolayers even when using high-viscosity epoxy embedding resins.
13. After polymerization, the specimen carriers can be removed from the resin block by fine forceps after careful trimming around the planchettes. Prior to ultrathin sectioning, it is helpful to select an ideally cultured area by placing the block on a microscope slide with the cells facing upward and focusing on the surface the block. Refer to original articles by Sawaguchi et al. (3, 4) for further details of the embedding of freeze-substituted cells processed by high-pressure freezing.
14. Grids coated with a plastic support film such as Formvar are essential due to the fragility of Lowicryl (especially Lowicryl K4M) resin sections. Epoxy resin sections should also be supported with Formvar films because the sections are subjected to repetitious washing in the process of immunolabeling and contrasting.
15. Sections are incubated by immersing the grid in a drop of the solution placed on a piece of parafilm with drop-sized depressions.
16. Sodium meta-periodate treatment reverse the deleterious effects of osmium tetroxide allowing some antibodies to recognize their antigens on osmicated, epoxy resin sections (13).
17. It is recommended to wash by immersing the grids in a jet of water or PBS to remove the non-specifically-bound antibody from the sections.
18. Monoclonal antibodies are usually used at concentration of 1–10 μg/mL, diluted with 5% normal serum/1% BSA/PBS. For polyclonal antibodies, twofold dilutions are examined in the first instance from 1/50 to 1/3,200. An optimal dilution and reaction time should be established experimentally.
19. The immunolabeled sections should be dried to stabilize the binding of gold particles.
20. The $KMnO_4$ stock solution has a purple-red color and can be kept for several months at room temperature. It should be noted that 0.1% $KMnO_4$ in higher concentrations of H_2SO_4 (>0.5 N) severely damages Lowicryl K4M ultrathin sections. On the other hand, lower concentrations of H_2SO_4 (<0.01 N) produce contaminants on the sections.

Prolonged incubation (>5 min) causes the loss of ultrathin sections on gold grids. Shorter incubations (<10 s) have no effect on the contrast.

Acknowledgments

The author would like to thank Professor Tatsuo Suganuma for expert advice and Soyuki Ide, Fumiyo Aoyama, and Yoshiteru Goto for their technical assistance.

References

1. Dahl, R. and Staehelin, L. A. (1989) High-pressure freezing for the preservation of biological structure: theory and practice. *J. Electron Microsc. Tech.* **13**, 165–174.
2. McDonald, K. L., Morphew, M., Verkade, P., and Müller-Reichert, T. (2007) Recent advances in high-pressure freezing: equipment- and specimen-loading methods. *Methods Mol. Biol.* **369**, 143–173.
3. Sawaguchi, A., McDonald, K. L., Karvar, S., and Forte, J. G. (2002) A new approach for high-pressure freezing of primary culture cells: the fine structure and stimulation-associated transformation of cultured rabbit gastric parietal cells. *J. Microsc.* **208**, 158–166.
4. Sawaguchi, A., Yao, X., Forte, J. G., and McDonald K. L. (2003) Direct attachment of cell suspensions to high pressure freezing specimen planchettes. *J. Microsc.* **212**, 13–20.
5. Sawaguchi, A., Ide, S., Goto, Y., Kawano, J., Oinuma, T., and Suganuma, T. (2001) A simple contrast enhancement by potassium permanganate oxidation for Lowicryl K4M ultrathin sections prepared by high pressure freezing/freeze substitution. *J. Microsc.* **201**, 77–83.
6. Kellenberger, E., Durrenberger, M., Villiger, W., Carlemalm, E., Wurtz, M. (1987) The efficiency of immunolabel on lowicryl sections compared to theoretical predictions. *J. Histochem. Cytochem.* **35**, 959–969.
7. Kramarcy, N. R. and Sealock, R. (1991) Commercial preparations of colloidal gold-antibody frequently contain free active antibody.*J. Histochem. Cytochem.* **39**, 37–39.
8. Slot, J. W. and Geuze, H. J. (1985) A new method of preparing gold probes for multiple-labeling cytochemistry. *Eur. J. Cell Biol.* **38**, 87–93.
9. Frens, G. (1973) Controlled nucleation for the regulation of the particle size in monodisperse gold suspensions. *Nature* **241**, 20–22.
10. De Mey, J., Moeremans, M., Geuens, R., Nuydens, R., and Brabander, M. D. (1981) High resolution light and electron microscopic localization of tubulin with the IGS (immuno gold staining) method. *Cell Biol. Int. Rep.* **5**, 889–899.
11. Griffiths, G., ed. (1993) Particle markers for immunoelectron microscopy. *Fine Structure Immunocytochemistry.* Springer-Verlag, Berlin, 279–306.
12. Kellenberger, E. (1991) The potential of cryofixation and freeze substitution: observations and theoretical considerations. *J. Microsc.* **161**, 183–203.
13. Bendayan, M. and Zollinger, M. (1983) Ultrastructural localization of antigenic sites on osmium-fixed tissues applying the protein A-gold technique. *J. Histochem. Cytochem.* **31**, 101–109.

Chapter 15

Immunoelectron Microscopy of Cryofixed Freeze-Substituted *Saccharomyces cerevisiae*

Jindriska Fiserova and Martin W. Goldberg

Abstract

Immunolabelling electron microscopy is a challenging technique with demands for perfect ultrastructural and antigen preservation. High-pressure freezing offers an ideal way to fix cellular structure. However, its use for immunolabelling has remained limited because of the low frequency of labelling due to loss of protein antigenicity or accessibility. Here we present a protocol for immunogold labelling of the yeast *Saccharomyces cerevisiae* that gives specific and multiple labelling while keeping the finest structural details. We use the protocol to reveal the organisation of individual nuclear pore complex proteins and the position of transport factors in the yeast *S. cerevisiae* in relation to actual transport events.

Key words: Yeast, nuclear, transport, transmission electron microscopy, freeze substitution, lowicryl, immunostaining.

1. Introduction

Immunoelectron microscopy requires preservation of both fine structural details and protein antigenicity. The classical approach to achieve this is the widely used Tokuyasu method (1). The sample is aldehyde-fixed at room temperature and frozen in sucrose/methyl cellulose. Sections are cut while frozen then thawed and immunolabelled. This avoids artefacts and epitope masking associated with dehydration and/or resin embedding (2, 3). Though a powerful method, there are drawbacks such as technical difficulties of dealing with cryosections, poorer chemical fixation of tissues compared to cryofixation methods and data interpretation (3). Superb preservation of ultrastructure is

S.D. Schwartzbach, T. Osafune (eds.), *Immunoelectron Microscopy*, Methods in Molecular Biology 657,
DOI 10.1007/978-1-60761-783-9_15, © Springer Science+Business Media, LLC 2010

without any doubt achieved by rapid freezing of the sample by high-pressure freezing (HPF) followed by low temperature fixation. Before immunolabelling, the frozen sample must be processed in a way that maintains antigenicity. This step is crucial and not always trivial in order to obtain fine structural detail and antigen preservation.

The literature covering theoretical aspects of HPF, freeze substitution and resin embedding is comprehensive (4–8). Briefly, by lowering the freezing point and suppressing the rate of ice crystal formation, HPF rapidly immobilises cellular contents in vitrified water circumventing disadvantages of slow chemical fixations (5). During freeze substitution, organic solvent (usually supplemented with chemical fixatives) substitutes for water in the specimen. Thus, fixation and sample dehydration occur at a freezing temperature which keeps ultrastructure preserved (5). Finally, infiltration and polymerisation of resins such as Lowicryl HM20 at freezing temperatures greatly enhance structural quality while still preserving some antigenicity (9). Interestingly, combined protocols of cryoimmunogold labelling (Tokuyasu method) with high-pressure freezing were recently published for yeast (10). Apparently, the "hybrid" method gives great membrane contrast and satisfactory structural preservation. On the other hand, it still requires cryosectioning expertise, specialised equipment for high-pressure freezing and twice as much time.

In this protocol, we present an efficient, specific and highly reproducible direct technique for immunogold labelling of high-pressure frozen yeast *Saccharomyces cerevisiae*. We have found that freeze substitution in glutaraldehyde, uranyl acetate and osmium tetroxide in acetone supplemented with a low concentration of water followed by embedding in Lowicryl HM20 simultaneously preserves structure and antigenicity in yeast as well as in mammalian cells.

2. Materials

2.1. Yeast Culture

1. Yeast cells, stored at −70°C.
2. YPD medium: 1% yeast extract, 2% peptone, 2% dextrose (all chemicals in **Section 2.1.** obtained from Sigma-Aldrich, St. Louis, MO).
3. YPD plates: Petri dishes with YPD medium containing 2% agar.
4. Synthetic dropout (SD) medium: 0.67% yeast nitrogen base, 2% dextrose, appropriate amino acid combination.
5. SD plates: Petri dishes with SD medium containing 2% agar.

6. Horizontal shaker (Scientific Laboratory Supplies, Nottingham, UK).

2.2. High-Pressure Freezing

1. Leica EM PACT high-pressure freezer (Leica Microsystems GmbH, Wetzlar, Germany).
2. Loading station (Leica Microsystems GmbH, Wetzlar, Germany) (*see* **Fig. 15.1A**).
3. Specimen carrier holder (Leica Microsystems GmbH, Wetzlar, Germany) (*see* **Fig. 15.1A**).
4. 200-μm deep specimen carriers (Leica Microsystems GmbH, Wetzlar, Germany) (*see* **Note 1** and **Fig. 15.1A**).
5. Flat specimen pod (Leica Microsystems GmbH, Wetzlar, Germany) (*see* **Fig. 15.1A, B**).
6. Torque wrench with 2-mm Allen key insert (Leica Microsystems GmbH, Wetzlar, Germany).
7. Loading device for flat specimen pod (Leica Microsystems GmbH, Wetzlar, Germany) (*see* **Fig. 15.1B**).
8. 25 L Dewar flask (VWR International Ltd, Lutterworth, UK) with liquid nitrogen (LN_2) (*see* **Note 2**).

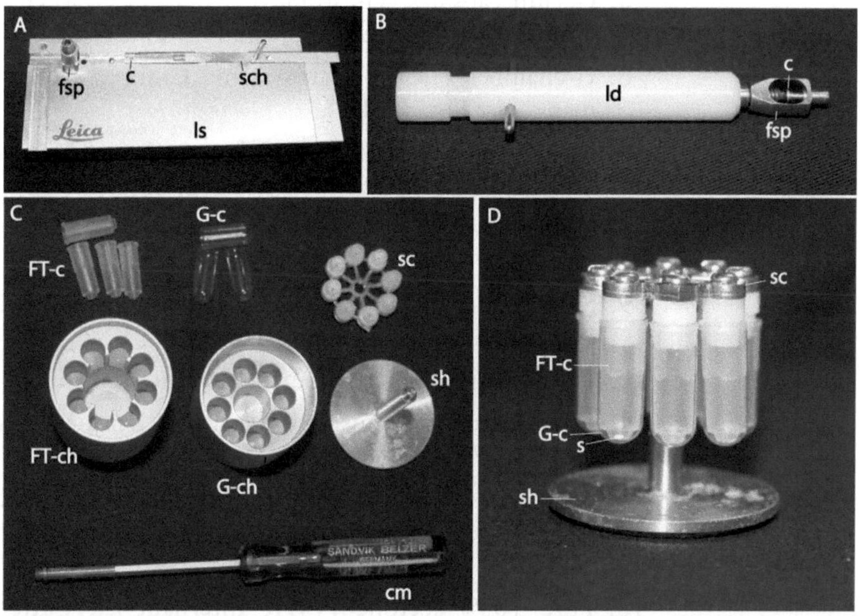

Fig. 15.1. High-pressure freezing and freeze substitution accessories. (**A**) Assembled loading station prior the high-pressure freezing of the sample by Leica EM PACT; loading station (*ls*), specimen carrier holder (*sch*), flat specimen pod (*fsp*), specimen carrier (*c*). (**B**) Loading device (*ld*) with flat specimen pod (*fsp*) loaded with specimen carrier (*c*). (**C**) Freeze substitution accessories: FT-capsules (*FT-c*), FT-chamber (*FT-ch*), G-capsules (*G-c*), G-chamber (*G-ch*), stem holder for spider cover (*sh*), spider cover (*sc*), cryomanipulator (cm). (**D**) FT-capsules (*FT-c*) containing specimen (*s*) inside G-capsules (*G-c*) attached to the spider cover (*sc*) on the top of the stem holder for spider cover (*sh*) during resin polymerisation.

9. 50 mL of methylcyclohexane (Sigma-Aldrich, St. Louis, MO).
10. Curved syringe needle Terumo 21G × $1^1/_2''$ (VWR International, Lutterworth, UK).
11. Wooden toothpicks (Plastico Ltd., Surrey, UK).
12. 25 mm diameter 0.2 or 0.45 μM Millipore® Filters MCE, MF (Fisher Scientific, Loughborough, UK).
13. Filter holder Swinnex™ 25 mm diameter (Fisher Scientific, Loughborough, UK).
14. 50 mL syringe with Lacer Luer (Greiner Bio-One, Stonehouse, UK).

2.3. Freeze Substitution and Resin Embedding

1. Leica EM AFS freeze substitution machine (Leica Microsystems GmbH, Wetzlar, Germany).
2. Two to three 25 L Dewar flasks (VWR International Ltd, Lutterworth, UK) filled with LN_2 (*see* **Note 2**).
3. FT-capsules with mesh bottom D 5×10 mm (Leica Microsystems GmbH, Wetzlar, Germany) (*see* **Fig. 15.1C**).
4. FT-chamber (Leica Microsystems GmbH, Wetzlar, Germany) (*see* **Fig. 15.1C**).
5. Gelatin capsules (G-capsules) size 1 (Leica Microsystems GmbH, Wetzlar, Germany) (*see* **Fig. 15.1C**).
6. G-chamber (Leica Microsystems GmbH, Wetzlar, Germany) (*see* **Fig. 15.1C**).
7. Cryomanipulator with M4 thread (Leica Microsystems GmbH, Wetzlar, Germany) (*see* **Fig. 15.1C**).
8. Spider cover (Leica Microsystems GmbH, Wetzlar, Germany) (*see* **Fig. 15.1C, D**).
9. Stem holder for spider cover (Leica Microsystems GmbH, Wetzlar, Germany) (*see* **Fig. 15.1C, D**).
10. Plastic Pasteur pipettes (Agar Scientific, Stansted, UK).
11. Long forceps with insulation coating (Leica Microsystems GmbH, Wetzlar, Germany).
12. Monostep™ Lowicryl HM 20 (nonpolar) (Polysciences, Inc., Eppelheim, Germany), pre-chilled to −25°C (*see* **Note 3**).
13. Pure 100% acetone (VWR International, Lutterworth, UK) pre-chilled to −25°C.
14. 2% osmium tetroxide (Agar Scientific, Stansted, UK) in acetone: dissolve 0.25 g of osmium tetroxide in 12.5 mL acetone (*see* **Note 4**).
15. Fixative: 0.2% uranyl acetate (Agar Scientific, Stansted, UK), 0.2% glutaraldehyde (Agar Scientific, Stansted, UK),

0.01% OsO_4, 5% H_2O in acetone. To prepare 50 mL of fixative solution, mix 250 μL of 2% osmium tetroxide, 200 μL of 50% glutaraldehyde (use glutaraldehyde up to 1 month old stored at 4°C), 2.5 mL distilled water and 45 mL acetone. Because uranyl acetate is not readily soluble in acetone, dissolve 0.1 g of uranyl acetate in 2 mL of methanol and add it immediately to the fixative (*see* **Note 5**). Aliquot fresh fixative in 1.5 mL cryovials and immediately freeze in LN_2. Fixative can be stored up to 3 months in LN_2.

2.4. Sectioning and Grid Immunolabelling

1. Leica EM UC6 ultramicrotome (Leica Microsystems GmbH, Wetzlar, Germany).
2. Formvar-coated 200-μm mesh nickel grids (Agar Scientific, Stansted, UK).
3. 10× PBS: Dissolve the following reagents in 750 mL of water: 80 g NaCl, 2.0 g KCl, 14.4 g Na_2HPO_4, 2.4 g KH_2PO_4. Adjust volume to 1 L with distilled H_2O and sterilise by autoclaving. Store at 4°C. Make 1× fresh as required. Adjust pH to 7.4 with NaOH, if necessary.
4. 1% BSA (Sigma-Aldrich, St. Louis, MO) in PBS, make fresh as required.
5. 0.1% glycine (Sigma-Aldrich, St. Louis, MO) in PBS, make fresh as required.
6. Primary antibody diluted in 1% BSA in PBS. The dilution depends on individual properties of the antibody and would usually range from 1:10 to 1:200 (*see* **Note 6**).
7. Secondary antibody conjugated to 5-nm gold particles (BB International, Cardiff, UK) diluted in PBS (*see* **Note** 7). Usually, 1:20 dilution is used for immunoTEM.
8. Double Distilled water.

2.5. Post-staining

1. 1% uranyl acetate in water. Use brown bottle to combine 1 g of uranyl acetate (*see* **Note 5**) in 100 mL of distilled water. Filter through 0.22 μm 25 mm diameter Polyethersulfone Membrane syringe filter (VWR International, Lutterworth, UK) before each use. Stable up to 1 month at 25°C.
2. Reynold's Lead citrate: 1.33 g of lead nitrate, 1.76 g of sodium citrate, 50 mL boiled distilled water, 1 M sodium hydroxide (all chemicals obtained from Agar Scientific, Stansted, UK). Prepare in a 50 mL scrupulously clean volumetric flask by combining about 30 mL boiled distilled water with the lead nitrate and agitate slightly to dissolve (*see* **Note 8**). Add the sodium citrate, shake for 1 min and then intermittently for 20 min. The solution should appear cloudy with no large particles. Add 8 mL of 1 M CO_2

free NaOH and invert slowly. The milky solution should turn clear. Bring to 50 mL volume with distilled water. Filter through 25 mm diameter Polyethersulfone Membrane syringe filter (VWR International, Lutterworth, UK) before each use. Stable up to 6 months at room temperature in a volumetric flask tightly stoppered with a plastic stopper to prevent the entry of CO_2. If apparent precipitate develops, make new solution.

3. Methods

3.1. Yeast Culture

1. Plate frozen yeast stock on YPD or SD selective plates (*see* **Note 9**). Allow colonies to form by growing for 3 days at 30°C.
2. The morning of the day before HPF, inoculate 10 mL of liquid YPD or SD media with a single yeast colony and grow on a horizontal shaker at 180 rpm for about 8–10 h at 30°C. Check the OD_{600} of the culture.
3. Assuming a culture doubling time is ~2 h, calculate the culture dilution required to inoculate 50 mL media so that after overnight growth, the culture will be mid-log phase with an OD_{600} value of 0.5–1, equivalent to $\sim 1\text{–}5 \times 10^7$ cells/mL. The dilution usually corresponds 1/100–1/200 of the final media volume (50 mL).
4. Inoculate 50 mL medium with the required dilution and grow overnight on a horizontal shaker at 180 rpm at 30°C.

3.2. High-Pressure Freezing

1. Prepare the Leica EM PACT high-pressure freezing unit and loading accessories according to the manufacturer's instructions.
2. Harvest 50 mL of yeast by gentle filtration through the 0.2 or 0.45 μM syringe filter. Place the yeast concentrated on the membrane filter on a fresh YPD plate to prevent drying (*see* **Note 10**).
3. Use wooden toothpicks to transfer ~2–3 mm^3 of the yeast from the membrane filter to the specimen carrier to slightly overload the specimen carrier. To remove the excess load, use the curved part of a syringe needle to align the yeast surface within the specimen carrier.
4. Load the specimen carrier into the flat specimen pod and tighten with the Torque wrench, attach the loading device and freeze in the Leica EM PACT high-pressure freezing unit according to manufacturer's instructions. The frozen

sample inside the specimen carrier can be stored indefinitely in LN_2.

3.3. Freeze Substitution and Resin Embedding

1. Program the Leica EM AFS freeze substitution unit for sample fixation and acetone washes as follows: T1: −90°C 49 h; S1: 5°C increment per hour up to −25°C; T2: −25°C 12 h; S2: -0°C 0 h; T3: −25°C 50 h (*see* **Note 11**). Allow the chamber to pre-chill for an hour.
2. Place the FT-capsules into cryovials, half-fill them with fixative and freeze in LN_2. Under LN_2, place the specimen carriers on the top of the frozen fixative inside the FT-capsules and close the lid (*see* **Note 12**). Place into the pre-chilled Leica EM AFS freeze substitution chamber and run the programme.
3. As soon as the temperature step T2 is finished, acetone washes can be performed. Place the FT-capsules into the FT-chamber filled with acetone chilled to −25°C and incubate for 15 min at −25°C. Repeat the acetone washing step twice by changing the acetone inside the FT-chamber after 15 min of incubation at −25°C.
4. Remove the specimen carriers from the FT-capsules. Ideally, by this stage the specimen becomes detached from the specimen carrier and can be seen inside the FT-capsule as a small yellowish spot. Otherwise, it needs to be scraped out of the specimen carrier with a pair of tweezers.
5. Infiltrate the specimen (now freely inside FT-capsule placed inside FT-chamber) with Lowicryl by filling the FT-chamber with 50% Lowicryl for 1 h at −25°C, 66% Lowicryl for 1 h at −25°C, 100% Lowicryl for 1 h at −25°C and 100% Lowicryl overnight at −25°C (*see* **Notes 13** and **14**).
6. Load G-chamber with G-capsules and place inside the substitution chamber of the Leica EM AFS. Fill the G-capsules with 100% Lowicryl and let chill down to −25°C. Place the FT-capsules with specimen inside the G-capsules and where needed refill with 100% Lowicryl. Close the G-capsules tightly with the spider cover and remove from the G-chamber using the cryomanipulator and place on top of the stem holder for the spider cover (*see* **Fig. 15.1D**).
7. Set the Leica EM AFS unit for resin embedding programme as follows: T1: −25°C 24 h; S1: 5°C increment per hour up to 25°C; T2: 25°C 24 100 h (*see* **Note 15**). Install UV lamp and start the polymerisation programme.

3.4. Sectioning and Grid Immunolabelling

1. When the polymerisation is finished, remove the spider cover with the attached G-capsules from the Leica EM AFS substitution chamber.

2. Remove polymerised resin blocks from G/FT-capsule covering using razor blade (*see* **Note 16**).
3. Trim the resin blocks and cut 60–70 nm sections using a diamond or glass knife.
4. Pick the sections up on Formvar-coated nickel grids (*see* **Notes 17–19**).
5. Rinse with 0.1% glycine in PBS three times for 1 min (*see* **Note 20**).
6. Block in 1% BSA in PBS four times for 1 min.
7. Incubate with primary antibody in a wet chamber using 5–10 μL droplets per grid for 1 h at room temperature or overnight at 4°C.
8. Rinse in PBS four times for 2 min.
9. Incubate with colloidal gold-conjugated secondary antibody in a wet chamber using a 5–10 μL antibody droplet per grid diluted 1:20–1:100 (*see* **Note 21**). Incubate for 1 h at room temperature.
10. Rinse in PBS three times for 5 s each.
11. Wash in PBS four times for 2 min each.
12. Wash in distilled water 10 times for 1 min each (*see* **Note 22**).

3.5. Post-staining

1. Float the grids for 10 min on a 20–50 μL droplet of 1% uranyl acetate (*see* **Note 23**).
2. Rinse by dipping in distilled water ~20 times.
3. Float the grids for 10 min on a 20–50 μL droplet of Reynold's lead citrate (*see* **Note 24**).
4. Rinse by dipping in water ~20 times.
5. Air dry the grids on a filter paper and observe with TEM. Cells fixed by this method exhibit nice membrane contrast (**Fig. 15.2A**) and individual organelles are distinct in the dense cytoplasm (**Fig. 15.2B**). The nuclear envelope (NE) covered by ribosomes is perforated by the nuclear pore complexes (NPCs), multiprotein gates that facilitate nucleocytoplasmic transport across the NE. Because NPCs are clearly visible (**Fig. 15.2C, D**), we use this method to resolve their structure and transport properties. Though extensively studied, the nature of the nucleocytoplasmic transport mechanism remains poorly understood. When the cryofixation freeze substitution technique is combined with immunostaining, individual proteins that are part of NPC or proteins undergoing transport can be detected (**Fig. 15.3**). The NPC protein called Nup116 contains GLFG domains that play

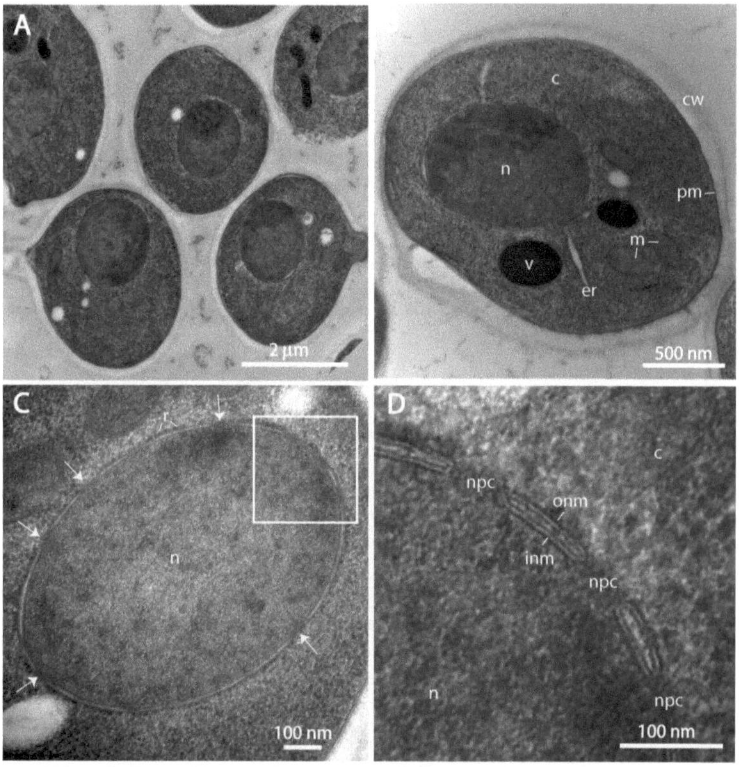

Fig. 15.2. High-pressure frozen and freeze-substituted yeast cells. (**A**) Overview of general yeast preservation. Individual yeast cells have round shapes, round nuclei and smooth membranes which are signs of good structural preservation during freezing, fixation and sectioning. (**B**) A well-preserved yeast cell shows no signs of freezing damage or stretching related to sectioning defects. Individual organelles such as the nucleus (*n*), vacuoles (*v*), endoplasmic reticulum (*er*) and mitochondria (*m*) are distinct inside the dense cytoplasm (*c*). The cell wall (*cw*) and plasma membrane (*pm*) are clearly seen. (**C**) High-magnification micrograph showing details of the nucleus. Nuclear pore complexes (marked by arrows) are clearly visible as gaps within the nuclear envelope covered by ribosomes (*r*). (**D**) High-magnification micrograph of the inset in (**C**) showing the detailed structure of the nuclear envelope and nuclear pore complex. The nuclear envelope has a smooth character and is interrupted by the nuclear pore complexes (*npc*). Outer and inner nuclear membranes (*onm* and *inm*) of the nuclear envelope are easily distinguished. Nuclear (*n*) and cytoplasmic (*c*) compartments are marked.

an important role in the translocation process. The localisation of the GLFG domains in two NPCs can be readily seen in **Fig. 15.3A**, whereas transport of GFP directed by the Spo12 nuclear localisation signal (11) through the nuclear pore complex is visualised in **Fig. 15.3B**. The cryofixation technique combined with immunostaining is thus a powerful technique for observation of nucleocytoplasmic transport of various cargoes via individual NPCs.

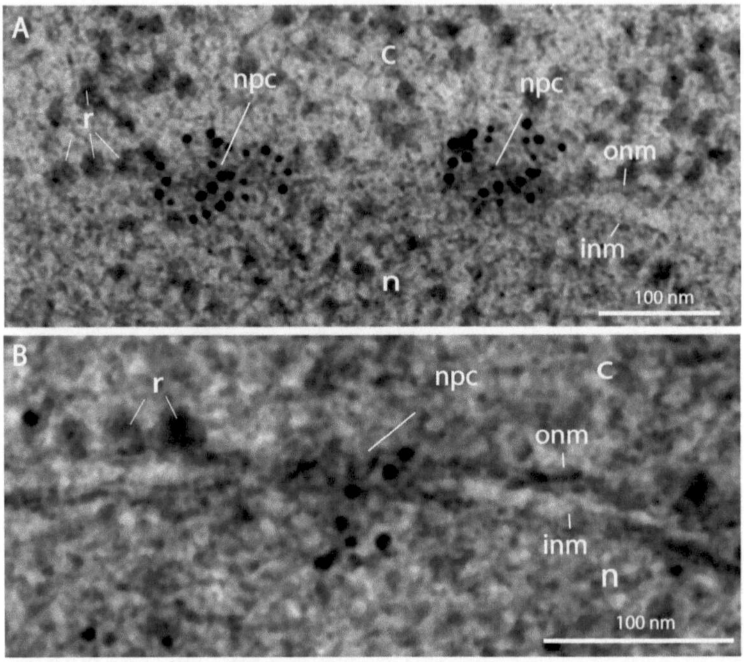

Fig. 15.3. Yeast sections immunolabelled after high-pressure freezing and freeze substitution. (**A**) Wild-type yeast immunolabelled with antibody against yeast Nup116GLFG and 5-nm gold-conjugated anti-rabbit secondary antibody (Agar Scientific, Stansted, UK). The nuclear envelope composed of inner and outer nuclear membranes (*inm* and *onm*) separates nuclear (*n*) and cytoplasmic (*c*) compartments. Ribosomes (*r*) attached to the outer nuclear membrane as well as cytoplasmic ribosomes are visible. Two nuclear pore complexes (*npc*) embedded in the nuclear envelope show very dense and specific labelling of GLFG domains of Nup116 at the periphery as well as inside the NPC centre. (**B**) Yeast carrying Spo12NLS-GFP, a GFP targeted to the nucleus by the nuclear localisation signal of Spo12, immunolabelled with anti-GFP primary antibody (Abcam, Cambridge, MA) and 5-nm gold-conjugated anti-rabbit secondary antibody (Agar Scientific, Stansted, UK). Immunolabel is seen inside the single nuclear pore complex (*npc*) and reflects transport of the Spo12NLS-GFP through the nuclear pore complex. Inner and outer nuclear membranes (*inm* and *onm*) that separate nuclear (*n*) and cytoplasmic (*c*) compartments can be clearly observed. Outer nuclear membrane is covered by ribosomes (*r*).

4. Notes

1. For high-pressure freezing we use flat specimen carriers (depth 200 μm) rather than the deeper carriers (depth 400 μm) to minimise freezing damage to the sample.
2. Caution: when working with LN_2 use a well-vented room, safety goggles and thermal protective gloves specially designed for cryogenic use.

3. Caution: Lowicryl is a methacrylic acid ester that can cause allergies, eye and skin irritation. It is very volatile. Work in a fume hood when possible, keep the area well vented and use protective clothing, glasses and 4H gloves.
4. Caution: osmium tetroxide in acetone is extremely volatile, toxic by inhalation and by skin contact. Work in a fume hood and wear gloves. 2% osmium tetroxide in acetone can be stored for several months in LN_2.
5. Caution: uranyl acetate contains trace amounts of U^{235}. Powder should not be inhaled and solutions should be discarded in special containers.
6. Antibody concentrations used in electron microscopy often tend to be up to 10× higher than those used for immunofluorescence observation. Test serial dilutions to find the best working concentration of your antibody.
7. The preferred size of the gold label conjugated to the secondary antibody depends mostly on the magnification used for the observation, in other words, it correlates with the demands for the resolution of the structural details. Smaller gold particles (1–5 nm) will obscure less of the sample and result in higher labelling sensitivity but they are difficult to see when an overview is required at low magnifications. The 5 and 10-nm gold particles can just be seen with 80,000 and 40,000 times magnification, respectively. Double labelling is possible when antibodies have been raised in different organisms. For instance 5-nm gold attached to an anti-mouse antibody can be used in conjunction with 10-nm gold attached to an anti-rabbit antibody if the primary antibodies were raised in mice and rabbits, respectively.
8. Caution: lead salts are extremely toxic. Wear gloves and work in a fume hood when weighing the powders.
9. Yeast is usually grown at 30°C on the complete medium, YPD. When amino acid selection is required, synthetic dropout (SD) media depleted of a certain amino acid or amino acid combination is used.
10. Before concentrating yeast, make sure all the instruments for high-pressure freezing are ready, e.g. toothpicks, 60° curved syringe needle, agar plates, loading station and Leica EM PACT. Once filtered, work as fast as possible to freeze yeast unstressed. The time from yeast concentration to freezing should be kept minimal, ideally about 30 s.
11. The best time to start the freeze substitution is Friday before 1 pm. This timing ensures the freeze substitution

runs over the weekend and allows sufficient time for acetone washes and Lowicryl infiltration the following Monday afternoon.

The Leica EM AFS unit is programmed as suggested to allow the 1-h period of chilling to −90°C, 48 h of incubation of the sample in fixative at −90°C, a warming up period up to −25°C taking 13 h and another incubation in fixative at −25°C for 12 h. In total, the sample is fixated for 73 h. So the acetone washes ideally follow 3 days and 1 h after the programme started and require the temperature of −25°C. The time to start acetone washes corresponds with the change from programme T2 to programme T3.

12. Fixative tends to evaporate from the FT-chamber over the 73-h period. To prevent the evaporation we use cryovials for the fixation step. We suggest scratching the label into the side of the cryovial to make sure it is not washed away during freeze substitution.
13. Oxygen strongly inhibits the polymerisation of methacrylate resins. Avoid harsh stirring when mixing Lowicryl HM20 resin.
14. The longer the 100% Lowicryl infiltration the better. We tested periods from several hours up to 3 days. The optimal results are achieved after at least 24 h of infiltration.
15. The resin hardness improves when UV light is applied for at least 24–72 h at 25°C. The longer the polymerisation, the better the resin quality.
16. When working with polymerised Lowicryl resin blocks, use protective gloves as remnants of unpolymerised Lowicryl can still be present.
17. During sectioning try to collect sections from the water surface as quickly as possible to minimise water extraction of the section surface.
18. Prior to immunolabelling, it is useful to immediately check for good structural preservation by post-staining one to two grids with uranyl acetate and Reynold's lead citrate. This avoids wasting time and immunoreagents on poorly preserved specimens. *See* **Section 3.5** for post-staining protocol.
19. For TEM structural observation, grids can be stored indefinitely. For immunolabelling, cut fresh sections to prevent loosing antigenicity.
20. During the immunostaining, grids are floated section-side down on the surface of a 20–50 μL droplet of the appropriate solution pipetted onto a parafilm strip. 5–10 μL droplets pipetted onto parafilm in a Petri dish layered with wet filter paper are used for antibody incubation.

21. Dilution of the secondary antibody for electron microscopy is usually between 1:20 and 1:100 depending on the secondary antibody. It is recommended to test serial dilutions to find the appropriate working concentration.
22. Sections can be post-stained immediately or allowed to dry and stored until processed further.
23. The final contrast of the sample is mostly achieved by binding of heavy metals like uranium or lead to the cellular protein structures during post-staining of the grid sections. Grids are usually floated section-side down on 20–50 μL droplets of solution pipetted onto the surface of a parafilm strip.
24. The lead citrate must be protected from contact with CO_2. To make a protective environment for post-staining with Reynold's citrate, place filter paper wetted with 1 M NaOH on a piece of parafilm in a Petri dish. Place several NaOH pellets around the periphery of the parafilm and keep covered. However, we rarely experience lead carbonate precipitation when using a carefully stored and filtered solution, working quickly and avoiding breathing on the lead citrate droplets. After use, the stain droplets of uranyl acetate and Reynold's citrate should be discarded in special containers.

Acknowledgements

We thank to Prof. S. Wente for kindly providing the antibodies and transgenic yeast strains. This work was funded by the Biotechnology and Biological Sciences Research Council, UK grant number BB/E015735/1.

References

1. Tokuyasu, K. T. (1973) A technique for ultracryotomy of cell suspensions and tissues. *J. Cell Biol.* **8**, 377–383.
2. van Donselaar, D., Posthuma, G., Zeuschner, D., Hmbel, B. M., and Slot, J. W. (2007) Immunogold labeling of cryosections from high-pressure frozen cells. *Traffic* **8**, 471–485.
3. Mobius, W. (2009) Cryopreparation of biological specimens for immunoelectron microscopy. *Ann. Anatomy*, doi: 10.1016/j.aanat.2008.11.004.
4. Mulholland, J. and Boststein D. (2002) Immunoelectronmicroscopy of aldehyde-fixed yeast cells. In *Methods in Enzymology. Guide to Yeast Genetics and Molecular and Cell Biology*, Part C, Guthrie, C. and Fink, G. R., eds. Elsevier, London, vol. 351, 50–81.
5. McDonald, K. (2007) Cryopreparation methods for electron microscopy of selected model systems. In *Methods in Cell Biology, Cellular Electron Microscopy*, McIntosh, J. R. ed. vol. **79**, 54–52.

6. McDonald, K. (1999) High pressure freezing for preservation of high resolution fine structure and antigenicity for immunolabelling. *Methods Mol. Biol.* **117**, 77–97.
7. Murray, S. (2008) High pressure freezing and freeze substitution of *Schizosaccharomyces pombe* and *Saccharomyces cerevisiae* for TEM. In *Methods in Cell Biology, Introduction to Electron Microscopy for Biologists*, Allen, T. D. ed. vol. 88, 4–17.
8. Studer, D., Humbel, B. M., and Chiquet, M. (2008) Electron microscopy of high pressure frozen samples: bridging the gap between cellular ultrastructure and atomic resolution. *Histochem. Cell Biol.* **130**, 877–889.
9. Giddings, T. H. (2003) Freeze-substitution protocols for improved visualisation of membranes in high-pressure frozen samples. *J. Microsc.* **212**, 53–61.
10. Griffith, J., Mari, M., Maziere, A. D., and Reggiori, F. (2008) A cryosectioning Procedure for ultrastructural analysis and the immunogold labelling of yeast *Saccharomyces Cerevisiae*. *Traffic* **9**, 1060–1072.
11. Terry, L. J. and Wente, S. R. (2007) Nuclear mRNA export requires specific FG nucleoporins for translocation through the nuclear pore complex. *J. Cell Biol.* **178**, 1121–1132.

Chapter 16

High-Resolution Molecular Localization by Freeze-Fracture Replica Labeling

Akikazu Fujita and Toyoshi Fujimoto

Abstract

The freeze-fracture technique splits the frozen lipid bilayer membrane into two halves and immobilizes membrane proteins and lipids by the vacuum evaporation of platinum and carbon. After a treatment by SDS to remove extramembrane materials, the specimen is subjected to immunogold labeling, which gives information on the two-dimensional distribution of membrane molecules and their relationship to various differentiated structures. In combination with rapid freezing, the freeze-fracture technique has an advantage over other methods using conventional chemical fixation because the distribution of lipids as well as proteins can be observed at the mesoscale in a wide area of the membrane.

Key words: Immunoelectron microscopy, freeze fracture, rapid freezing, SDS-FRL, physical fixation.

1. Introduction

The freeze-fracture technique was developed in the 1960s and 1970s and has provided valuable information on the ultrastructural architecture in the cellular membrane (1). The two-dimensional information obtained by the technique cannot be easily obtained by conventional electron microscopy using ultrathin sections. Notwithstanding this advantage, specimens obtained by the classical freeze-fracture technique could not be used for immunocytochemistry and therefore did not give any information about the molecular composition of the observed structure. The technique was dramatically improved when the late Dr. Kazushi Fujimoto developed a new method to treat

S.D. Schwartzbach, T. Osafune (eds.), *Immunoelectron Microscopy*, Methods in Molecular Biology 657,
DOI 10.1007/978-1-60761-783-9_16, © Springer Science+Business Media, LLC 2010

freeze-fracture replicas and demonstrated that proteins retained in the replica can be specifically labeled by antibodies (2). He showed that treatment with sodium dodecyl sulfate (SDS) removed extramembrane materials but not molecules cast in the replica, which remained a viable substrate for various probes. The technique named SDS-digested freeze-fracture replica labeling (SDS-FRL) by Dr. K. Fujimoto (3) and freeze-fracture replica immunolabeling technique (FRIL) by other groups (4, 5) made it possible to determine the molecular composition of fine structures delineated by freeze-fracture electron microscopy.

SDS-FRL has been applied successfully to localize various proteins and lipids. Both cultured cells and tissues in vivo, and both the plasma membrane and other cellular structures, have been examined. The two-dimensional distribution of the labeling can be analyzed objectively by statistical methods (6, 7). Lipid droplets, which have a characteristic shape, have been studied by SDS-FRL (5). Other intracellular organelles are often difficult to identify by their morphology alone in the freeze-fracture replica. Therefore, labeling of organelle-specific markers may be necessary for unambiguous identification.

One of the major advantages of SDS-FRL is that it can immobilize membrane molecules without chemical fixation. The process by which the thin layer of platinum/carbon evaporated in the vacuum stabilizes both proteins and lipids is called "physical fixation." To take full advantage of this process, it is desirable that live cells or tissues are frozen rapidly without any pretreatment. Sophisticated rapid-freezing methods are available for this purpose (1), but we will introduce an easy-to-use rapid-freezing method for cultured cells that does not require expensive instrumentation (8). The step-by-step method to label freeze-fracture replicas will also be described.

2. Materials

2.1. Cell Culture

1. Gold foil: A piece of gold foil (20 μm in thickness, The Nilaco Corp., Tokyo, Japan) cut into an asymmetrical trapezoidal shape (approximate size: upper side 1 mm; lower side and height 2 mm). The asymmetrical shape helps in easily discerning the cell side in subsequent procedures (**Fig. 16.1A, B-a**). A slit is cut in the gold foil when necessary (**Fig. 16.1A**) (*see* **Note 1**). The gold foil pieces are treated with concentrated hydrochloric acid for 20 min, rinsed extensively with distilled water, and immersed in acetone for 20 min. The pieces are heat-sterilized (180°C, 1 h) before being used for cell culturing.

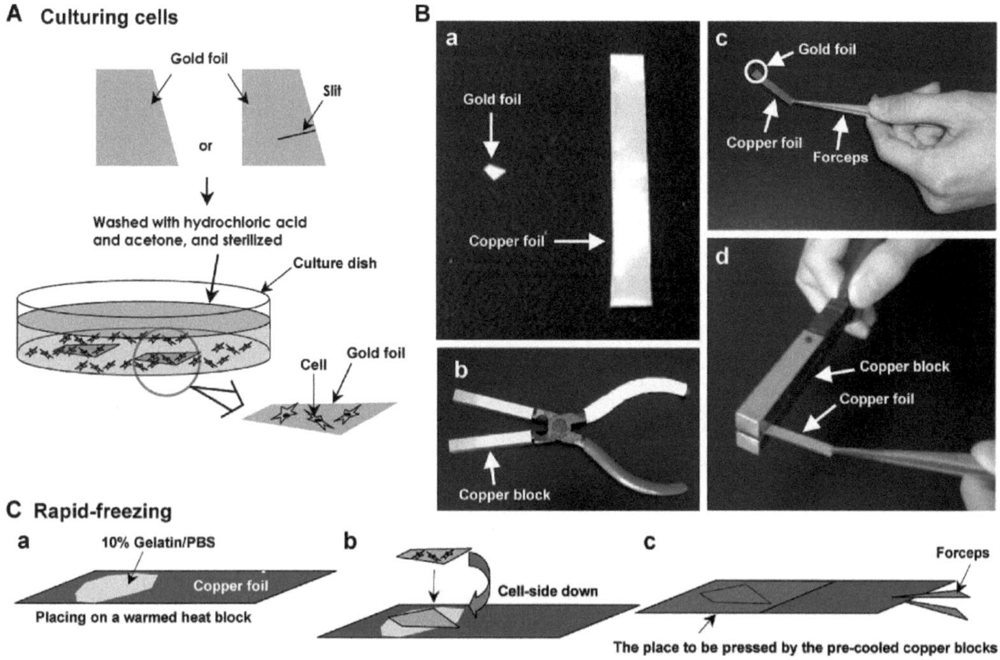

Fig. 16.1. Rapid freezing of monolayer cultured cells. (**A**) Cells are cultured on pieces of gold foil (20 μm in thickness), which are cut in a trapezoidal shape. (**B**) (*a*) Gold and copper foils. (*b*) Gold-plated copper blocks attached to a plier. (*c*) The copper foil is held by forceps at an edge distant from the gold foil. (*d*) The gold foil portion is pressed between the cold copper blocks attached to a plier. (**C**) Scheme for the rapid-freezing procedure. Cells grown on a piece of gold foil were inverted onto prewarmed gelatin/PBS on copper foil (20 μm in thickness) (*a, b*). The corner of the copper foil is held with forceps, and the cell sandwich is rapidly pressed between precooled gold-plated copper blocks (*c*). The blocks are kept in liquid nitrogen and are taken out immediately before use.

2. Culture medium: Add 10% fetal bovine serum (FBS, JRH Biosciences, KS) to Dulbecco's Modified Eagle's Medium (DMEM) (Invitrogen, Carlsbad, CA).
3. Phosphate-buffered saline (PBS): 137 mM NaCl, 2.7 mM KCl, 10 mM Na_2HPO_4, 10 mM KH_2PO_4.
4. Trypsin/EDTA solution for cell dispersion: 0.05% trypsin and 0.02% EDTA in PBS. Filter sterilize.
5. Plastic tissue culture dishes (35 mm).

2.2. Rapid Freezing

1. Gold-plated copper blocks attached to two arms of a fine plier (**Fig. 16.1B-b**; Dosaka EM, Kyoto, Japan).
2. 10% Gelatin/PBS solution: Dissolve gelatin powder in PBS for a final concentration of 10% by heating and keep at 4°C. Heat to 37°C prior to use.
3. Copper foil: Copper foil (20–40 μm in thickness; The Nilaco Corp., Tokyo, Japan.) cut into a thin rectangular shape (**Fig. 16.1B-a**; the foil shown in the figure is about 5 mm×35 mm).
4. Small cryogenic vials (Corning Inc., Corning, NY).

2.3. Freeze Fracture

1. Balzers BAF400T (Bal-Tec, Liechtenstein) or any other freeze-fracture apparatus.
2. SDS digestion buffer: 2.5% (w/v) SDS, 15 mM Tris–HCl, pH 8.0.
3. Glycerol.
4. Ceramic dishes with dimples.
5. Stereomicroscope.

2.4. Immunolabeling

1. PBS.
2. 0.1% Tween 20 in PBS.
3. Blocking buffer: 3% bovine serum albumin (BSA) in PBS.
4. Antibody dilution buffer: 1% BSA in PBS.
5. Washing buffer: 0.1% BSA in PBS.
6. Primary antibodies.
7. Colloidal gold-conjugated reagents: colloidal gold (5–15 nm)-conjugated secondary antibody (BioCell, UK) or colloidal gold-conjugated protein A (Utrecht University, Utrecht, The Netherlands) (*see* **Note 2**).
8. Formvar-coated, 50-mesh copper grid (Nisshin EM, Tokyo, Japan).
9. Multiwell ceramic plate.
10. Stereomicroscope.
11. Parafilm.
12. Moisture chamber.

3. Methods

In the conventional freeze-fracture method, cells are first fixed by chemical fixatives, such as formaldehyde and glutaraldehyde. They are then infiltrated with glycerol for cryoprotection and frozen by dipping into liquid propane, liquid nitrogen thrush, or other appropriate cryogens. The frozen specimen is set on a stage with a temperature control and cracked with a knife. The temperature is usually maintained at around −100°C in a vacuum lower than $\sim 1 \times 10^{-5}$ mmHg. Because the fracture plane frequently passes through the hydrophobic center of the lipid bilayer, the membrane is split into outer and inner leaflets. The two surfaces of the membrane that are revealed by the splitting are called the exoplasmic face (E face) and the protoplasmic face (P face) (**Fig. 16.2B**). The true surface of the membrane is not revealed by freeze

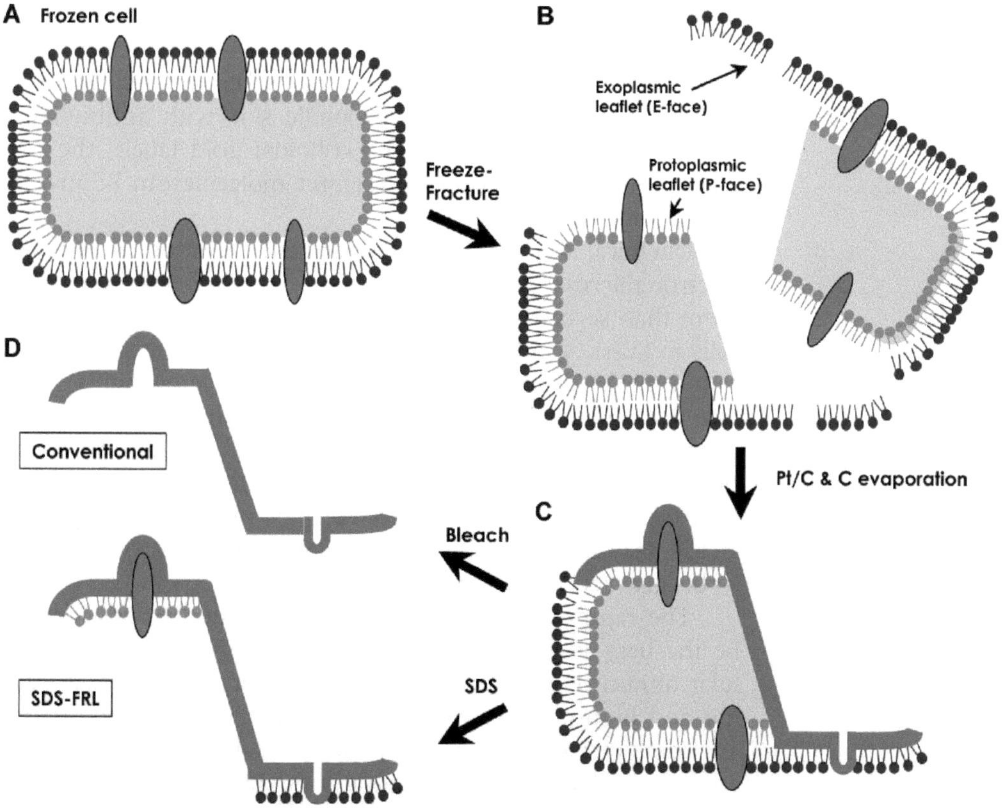

Fig. 16.2. Conventional freeze-fracture and SDS-FRL. (**A, B**) Fracturing of frozen specimens splits the plasma membrane into two halves and reveals the inner protoplasmic (P face) and outer exoplasmic (E face) face. (**C**) The surfaces revealed by freeze fracturing are coated by vacuum evaporation of platinum/carbon (Pt/C) and carbon (**C**). (**D**) In the conventional freeze-fracture method, the deposition of Pt/C (2 nm) is followed by that of C (20 nm), and the replica is thawed and treated with household bleach to digest all attached biological materials. In SDS-FRL, the order of Pt/C and C vacuum evaporation may need to be changed for better immunoreactivity (*see* **Note 5**). Moreover, the obtained replica is treated with SDS to digest extramembrane materials alone. Membrane molecules cast by the replica remain and are subjected to immunolabeling.

fracture. The P and E faces are coated with platinum/carbon (Pt/C) and carbon (C) by vacuum evaporation; the coating creates thin replicas (**Fig. 16.2C**). After thawing, the replica is removed from the specimen by the digestion of biological materials with household bleach (**Fig. 16.2D**). The replica is then rinsed, picked up on Formvar-coated EM grids, and examined with a standard transmission electron microscope (EM).

In SDS-FRL, the general procedure is the same as in the conventional method. Specimens are usually frozen rapidly without any prior fixation. After freeze-fracture and replica formation, the samples are treated with SDS, instead of household bleach, to remove only extramembrane materials (**Fig. 16.2D**). The replicas are incubated with the primary antibody, then with colloidal gold-conjugated secondary reagent, similar to other immunoelectron microscopic techniques. Because the cytoplasmic and exoplasmic leaflets of the membrane are cast by the Pt/C replica from the hydrophobic interface, membrane molecules are stabilized and can be accessed from the hydrophilic side with antibodies or other probes. By observing the colloidal gold labels, the two-dimensional distribution of the target molecule can be mapped on the wide membrane plane.

Chemical fixation has been used in conventional immuno-electron microscopic techniques. However, chemical fixatives take more than several seconds to reach and react with molecules in cells and tissues and may sometimes induce an artificial redistribution of molecules by cross-linking. Moreover, most lipids do not react with aldehyde fixatives. Therefore, when probes are applied after aldehyde fixation, the bound probes may induce a redistribution of the lipids (9). Glycerol that is used for the cryoprotection of fixed samples may also affect the membrane structure. These factors may cause artifacts when the molecular distribution needs to be analyzed at the nanometer level.

The rapid freezing of live cells and tissues is considered to be the best way to preserve molecular distribution and the cellular ultrastructure. Various instruments for metal-contact freezing and high-pressure freezing have been developed and some are commercially available, but they are relatively expensive and are not available in many laboratories. Liquid helium used for metal-contact freezing is also costly. To circumvent this problem, we developed a simple rapid-freezing method that uses two metal blocks cooled by liquid nitrogen (8). Although this method can be applied only to culture cells growing in a monolayer, it does not require any expensive equipment and is easy to use.

Samples that are fixed chemically can be used for proteins localized to specific structures (6, 10). The fixation needs to be weak so that cross-linking can be disrupted by the subsequent SDS treatment. For chemically fixed samples, glycerol infiltration and conventional freezing methods may be used.

3.1. Sample Preparation for Rapid Freezing

1. Put sterilized gold foil pieces cut in a trapezoidal shape into 35-mm plastic tissue culture dishes (**Fig. 16.1A**) containing the culture medium.
2. Disperse cultured cells with the trypsin/EDTA solution and seed them onto gold foils in culture dishes.
3. Grow cells to 70–100% confluency (*see* **Note 3**).

3.2. Rapid Freezing

1. Smear a small amount (1–2 μl) of prewarmed 10% gelatin in PBS on the copper foil which is kept on a warm heat block.
2. Invert a piece of gold foil with the cell side down (**Fig. 16.1C**) on the gelatin area of the copper foil (*see* **Note 4**).

3. Cool the two gold-plated copper blocks attached to a plier in liquid nitrogen and take them out immediately before use.
4. Holding the corner of the copper foil with forceps, rapidly press the gold-cell-copper sandwich between the cooled copper blocks (**Fig. 16.1B-c,d**). After pressing, plunge the sandwich in liquid nitrogen.
5. Keep the frozen cell sandwich in liquid nitrogen until further processing. For long-term storage, put sandwich specimens in a small cryogenic vial and store them in liquid nitrogen dewars.

3.3. Freeze-Fracture Replica Formation

1. Mount the frozen specimen on a Balzers specimen table (Bal-Tec) by clamping an edge of the copper foil to the table (**Fig. 16.3B-a**). The unnecessary portion of the copper foil is easily cut by ordinary scissors. All of the operations are carried out in liquid nitrogen.

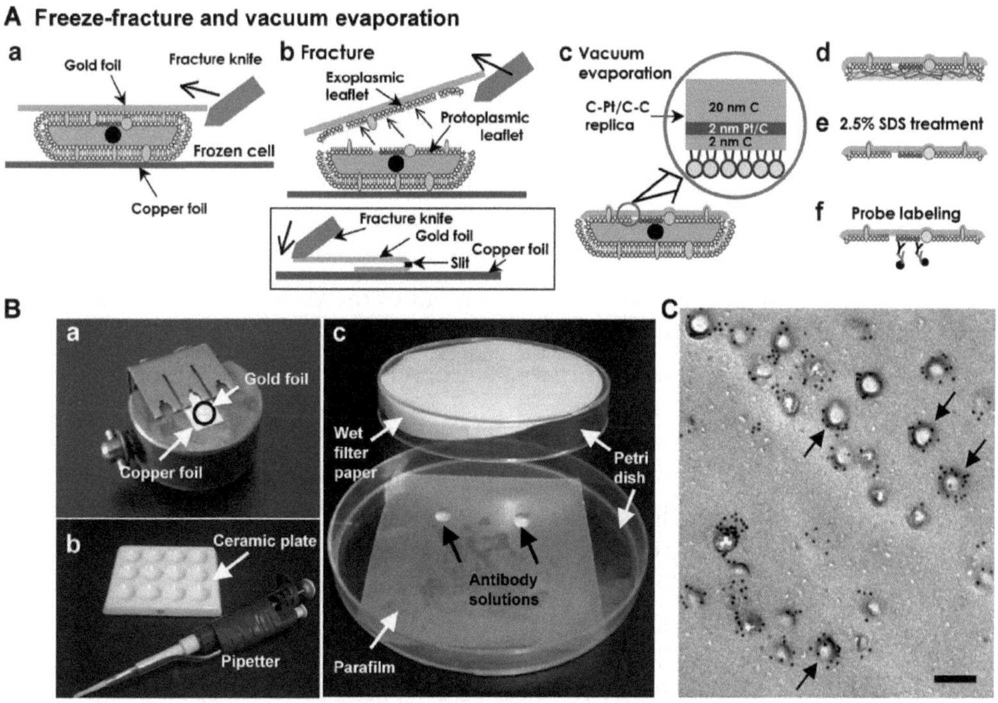

Fig. 16.3. SDS-FRL. (**A**) The scheme of the SDS-FRL procedure for monolayer cultured cells. After freeze-fracture, the samples left on the copper foil and the gold foil reveal the P and E faces of the basal plasma membrane, respectively. To make the replica on the gold foil, the foil needs to be bent sharply by the fracture knife as depicted in the box. (**B**) (*a*) A frozen specimen, consisting of a copper foil, a gold foil, and sandwiched cells, is fixed tightly by a clamp of a Balzers specimen table. (*b*) The washing, blocking, and picking up of replicas are done in wells of a ceramic plate. (*c*) The antibody incubation of replicas is done in drops made on a Parafilm sheet. As little as 20 μL of the solution is sufficient (*black arrows*). A moisture chamber made in a plastic Petri dish (100 mm diameter) is easy to use. (**C**) Immunogold labeling of caveolin-1 on the P face of human fibroblasts. Replicas are labeled with rabbit anti-caveolin-1 antibody, followed by 10-nm colloidal gold-conjugated anti-rabbit IgG antibody. Caveolae are observed as indentations of 70–100 nm in diameter (*arrows*). Scale bar, 100 nm.

2. Transfer the specimen table to the precooled specimen stage of the freeze-fracture apparatus and warm to −95°C.
3. Precool knife to liquid nitrogen temperature.
4. Apply vacuum to the freeze-fracture apparatus. When the vacuum is below $\sim 5\times10^{-6}$ mmHg, fracture the specimen sandwich by inserting the knife between the gold and copper foils so that an edge of the gold foil is lifted from the copper foil (**Fig. 16.3A-a,b**). In most cases, the specimen left on the copper foil exposes the P face of the basal plasma membrane. The gold foil may fly away by the above procedure. If it is necessary to observe the E face of the basal plasma membrane, bend the gold foil in the middle by using the slit and leave the gold foil on the copper foil with the cell side facing upward (**Fig. 16.3A-b, box**). For these operations, it is easier to move the knife by manual control.
5. Shadow the freeze-fractured specimen by electron-beam evaporation of Pt/C and C. Shadow first with 2 nm of carbon (C) at an angle of 90° to the specimen surface, then with 1–2 nm of Pt/C at an angle of 30–45°, and finally with 10–20 nm of C at an angle of 90° (**Fig. 16.3A-c**). This three-step procedure, hereafter called C-Pt/C-C, is favored over the conventional two-step procedure, Pt/C-C, because many molecules are labeled better in the C-Pt/C-C replica than in the Pt/C-C replica (*see* **Note 5**). The thickness of the evaporated material can be adjusted by using a crystal thickness monitor.
5. Take the specimen out of the vacuum chamber and thaw.
6. With a stereomicroscope, locate the area where the gold foil is torn away by observing its imprint left on the gelatin-smeared area of the copper foil. Shave the gelatin layer around but not within the imprinted area on the copper foil using a needle or a forceps tip (*see* **Note 4**).
7. Rinse the specimen briefly in PBS to wash off the shaved replica debris from outside the imprinted area of the copper foil by immersing the whole specimen in the small wells of a ceramic plate (**Fig. 16.3B-b**).
8. Transfer the copper foil to wells containing SDS digestion buffer to release the replica of the cell specimen (*see* **Note 6**). The replica should be easily released from the copper foil.
9. Transfer the released replicas to a 1.5-mL plastic tube using a pipette and incubate at 70°C for more than 60 min to dissolve the extramembrane materials (**Fig. 16.2D**) (*see* **Note 7**).

10. For long-term storage, add glycerol to a final concentration of 50% and store the specimen at –20°C until use (*see* **Note 8**).

3.4. Immunolabeling and EM Observation

1. Under a stereomicroscope, rinse the replicas five to six times by transferring replicas to new wells containing 0.1% Tween 20 in PBS using a pipette with a 10 μL tip (*see* **Note 9**). Five to ten small replica fragments are taken up into 1 μL of solution.
2. Wash the replicas in each well by gently stirring the solution by drawing into and releasing from a pipette.
3. Transfer the replicas to a well containing blocking buffer and incubate for more than 10 min at room temperature.
4. Dilute the primary antibody with antibody dilution buffer to an appropriate concentration (in most cases, 10–20 μg/mL) and place as drops (ca. 20 μL) on a sheet of Parafilm in a moisture chamber (**Fig. 16.3B-c**) (*see* **Note 10**). The primary antibody preincubated with the excess antigen, non-immune IgG, or the dilution buffer by itself may be used for controls.
5. Rinse the replicas once with washing buffer, transfer them to a drop of the primary antibody solution or control solution, and incubate overnight at 4°C or for 1 h at 37°C.
6. Rinse the replicas with antibody dilution buffer five or six times by transferring replicas to new buffer containing wells.
7. After rinsing, place the replicas in a 20 μL drop of colloidal gold-conjugated secondary antibody diluted 1:30 or colloidal gold-conjugated protein A diluted 1:70 and incubate for 30 min at 37°C.
8. Rinse the replicas 5–6 times with washing buffer by transferring the replicas to new buffer containing wells.
9. Just prior to mounting replicas on EM grids, rinse the replicas three times with distilled water by transferring to wells containing distilled water (*see* **Note 11**).
10. Pick up the replica onto a Formvar-coated EM grid in the third well containing distilled water.
11. After drying, store the grids in an appropriate grid case until the replicas can be viewed in the transmission EM. **Figure 16.3C** shows the distribution of caveolin-1 on the P face of human fibroblasts prepared by SDS-FRL and labeled with anti-caveolin-1 antibody followed by 10-nm colloidal gold-conjugated anti-rabbit IgG.

4. Notes

1. The slit is cut so that the foil can be bent easily by the knife of the freeze-fracture apparatus. By bending the gold foil, the freeze-fractured cell specimen on the gold foil faces upward and is coated by the vacuum evaporation (*see* **Section 3.3, Fig. 16.3A-b**). The replica made on the gold foil is complementary to that on the copper foil and useful to label molecules in the outer leaflet of the basal cell membrane.
2. The colloidal gold particles need to be larger than 5 nm to be clearly visible on the replica. We usually use 10-nm colloidal gold, because it is sometimes difficult to distinguish 5-nm colloidal gold from the intramembrane particles that are thought to represent transmembrane proteins. For double labeling, a combination of 10 and 15-nm colloidal gold particles may be used.
3. Because cells on the gold foil cannot be observed with a conventional light microscope, the cell density needs to be estimated from that of the surrounding areas in the plastic dish. Most cells grow very well on gold foil. If it is necessary to confirm that cells grow on the gold foil, the cell nucleus can be stained by 4′,6-diamidino-2-phenylindole (DAPI) and observed by fluorescence microscopy.
4. The gelatin is used for the following reasons. First, it facilitates the release of replicas from the copper foil, because gelatin is quickly dissolved by the SDS digestion buffer treatment. Without the gelatin smear, the replica often adheres to the copper foil tightly and needs to be scraped mechanically. Second, by shaving off the gelatin area from the outside of the gold foil before the SDS digestion step, most of the replica made on non-cellular areas can be excluded. This procedure increases the probability that the recovered replica contains the cellular specimen.
5. Many proteins and lipids are labeled with more intensity in C-Pt/C-C or C-Pt/C replicas than in conventional Pt/C-C replicas (6, 7, 11). By evaporating C before Pt/C, membrane molecules in the replica may retain some flexibility and react well with probes. Although evaporation of C before Pt/C makes small structures, such as intramembrane particles, less visible, this disadvantage is minimized by making the first C layer very thin (i.e., 1–2 nm).

6. In contrast to the replica made on the gelatin area, the replica made on the gold foil is not released easily by SDS treatment. We usually detach it physically using a pipette tip.
7. The replica must be heated in the SDS digestion buffer quickly to deactivate proteases and other degradation enzymes. For specimens that are difficult to digest, stronger treatments can be applied; extension of the SDS digestion to overnight, addition of stirring by a magnetic bar, and even autoclaving (121 °C for 15 min) are possible options. Through the treatment process, the replica may break up into fine fragments, which are barely visible without a stereomicroscope. To prevent this fragmentation, a method to coat the replica by the Lexan plastic was developed and has been applied to the grid-mapped freeze-fracture analysis of complex brain tissues (4).
8. To avoid repeated freeze thawing, the replicas are kept in a 50% glycerol solution, which does not freeze at –20°C. Replicas stored in this manner are reactive to various probes for many years.
9. Pipette tips with smooth edges are preferable, because replicas tend to stick to the tip when its edge is rough.
10. It is handy to make a moisture chamber from a plastic culture dish. Place a sheet of wet filter paper on the bottom part of the dish and use the lid part to place parafilm for antibodies (**Fig. 16.3B-c**).
11. This rinsing step needs to be done individually for each replica, because replicas tend to adhere tightly to the vessel wall or the pipette tip when placed in a solution without any protein or detergent.

Acknowledgments

The authors are thankful to Dr. Jinglei Cheng, Mr. Takehiro Yamada, Mr. Kousuke Chiba, and Mr. Nobuto Tamaki for their excellent technical help in the freeze-fracture experiments. This work was supported by Grants-in-Aid for Scientific Research and the Global COE Program "Integrated Molecular Medicine for Neuronal and Neoplastic Disorders" of the Ministry of Education, Culture, Sports, Science, and Technology of the Japanese Government.

Reference

1. Severs, N. J. and Shotton, D. M., ed. (1995) *Rapid Freezing, Freeze Fracture, and Deep Etching*. Wiley-Liss, New York.
2. Fujimoto, K. (1995) Freeze-fracture replica electron microscopy combined with SDS digestion for cytochemical labeling of integral membrane proteins. Application to the immunogold labeling of intercellular junctional complexes. *J. Cell Sci.* **108**, 3443–3449.
3. Fujimoto, K. (1997) SDS-digested freeze-fracture replica labeling electron microscopy to study the two-dimensional distribution of integral membrane proteins and phospholipids in biomembranes: practical procedure, interpretation and application. *Histochem. Cell Biol.* **107**, 87–96.
4. Rash, J. E. and Yasumura, T. (1999) Direct immunogold labeling of connexins and aquaporin-4 in freeze-fracture replicas of liver, brain, and spinal cord: factors limiting quantitative analysis. *Cell Tissue Res.* **296**, 307–321.
5. Robenek, H., Buers, I., Hofnagel, O., Robenek, M. J., Troyer, D., and Severs, N. J. (2009) Compartmentalization of proteins in lipid droplet biogenesis. *Biochim. Biophys. Acta.* **1791**, 408–418.
6. Hagiwara, A., Fukazawa, Y., Deguchi-Tawarada, M., Ohtsuka, T., and Shigemoto, R. (2005) Differential distribution of release-related proteins in the hippocampal CA3 area as revealed by freeze-fracture replica labeling. *J. Comp. Neurol.* **489**, 195–216.
7. Fujita, A., Cheng, J., Hirakawa, M., Furukawa, K., Kusunoki, S., and Fujimoto, T. (2007) Gangliosides GM1 and GM3 in the living cell membrane form clusters susceptible to cholesterol depletion and chilling. *Mol. Biol. Cell.* **18**, 2112–2122.
8. Fujimoto, T. and Fujimoto, K. (1997) Metal sandwich method to quick-freeze monolayer cultured cells for freeze-fracture. *J. Histochem. Cytochem.* **45**, 595–598.
9. Kusumi, A. and Suzuki, K. (2005) Toward understanding the dynamics of membrane-raft-based molecular interactions. *Biochim. Biophys. Acta* **1746**, 234–251.
10. Masugi-Tokita, M. and Shigemoto, R. (2007) High-resolution quantitative visualization of glutamate and GABA receptors at central synapses. *Curr. Opin. Neurobiol.* **17**, 387–393.
11. Schlormann, W., John, M., Steiniger, F., Westermann, M., and Richter, W. (2007) Improved antigen retrieval in freeze-fracture cytochemistry by evaporation of carbon as first replication layer. *Histochem. Cell Biol.* **127**, 633–639.

Chapter 17

Pre-embedding Electron Microscopy Methods for Glycan Localization in Chemically Fixed Mammalian Tissue Using Horseradish Peroxidase-Conjugated Lectin

Yoshihiro Akimoto and Hayato Kawakami

Abstract

In histochemistry and cytochemistry, horseradish peroxidase-labeled lectins are often used as probes for the localization of carbohydrates in cells and tissues. In transmission electron microscopy, the most commonly used procedure for detection of carbohydrates is lectin–gold labeling. Horseradish peroxidase catalyzes the formation of insoluble polymerized diaminobenzidine which on exposure to osmium tetroxide forms osmium black, a compound visible in the electron microscope, making horseradish peroxidase an alternative to the more frequently used colloidal gold. This chapter describes a pre-embedding method for carbohydrate localization in which tissue sections are incubated with horseradish peroxidase-conjugated lectin prior to embedding in resin.

Key words: Pre-embedding method, horseradish peroxidase (HRP), glycans, lectin, electron microscopy.

1. Introduction

Glycans are examined histochemically by using carbohydrate-specific antibodies and lectins. Lectins are carbohydrate-binding proteins other than enzymes or antibodies. Various lectins have been extracted from plants, animals, and microorganisms. Compared with antibodies, lectins have useful specificities for complex glycans and are less expensive. Changes in glycans during development or under pathological conditions can be observed histochemically by lectins (1). Lectins are used as markers for malignant transformation of tumors and differentiation of a specific kind of cell or tissue (2). Researches on lectins have been reviewed in several books (2, 3).

S.D. Schwartzbach, T. Osafune (eds.), *Immunoelectron Microscopy*, Methods in Molecular Biology 657,
DOI 10.1007/978-1-60761-783-9_17, © Springer Science+Business Media, LLC 2010

For transmission electron microscopy of lectin histochemistry, pre-embedding and post-embedding methods are used (4). In the pre-embedding method, the binding of lectin to sugar residues is performed before the tissue is embedded in resin. In the post-embedding method, ultrathin sections from resin-embedded specimens are prepared and then incubated with lectin. In the pre-embedding method, HRP is commonly used as a labeling marker, while in the post-embedding method colloidal gold is used as it is more difficult for colloidal gold conjugates to penetrate into cells or tissues than HRP conjugates because of their larger size.

In the HRP-labeling method, the localization of the desired sugar residue is revealed by the HRP-catalyzed formation of insoluble polymerized diaminobenzidine (DAB) at the site of HRP-labeled lectin binding. By binding osmium tetroxide (O_sO_4), the polymerized DAB is converted to the product osmium black which can be observed under the electron microscope. The merits of the pre-embedding HRP-labeling method are that 1) sensitivity is much better than by post-embedding methods, as the binding occurs before embedding in resin; 2) the electron microscopic image can be compared with the light microscopic image in the same section or serial sections; 3) the electron microscopic image at low magnification can be observed with the HRP-labeling method; and 4) membrane structure is well preserved. The limitations of the pre-embedding HRP-labeling method is that 1) the results obtained are qualitative, not quantitative, whereas the post-embedding colloidal gold-labeling method is quantitative, and 2) penetration of lectin conjugates into tissue is sometimes problematic (5).

There are two methods for HRP detection of lectin binding. In the direct method, HRP is directly conjugated to the lectin. In the indirect method, a biotinylated lectin is used and the lectin is detected with HRP-conjugated streptavidin. The direct method is more convenient and gives less nonspecific binding than the indirect method. However, the indirect method employing biotinylated lectins is also widely used as it has high sensitivity. This chapter describes the localization of glycans by the pre-embedding labeling method using the HRP technique at the electron microscopic level.

2. Materials

2.1. Tissue Fixation

1. Phosphate-buffered saline (PBS): Dissolve 8.00 g NaCl, 0.20 g KCl, 0.24 g KH_2PO_4, 1.44 g Na_2HPO_4 in 1 l of distilled water.

2. 2.5% (v/v) glutaraldehyde–PBS: Dilute one volume of 25% (v/v) glutaraldehyde solution (EM grade, TAAB, Berkshire, United Kingdom) with nine volumes of PBS.

2.2. Sectioning Frozen Fixed Tissue

1. 10%, 15%, 20% (w/v) sucrose–PBS.
2. OCT compound (Sakura Finetek, Torrance, CA, USA).
3. Cryo-mold (Sakura Finetek).
4. Dry ice ethanol.
5. Cryostat (Leica, Wetzlar, Germany).
6. Poly-L-lysine-coated glass slides (Polysciences, Warrington, PA, USA).
7. PAP pen (Invitrogen, Carlsbad, CA, USA).

2.3. Glycans Detection

1. Moisture chamber.
2. Five percent (w/v) bovine serum albumin (BSA, Fraction V; Sigma, Saint Louis, MO, USA)–PBS.
3. HRP-conjugated lectins (EY Laboratories, San Mateo, CA, USA).
4. 0.2 M Phosphate buffer (pH 7.4): Prepare by using 0.2 M NaH_2PO_4 to adjust the pH of a 0.2 M Na_2HPO_4 solution to 7.4.
5. 3,3′-Diaminobenzidine–4HCl (DAB; Sigma).
6. 0.05% (w/v) DAB–PBS: Prepare a fresh solution by dissolving 50 mg DAB in 100 ml PBS. Mix rapidly on a magnetic stirring plate and keep in dark.
7. 0.05% (w/v) DAB–0.005% (w/v) H_2O_2–PBS: Prepare a fresh solution by dissolving 50 mg DAB in 100 ml PBS. Add 16.7 μl 30% hydrogen peroxide just before start of reaction.

2.4. Post-fixation of Tissue

1. One percent (v/v) glutaraldehyde–PBS: Dilute one volume of 25% (v/v) glutaraldehyde solution (EM grade, TAAB, Berkshire, United Kingdom) with 24 volumes of PBS.
2. One percent (w/v) OsO_4–0.1 M phosphate buffer (pH 7.4): Mix equal volumes of 2% (w/v) OsO_4 in distilled water and 0.2 M phosphate buffer (pH 7.4).
3. 50, 70, 90, and 100% (v/v) ethanol-distilled water.
4. Gelatin capsule (SPI Supplies/Structure Probe, West Chester, PA, USA).
5. Epon 812 (plastic embedding resin; TAAB).

2.5. Sectioning and Post-staining

1. Ultramicrotome (Ultracut UCT; Leica).
2. Four percent (w/v) lead citrate in distilled water.

3. Single-edge razor blade.
4. Scotch tape.
5. Light microscope.

3. Methods

3.1. Tissue Fixation

1. Slice animal tissue into small pieces (size 5 mm × 5 mm × 2 mm) with razor blade as quickly as possible (*see* **Note 1**).
2. Fix tissue with cold 2.5% glutaraldehyde–PBS at 4°C by gently shaking for 1 h.

3.2. Sectioning Frozen Fixed Tissue

1. Wash tissue three times for 10 min with cold PBS.
2. Cryoprotect the tissue specimens by immersing them sequentially in 10, 15, and 20% sucrose–PBS at 4°C each for a minimum of 4 h and a maximum of 12 h.
3. Mount the tissue specimens in OCT compound using cryomolds and freeze using dry ice ethanol.
4. Cut the specimens into 10-μm-thick sections with a cryostat.
5. Place the sections on poly-L-lysine-coated glass slides.
6. Air-dry the sections with cold air (*see* **Note 2**) for 30 min to attach the sections firmly to the glass slides.
7. Draw a circle around the section with a PAP pen (*see* **Note 3**).
8. Immerse sections in 0.6% H_2O_2–PBS at room temperature for 15 min to eliminate endogenous peroxidase activity (*see* **Note 4**).
9. Wash the sections three times for 5 min with cold PBS.

3.3. Glycan Detection (see Note 5)

1. Incubate the sections with 5% BSA–PBS (*see* **Note 6**) at room temperature for 1 h to block non-specific binding of lectin.
2. Rinse the sections with PBS.
3. Incubate the sections with HRP-conjugated lectin (5~50 μg/ml PBS) (*see* **Note** 7) for 24 h in a moisture chamber at 4°C (*see* **Note 8**).
4. Wash the sections six times for 5 min with cold PBS.
5. Incubate the sections in 0.05% DAB solution for 10 min at room temperature.
6. Remove the sections from the DAB solution. Initiate the HRP reaction with freshly prepared 0.05% DAB–0.005%

H_2O_2–PBS and incubate the sections at room temperature for 2–10 min monitoring the reaction under a microscope. React sections until they become dark (*see* **Note 9**).

7. When the sections become dark, stop the reaction by removing the DAB solution and wash the sections three times for 5 min with cold PBS.

3.4. Post-Fixation of Tissues

1. Fix the labeled sections by incubation for 10 min at 4°C with 1% glutaraldehyde–PBS.
2. Wash the sections three times for 5 min with cold PBS.
3. Place the glass slides on ice in a fume hood and osmicate them by incubation for 1 h at 4°C with 1% O_sO_4–0.1 M phosphate buffer (pH 7.4).
4. Wash the sections three times for 5 min with cold distilled water.

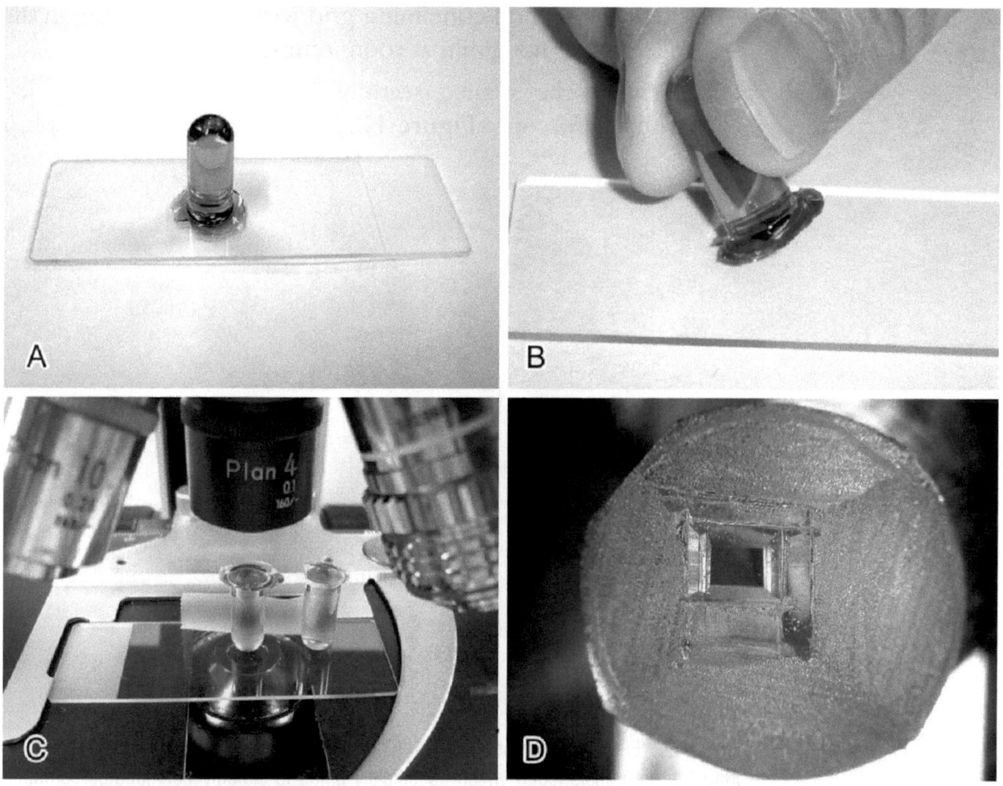

Fig. 17.1. (**A**) Embedding in resin. The gelatin capsule containing the embedding medium is inverted over the tissue section and the resin is polymerized at 60°C for 24 h. (**B**) The gelatin capsule containing Epon can be snapped off the surface of the glass slide by heating the slide to 170°C on a hot plate. (**C**) The inverted Epon block containing the resin-embedded section can be attached to a glass slide with Scotch tape and observed under the light microscope. (**D**) The trimmed resin block.

5. Dehydrate the sections by passage through a graded series of ethanol (50, 70, 90, and 100%) incubating for 5 min in each concentration.
6. Fill a gelatin capsule with Epon812 and before the section has dried, invert the resin-filled capsule over the tissue section on the glass slide as shown in **Fig. 17.1A**.
7. Polymerize the resin at 60°C for 24 h.
8. Detach the Epon-embedded sections from the glass slide by heating the slide to 170°C on a hot plate until the gelatin capsule containing resin can be snapped off the surface of the slide (**Fig. 17.1B**).

3.5. Sectioning and Post-staining

1. Observe the section under the light microscope attaching the inverted Epon block to a glass slide using Scotch tape and determine the trimming area (**Fig. 17.1C**).
2. Trim the resin block under a binocular microscope with a single-edge razor blade (**Fig. 17.1D**).
3. Cut ultrathin sections with ultramicrotome.
4. Stain the section containing grid with 4% lead citrate in distilled water for 4 min at room temperature.
5. Observe the stained sections under a transmission electron microscope. **Figure 17.2** is an electron micrograph of

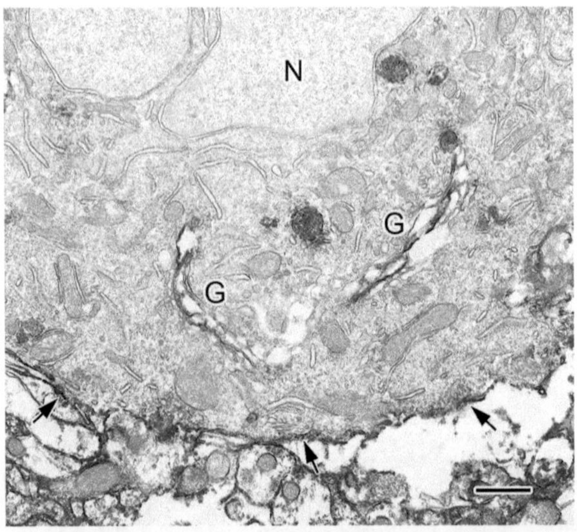

Fig. 17.2. Electron micrograph of Purkinje cells from a 9-week-old rat stained with *Maackia amurensis* lectin which specifically binds to sialoglycoconjugates containing the Siaα2-3Gal group. The stained plasma membranes of the Purkinje cells are indicated by *arrows*. Positive staining is also observed in the Golgi apparatus (G) and the plasma membrane of cell processes around the Purkinje cell. N, nucleus. Bar: 1 μm. (Reproduced from Sasaki et al. (6) with permission from The Histochemical Society, Inc.)

Purkinje cells from a rat cerebellum stained with *Maackia amurensis* lectin (6). An electron-dense reaction product (osmium black) is seen in the plasma membrane (arrows) and the Golgi apparatus (G) of the Purkinje cell. It shows that sialoglycoconjugates containing the Siaα2-3Gal group are present in the plasma membrane and Golgi apparatus of Purkinje cells.

4. Notes

1. If possible, the tissue should be fixed by perfusion with cold 2.5% glutaraldehyde–PBS prior to removal from the animal.
2. For better preservation of tissue structure in some tissues such as testes and ovaries, frozen sections attached to glass slides are rapidly transferred to PBS without air-drying (7).
3. Drawing a water-repellent circle around the section with a PAP pen prevents streaming of antibody and other solutions from the section and reduces the amount of antibody solution needed. This allows all incubations to be done with small solution volumes placed in this circle. However, when there are many slides to process, in order to keep the incubation times constant, washing should be done with a washing bottle and the DAB reaction should be done in a staining jar by putting all slides in a cage.
4. This step can be omitted if there is no endogenous peroxidase activity in the tissue.
5. Both negative and positive histochemical control experiments should be carried out to confirm the specificity of the lectin-binding reaction. Negative controls should include (1) the addition of the appropriate inhibitory sugars (0.1 M) specific for each lectin to the HRP-conjugated lectin–PBS to compete with the target for lectin binding; (2) omission of the lectin; and (3) destruction of target glycans by incubation with the appropriate glycosidases. The positive control should include incubation of the cells or tissues with a lectin whose binding to the cell or the tissue has been established experimentally.
6. Do not use normal serum for blocking as glycoconjugates present in serum competitively inhibit specific binding of lectin.
7. As each lectin has a different affinity for sugar residues, the optimal lectin concentration to use should be determined by

light microscopic observation before electron microscopic experiments are performed.

8. For the indirect method of HRP detection of lectin binding, incubate the sections with biotinylated lectin (5–50 μg/ml PBS) for 24 h in a moisture chamber at 4°C. Wash the sections six times for 5 min with cold PBS. Incubate the sections with HRP–streptavidin (0.5 μg/ml PBS) for 24 h at room temperature. Then proceed to **Section 3.3**, Step 4.
9. When the DAB reaction product diffuses away from the primary site of reaction because of overreaction for DAB deposition, samples should be reacted for a shorter period time.

References

1. Brooks, S. A., Leathem, A. J. C., and Schumacher, U. (1997) Lectin histochemistry: A concise practical handbook. BIOS Scientific Publishers in association with the Royal Microscopical Society – (Royal Microscopical Society microscopy handbooks; 36).
2. Sharon, N. and Lis, H.(2003) *Lectins*, 2nd edn. Kluwer Academic Publishers, Dordrecht, The Netherlands
3. Varki, A., Cummings, R. D., Esko, J. D., Freeze, H. H., Hart, G. W., and Etzler, M. E. (2009) *Essentials of Glycobiology*, 2nd edn. Cold Spring Harbor Laboratory Press, New York.
4. Kawakami, H. and Hirano, H. (1993) Lectin histochemistry and cytochemistry. In *Electron Microscopic Cytochemistry and Immunocytochemistry in Biomedicine*, Ogawa, K. and Barka, T., eds. CRC Press, Boca Raton, FL, 319–332.
5. Oliver, C. (1994) Pre-embedding labeling methods. In *Immunocytochemical Methods and Protocols*, Javois, L. C., ed. *Methods Mol. Biol.* **34**, 315–319.
6. Sasaki, T., Akimoto, Y., Sato, Y., Kawakami, H., Hirano, H., and Endo, T. (2002) Distribution of sialoglycoconjugates in the rat cerebellum and its change with aging. *J. Histochem. Cytochem.* **50**, 1179–1186.
7. Kanai, Y., Kawakami, H., Takata, K., Kurohmaru, M., Hayashi, Y., Nishida, T., and Hirano, H. (1990) Localization of Forssman glycolipid and GM1 ganglioside intracellularly and on the surface of germ cells during fetal testicular and ovarian development of mice. *Histochemistry* **94**, 561–568.

Chapter 18

Pre-embedding Nanogold Silver and Gold Intensification

Akitsugu Yamamoto and Ryuichi Masaki

Abstract

Pre-embedding nanogold silver and gold intensification methods involve immunoreactions with nanogold-labeled antibodies and intensification of the nanogold particles before embedding and ultrathin sectioning. These highly sensitive methods show good resolution and ultrastructural preservation. They also are useful for simultaneous observation of immunolabeled cells under light and electron microscopes and for 3D immunoelectron microscopic analyses. Silver intensification is usually superior for immunolabeling. On the other hand, ultrastructural preservation is better when gold intensification is used. In this chapter, we introduce pre-embedding nanogold silver and gold intensification procedures for use primarily with cultured cells.

Key words: Silver intensification, gold intensification, nanogold, pre-embedding immunoelectron microscopy, cultured cells.

1. Introduction

Determining the localization of a protein is crucial for understanding its function in cells and tissues. Pre-embedding nanogold silver and gold intensification methods have been widely used for this purpose (1, 2). These pre-embedding methods involve immunolabeling samples before resin-embedded sections are prepared, such that antibodies can bind throughout the samples. As a result, these methods are highly sensitive and result in marked ultrastructural preservation. Additionally, it is easy to observe the same immunolabeled cells under a light microscope and an electron microscope. Currently, nanoscale 3D analyses that employ reconstruction of serial ultrathin sections or electron tomography are important tools in studies of cell biology. Nanogold silver and gold intensification methods are amenable to 3D analyses. In addition, the reaction product does not diffuse

S.D. Schwartzbach, T. Osafune (eds.), *Immunoelectron Microscopy*, Methods in Molecular Biology 657,
DOI 10.1007/978-1-60761-783-9_18, © Springer Science+Business Media, LLC 2010

following nanogold intensification resulting in high resolution when determining the localization of proteins and facilitating quantitative analysis. Importantly, conventional epoxy resin embedding can be used with these methods and specific machines or techniques are not required.

Conventional gold particles (3–50 nm in diameter) are, in many cases, not useful for pre-embedding immunoelectron microscopy because they do not permeate well into cells or organelle compartments without compromising ultrastructural details. Using smaller gold particles (nanogold, ~1 nm in diameter), therefore, is essential. Because it is not easy to observe these small particles on ultrathin sections of cells, nanogold particles are intensified by increasing the diameter to 10–100 nm with silver or gold deposited on their surface (3).

Silver intensification procedures usually result in heavier labeling with smaller and more homogenous particles than those observed following gold intensification. On the other hand, conventional OsO_4 post-fixation can be used with gold intensification, but not silver intensification, and gold intensification results in better ultrastructural preservation of cells. Thus, to obtain clear ultrastructural details, silver intensification generally requires stronger fixation conditions than those used for gold intensification. Silver intensification samples are usually processed in the dark room, whereas gold intensification samples can be processed under light conditions. Some improvements have been made to both intensification methods. For instance, He et al. (4) have used gold intensification to produce small and homogenous particles that are comparable to those obtained with silver intensification. In both procedures, it is important to use highly specific antibodies. Primary antibodies should be checked using immunofluorescence microscopy, and the antibodies should be titrated to determine the optimal concentration. In this chapter, we introduce pre-embedding nanogold silver and gold intensification procedures to be used primarily with culture cells.

2. Materials

2.1. Fixation and Permeabilization of Cultured Cells

1. Cell desk LF1 low autofluorescence 13.5-mm diameter plastic coverslips (Sumitomo Bakelite, Tokyo, Japan) (*see* **Note 1**).
2. Milli-Q water produced by a Milli-Q system (Millipore Co., Billerica, MA). Use Milli-Q water in all the solutions.
3. Phosphate-buffered saline (PBS): Prepare 1.0 L of 25× stock solution with 200 g of NaCl, 36.3 g of Na_2HPO_4• $12H_2O$, 2.5 g of KH_2PO_4, and 2.5 g of KCl. Autoclave

the solution before storing it at room temperature. Prepare a working solution of PBS by diluting one part of stock solution with 24 parts Milli-Q water.

4. 0.1 M sodium phosphate buffer, pH 7.4 (PB): Prepare 1.0 L of 3× (0.3 M) stock solution with 8.9 g of NaH_2PO_4•$2H_2O$ and 87.0 g of Na_2HPO_4•$12H_2O$, and autoclave the solution before storing it at room temperature. Prepare a working PB solution by diluting one part of the stock solution with two parts Milli-Q water and adjust the pH if necessary with NaOH.
5. 10% (w/v) paraformaldehyde solution: Add ~45 mL of Milli-Q water to 5 g of paraformaldehyde powder (Nakalai Tesque, Kyoto, Japan) and heat to 60°C. Add a few milliliters of 1.0 M NaOH drop wise while stirring in a draft chamber until the solution clears. Adjust the volume of the solution to 50 mL with Milli-Q water. This solution should be used within 24 h (*see* **Note 2**).
6. 4% (w/v) paraformaldehyde solution: Mix 4.0 mL of 10% (w/v) paraformaldehyde solution, 3.3 mL of PB (3×), and 2.7 mL of Milli-Q water (10 mL total). This solution should be used within 24 h.

2.2. Immunogold Labeling

1. Saponin (Nakalai Tesque, Kyoto, Japan)
2. Bovine serum albumin (BSA, Sigma-Aldrich, St Louis, MO)
3. Normal goat serum (Vector Laboratories, Burlingame, CA)
4. Cold water fish skin gelatin (Sigma-Aldrich, St Louis, MO)
5. Blocking solution A: PB containing 0.005% (w/v) saponin, 10% (w/v) BSA, 10% (v/v) normal goat serum, and 0.1% (w/v) cold water fish skin gelatin
6. Blocking solution B: PB containing 0.1% (w/v) saponin, 10% (w/v) BSA, 10% (v/v) normal goat serum, and 0.1% (w/v) cold water fish skin gelatin
7. Primary antibody to protein of interest
8. Nanogold Fab' conjugate (Nanoprobes, Yaphank, NY): The IgG Fab' fragment from goat anti-mouse IgG, anti-rabbit IgG, or anti-rat IgG covalently conjugated to 1.4-nm colloidal gold particles (*see* **Note 3**)
9. Glutaraldehyde: EM grade, 25% (w/w) aqueous solution (Electron Microscopy Sciences, Hatfield, PA) (*see* **Note 2**)

2.3. Nanogold Silver Intensification Procedure

1. 50 mM Hepes-NaOH, pH 5.8
2. HQ silver enhancement kit (Nanoprobes, Yaphank, NY)
3. OsO_4 (TAAB Laboratories Equipment, Berks, UK): The stock 2% (w/v) aqueous solution should be kept separate from other reagents (*see* **Note 2**)

4. 0.5% (w/v) OsO_4 in PB: Mix 2.5 mL of 2% (w/v) aqueous OsO_4 solution, 3.3 mL of PB (3×), and 4.2 mL of Milli-Q water (10 mL total), and cool the solution on ice.

2.4. Nanogold Gold Intensification Procedure

1. PBS containing 50 mM glycine
2. PBS containing 1% (w/v) BSA
3. GoldEnhance EM kit (Nanoprobes, Yaphank, NY)
4. 1 % (w/v) aqueous sodium thiosulfate solution. Prepare this solution immediately before use.
5. 1% (w/v) OsO_4 and 1.5% (w/v) potassium ferrocyanide in PB: Dissolve ferrocyanide powder for a final concentration of 1.5% (w/v) to 1% (w/v) OsO_4 in PB immediately before use.

2.5. Embedding in Epoxy Resin and Ultrathin Sectioning

1. A graded series of ethanol solutions: 30, 50, 70, 90, and 100% (v/v) ethanol.
2. Epoxy resin mixture (TAAB Laboratories Equipment, Berks, UK): Mix 24.5 mL of epon812, 15.0 mL of dodecenyl succinic anhydride (DDSA), 13.5 mL of methyl nadic anhydride (MNA), and 0.75 mL of 2,4,6-tri(dimethylaminomethyl)phenol (DNP30) (*see* **Note 4**)
3. Silicon rubber flat embedding mold (Dosaka EM Co., Kyoto, Japan)
4. Diamond Knife (Diatome AG, Biel, Switzerland)
5. Ultramicrotome (Ultracut N, Reichert-Jung Optische Werke, Vienna, Austria)
6. Bioden mesh cement (0.25% neoprene solution in toluene, Ohkenshoji Co. LTD, Tokyo, Japan)
7. Veco specimen grid (200 mesh, copper) (Electron Microscopy Sciences, Hatfield, PA). The EM grids are coated with Bioden mesh cement by placing the grids on filter paper and dropping the mesh cement onto the grids
8. 2% (w/v) aqueous uranyl acetate solution
9. Lead staining solution (Sigma, St Louis, MO)

3. Methods

3.1. Fixation and Permeabilization of Cultured Cells

1. Cells are cultured on LF1 plastic coverslips, glass coverslips, or CELLocate in a 24-well plate containing 0.5 mL of culture medium in each well.
2. The culture medium is removed, cells in each well are washed with 1 mL of PBS for a few seconds and then fixed

with 0.5 mL of 4% (w/v) paraformaldehyde for 2 h at room temperature (*see* **Note 5**).

3. The cells are washed three times for 5 min each in 1 mL of PB.
4. The fixed cells in a 24-well plate are preserved in PB in a refrigerator before use. The cells should be kept from drying.

3.2. Immunogold Labeling

1. The fixed cells are permeabilized with 0.5 mL of PB containing 0.25% (w/v) saponin for 30 min at room temperature and washed once with 1 mL of PB for 3 min (*see* **Note 6**).
2. The fixed and permeabilized cells are incubated with 0.5 mL of blocking solution A (for silver intensification) or blocking solution B (for gold intensification) for 30 min (*see* **Note 7**).
3. The blocking solution is removed and replaced with 0.2 mL of primary antibody in blocking solution A (silver intensification) or blocking solution B (gold intensification) at a concentration similar to that used for immunofluorescence. Samples are incubated in a moist chamber overnight in a refrigerator.
4. The primary antibody is removed and the cells are washed six times for 10 min with 1 mL of PB containing 0.005% (w/v) saponin.
5. The cells are incubated in a moist chamber with 0.2 mL of nanogold Fab' conjugates (1:500 dilution) in blocking solution A (silver intensification) or blocking solution B (gold intensification) as a secondary antibody for 2 h at room temperature.
6. The secondary antibody is removed and the cells are washed six times for 10 min with 1 mL of PB containing 0.005% (w/v) saponin followed by a 10-min wash in 1 mL of PB.
7. The immunogold-labeled cells are fixed with 0.5 mL of 1% (w/v) glutaraldehyde in PB for 10 min at room temperature (*see* **Note 8**).

3.3. Nanogold Silver Intensification Procedure

1. The immunogold-labeled cells are washed three times for 3 min in 1 mL of 50 mM Hepes-NaOH (pH 5.8) followed by a 1-min wash in 1 mL of Milli-Q water.
2. Silver intensification solution is made immediately before use by mixing equal volumes of the three reagents from the HQ silver enhancement kit in a dark room (*see* **Note 9**).
3. The immunogold-labeled cells are incubated in ~0.2 mL of silver intensification solution at 20°C for 6 min in the dark (*see* **Note 10**).

4. The silver intensification solution is removed, and the immunogold-labeled cells are washed three times for 1 min each in 1 mL of Milli-Q water in the dark.
5. The immunogold-labeled cells are taken out of the dark room and post-fixed in 0.5 mL of 0.5% (w/v) OsO_4 at 4°C for 90 min in a draft chamber (*see* **Note 11**).
6. The immunogold-labeled cells are washed with 1 mL of Milli-Q water three times for 1 min (*see* **Note 12**)

3.4. Nanogold Gold Intensification Procedure

1. The immunogold-labeled cells are washed three times for 5 min in 1 mL of PBS containing 50 mM glycine, followed by three 5-min washes in 1 mL of PBS containing 1% (w/v) BSA and three 5-min washes in 1 mL of Milli-Q water.
2. Gold intensification solution is prepared immediately before use by mixing one part of solutions A and five parts of solution B in an Eppendorf microcentrifuge tube. After 5 min, one part of solutions C and D is added to the Eppendorf tube and the solution is mixed (*see* **Note 13**).
3. The immunogold-labeled cells are incubated in 0.1 mL of the gold intensification solution at room temperature for 2 min (*see* **Note 13**).
4. The gold intensification solution is removed; the immunogold-labeled cells are incubated in 0.5 mL of freshly prepared 1% (w/v) aqueous sodium thiosulfate solution for few seconds and washed in 1 mL of Milli-Q water three times for 1 min.
5. The immunogold-labeled cells are post-fixed in 0.5 mL of 1% (w/v) OsO_4 and 1.5% (w/v) potassium ferrocyanide in PB for 60 min at room temperature in a draft chamber (*see* **Note 14**).
6. The immunogold-labeled cells are washed in 1 mL of Milli-Q water three times for 1 min (*see* **Note 12**)

3.5. Embedding in Epoxy Resin and Ultrathin Sectioning

1. The immunogold-labeled cells are dehydrated at room temperature in 1 mL of a graded series of ethanol solutions: 30, 50, 70, and 90% (v/v) ethanol, and twice in 100% ethanol for 10 min
2. The immunogold-labeled cells are incubated in 0.5 mL of a 50% (v/v) epoxy resin ethanol mixture for 30 min at room temperature.
3. The immunogold-labeled cells are incubated in 0.5 mL of epoxy resin mixture two times for 1 h at room temperature.
4. The cavities in a silicon rubber flat embedding mold are filled with epoxy resin mixture. The bent edge of a LF1 coverslip is cut off using scissors and the cavity is covered with the

coverslip with the cell side down. The samples are then incubated for 2 days at 60°C (*see* **Note 15**).

5. After the epoxy resin polymerizes, the LF1 coverslip is removed from the epoxy resin block using tweezers (*see* **Note 16**).
6. The cells on the newly exposed epoxy resin block surface are observed under a light microscope and the positions of interest are marked with a marker. In the case of silver intensification, the brown reaction products are often but not always visible under a light microscope as seen in **Fig. 18.1A** (*see* **Note 17**).

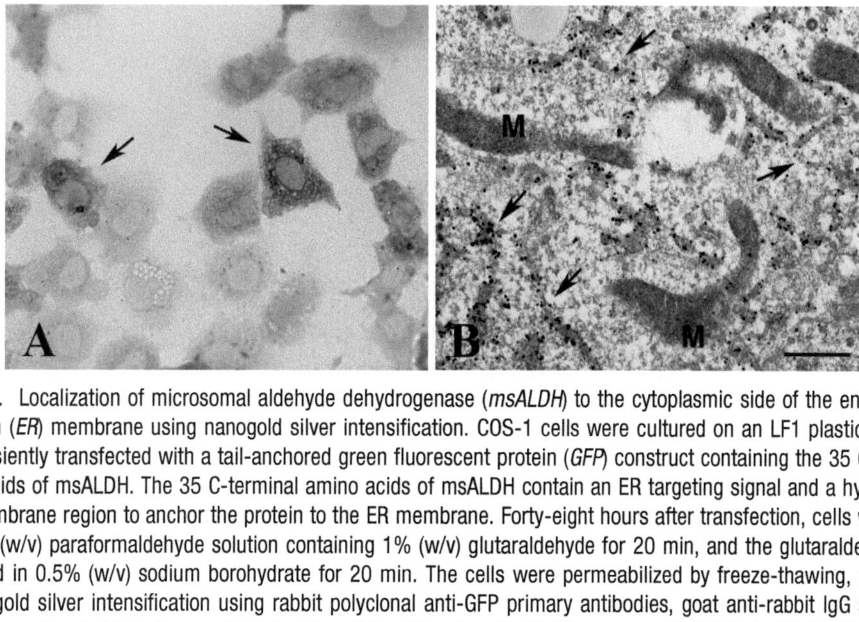

Fig. 18.1. Localization of microsomal aldehyde dehydrogenase (*msALDH*) to the cytoplasmic side of the endoplasmic reticulum (*ER*) membrane using nanogold silver intensification. COS-1 cells were cultured on an LF1 plastic coverslip and transiently transfected with a tail-anchored green fluorescent protein (*GFP*) construct containing the 35 C-terminal amino acids of msALDH. The 35 C-terminal amino acids of msALDH contain an ER targeting signal and a hydrophobic transmembrane region to anchor the protein to the ER membrane. Forty-eight hours after transfection, cells were fixed with 4% (w/v) paraformaldehyde solution containing 1% (w/v) glutaraldehyde for 20 min, and the glutaraldehyde was quenched in 0.5% (w/v) sodium borohydrate for 20 min. The cells were permeabilized by freeze-thawing, processed for nanogold silver intensification using rabbit polyclonal anti-GFP primary antibodies, goat anti-rabbit IgG secondary antibodies conjugated to 1.4-nm colloidal gold particles (1) and embedded in epoxy resin. (**A**) Embedded cells observed under a light microscope showing silver-intensified nanogold label in some cells indicative of strong expression of the GFP-msALDH chimera (arrows). (**B**) An electron micrograph of a COS-1 cell expressing GFP-msALDH. Numerous 10–20-nm silver-intensified nanogold particles can be seen bound to the cytoplasmic side of the ER membranes. Arrows indicate the ER. M: mitochondria. Bar: 1 μm.

7. 80 nm thick ultrathin sections of the marked region are cut horizontally to the cell layer using an ultramicrotome equipped with a diamond knife and picked up on Bioden mesh cement-coated copper EM grids.
8. The sections are stained with 2% (w/v) uranyl acetate solution for 1 h and briefly washed three times in Milli-Q water. The sections then are stained with lead staining solution for 1 min and washed briefly three times in Milli-Q water.
9. The sections are observed under an electron microscope. **Figure 18.1B** shows an electron micrograph immunolocalizing GFP-msALDH to the cytoplasmic side of the ER

membrane. The 10–20-nm particles seen bound to the cytoplasmic side of the ER membrane were formed by silver intensification of the 1.4-nm colloidal gold particles conjugated to the secondary antibody. **Figure 18.2** shows an electron micrograph immunolocalizing heat shock protein 47 to the ER lumen. Gold intensification of the 1.4-nm colloidal gold particles conjugated to the secondary antibody produced the 10–20-nm particles seen in the luminal space of the ER.

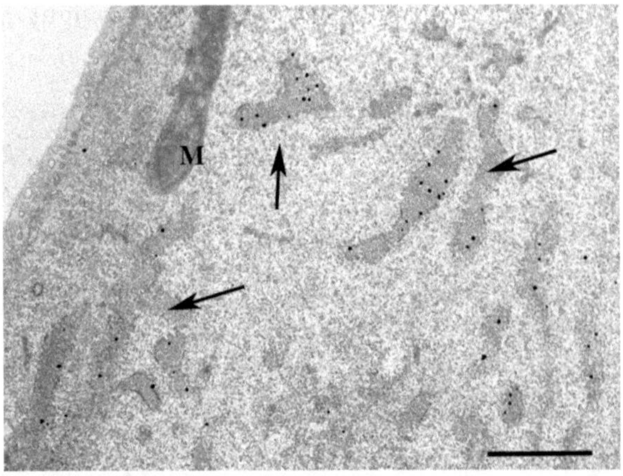

Fig. 18.2. Localization of heat shock protein 47 (HSP47) to the ER lumen by nanogold gold-intensified immunoelectron microscopy. NIH3T3 cells were cultured on an LF1 plastic coverslip, fixed with 4% (w/v) paraformaldehyde for 2 h, permeabilized in 0.25% (w/v) saponin solution, and processed for nanogold gold intensification procedure using mouse monoclonal anti-HSP47 primary antibody and goat anti-mouse IgG secondary antibody conjugated to 1.4-nm colloidal gold particles. Gold-intensified nanogold particles were observed in the luminal space of the ER. *Arrows* indicate the ER. M: mitochondria. Bar: 1 μm.

4. Notes

1. LF1 coverslips can be easily removed from the epoxy resin block after embedding making them an ideal surface for cell growth. The coverslips are packed in a sterilized 24-well plate. Remove the black spacer on the coverslips in a clean ventilator. LF1 coverslips have a bent edge and can easily be removed from 24-well plates using a tweezers. The coverslips can be placed in a 24-well plate, cell side up, containing the fixation, wash, immunolabeling, intensification, and dehydration solutions. CELLocate can be used when it is necessary to identify the positions of cells of interest, such as microinjected cells. Glass coverslips with grids can also be obtained from Matsunami Glass Industries (Osaka, Japan).

2. Paraformaldehyde and other chemicals used for fixation, post-fixation, and staining are hazardous. Solutions should be prepared and used in a hood. Protective eyeware and gloves should be worn.
3. IgG nanogold conjugate is also available (use at 1:100 dilution), but immunogold labeling may be relatively weak because cells are less permeable to IgG conjugates than Fab' conjugates.
4. Epoxy resin mixture can be stored in syringes sealed with Parafilm for 1 month in a freezer.
5. If this fixative results in unsatisfactory immunolabeling or ultrastructural preservation, change the composition of the fixative. Adding glutaraldehyde to the paraformaldehyde solution may improve preservation of ultrastructural details. Fix cells in 4% (w/v) paraformaldehyde solution containing 0.05–1% (w/v) glutaraldehyde for 20–30 min. Quench the glutaraldehyde with freshly prepared 0.5% (w/v) sodium borohydrate dissolved in PB for 20 min at room temperature followed by three 3-min washes with PB. Higher concentrations of glutaraldehyde are likely to denature antigens decreasing immunolabeling. The optimal fixation condition is identified using various concentrations of glutaraldehyde. On the other hand, when immunolabeling is weak, the 2-h fixation with 4% (w/v) paraformaldehyde solution may be too long. Reduce the fixation step to 20–30 min, which is similar to the duration used in immunofluorescence procedures. If the antibody works well for immunofluorescence microscopy, significant labeling will also be obtained with the nanogold silver and gold intensification methods. Shorter fixations occasionally result in poor ultrastructural preservation, however, especially for silver intensification.
6. Saponin treatment results in effective and stable permeabilization. For silver intensification, good permeabilization can also be achieved by freeze-thawing, and the structural integrity of the cells is usually better than that obtained with saponin. For freeze-thawing, the cells are soaked in PB containing 14% (v/v) glycerol and 35% (w/v) sucrose for 15 s, frozen in liquid nitrogen, and thawed. Samples are then washed three times for 3 min with PB.
7. Blocking solution A is used for immunolabeling with silver intensification and blocking solution B is used for immunolabeling with gold intensification because better labeling is obtained using blocking solution B for gold intensification. If the labeling inside the organelles, such as the luminal side of the endoplasmic reticulum, is low, substitute blocking solution B [0.1% (w/v) saponin] for blocking solution A [0.005% (w/v) saponin] when using silver intensification.

8. PBS is not recommended as a fixative dilution buffer or as a washing solution because Cl^- inhibits the silver intensification reaction.
9. The silver intensification solution is viscous. Use a Pasteur pipette with a large internal diameter. The silver intensification process is light sensitive and all steps should be performed in a darkroom using a red safelight.
10. The silver intensification reaction is temperature dependent and does not proceed efficiently at temperature below 20°C.
11. Higher OsO_4 concentrations dissolve the silver deposited on the nanogold particles.
12. The immunogold-labeled cells can be stored in a refrigerator for a few days before embedding in epoxy resin.
13. To obtain bigger intensification particles, mix one part of solution A and one part of solution B. Five minutes later add one part each of solutions C and D. The manufacturer's instructions mention solution B as a stabilizing reagent for the gold intensification reaction. Bigger intensification particles also are obtained with longer incubations (e.g., 3–5 min).
14. Gold deposits are resistant to 1% (w/v) OsO_4. Potassium ferrocyanide enhances the contrast of the membranes and clarifies the cellular ultrastructure.
15. Cells on glass coverslips and CELLocate coverslips are placed onto a drop of epoxy resin on a glass slide with the cell side up. Embedding capsules (truncated pyramid, 8 mm in diameter, TAAB Laboratories Equipment, Berks, UK) are filled with epoxy resin mixture and placed upside-down on the coverslips.
16. Cells on glass coverslips and CELLocate are removed by heating the epoxy resin block in the embedding capsule for several seconds over the flame of an alcohol lamp and twisting the embedding capsule with a pliers.
17. In the case of gold intensification, the reaction product is hardly visible under a light microscope. When big intensification particles are formed (*see* **Note 13**), the blue reaction products are sometime visible under a light microscope.

Acknowledgments

The authors would like to thank Yuni Abe, Miho Ueda, Eri Toyama, and Yumiko Kosuga for technical assistance.

References

1. Masaki, R., Kameyama, K., and Yamamoto, A. (2003) Post-translational targeting of a tail-anchored green fluorescent protein to the endoplasmic reticulum. *J. Biochem.* **134**, 415–426.
2. Wakana, Y., Takai, S., Nakajima, K., Tani, K., Yamamoto, A., Watson, P., Stephens, D. J., Hauri, H. P., and Tagaya, M. (2008) Bap31 is an itinerant protein that moves between the peripheral endoplasmic reticulum (ER) and a juxtanuclear compartment related to ER-associated degradation. *Mol. Biol. Cell* **19**, 1825–1836.
3. Lah, J. J., Hayes, D. M., and Burry, R. W. (1990) A neutral pH silver development method for the visualization of 1-nanometer gold particles in pre-embedding electron microscopic immunocytochemistry. *J. Histochem. Cytochem.* **38**, 503–508.
4. He, W., Kivork, C., Machinani, S., Morphew, M. K., Gail, A. M., Tesar, D. B., Tiangco, N. E., McIntosh, J. R., and Björkman, P. J.(2007) A freeze substitution fixation-based gold enlarging technique for EM studies of endocytosed nanogold-labeled molecules. *J. Struct. Biol.* **160**, 103–113.

Chapter 19

The Post-embedding Method for Immunoelectron Microscopy of Mammalian Tissues: A Standardized Procedure Based on Heat-Induced Antigen Retrieval

Shuji Yamashita

Abstract

We describe a standardized method of fixation, antigen retrieval, and image contrasting for post-embedding immunoelectron microscopy. Tissues are fixed with formaldehyde solutions containing Ca^{2+} and Mg^{2+} ions at pH 7.4 and then at pH 8.5. After dehydration with dimethylformamide, the specimens are embedded in LR-White resin. For antigen retrieval, ultrathin sections are heated in 20 mM Tris–HCl buffer (pH 9.0) for 1 h at 95°C. After immunogold labeling, the sections are treated with a mixture of tannic acid and glutaraldehyde, with OsO_4 solution, and then double-stained with uranyl acetate and lead citrate. The standardized method yields strong and reproducible immunoreactions for many antigens showing excellent image contrast without destruction of fine structures.

Key words: Immunoelectron microscopy, post-embedding method, universal fixatives, LR-White, heat-induced antigen retrieval, mammalian tissue, electron staining with tannic acid.

1. Introduction

Three main methods have been used for immunoelectron microscopy, i.e., the pre-embedding method, the post-embedding method, and cryoultramicrotomy. In the pre-embedding methods, antibodies are unable to penetrate into the cell organelles evenly and immunoreactions are negative in a deep portion of tissue sections. Since immunoreactions occur on the surface of the ultrathin sections in the post-embedding method, the post-embedding method gives reproducible and reliable immunostaining results (1). The colloidal gold conjugates

S.D. Schwartzbach, T. Osafune (eds.), *Immunoelectron Microscopy*, Methods in Molecular Biology 657,
DOI 10.1007/978-1-60761-783-9_19, © Springer Science+Business Media, LLC 2010

widely used for the post-embedding method facilitate observation of immunoreactions with an electron microscope and quantitative analysis of antigen concentrations. Multiple labeling with colloidal gold conjugates of different sizes is easily performed in the post-embedding method (2). Although the post-embedding method has these advantages, it has been used for a limited number of antigens because antigenicity is frequently lost during the dehydration and embedding procedures. Suppression of immunoreaction in resin-embedded tissues may be due to intra- and inter-crosslinks within aldehyde-treated proteins and the destruction of secondary and tertiary protein structures during dehydration (3). Furthermore, specimens embedded in acrylic resins without osmication show poor preservation of membranes and low contrast for cell organelles.

Fixation is one of the most important aspects of sample preparation for immunoelectron microscopy because it affects the strength of the immunoreaction and the preservation of fine cellular structures. In general, since fixatives that provide good morphology crosslink macromolecules rapidly and tightly, form a gel-like structure in the tissues, and directly modify epitopes, these fixatives severely inhibit immunoreactions. Due to these contradictory effects, many kinds of fixatives, fixation conditions, and procedures have been employed to localize each antigen. A mixture of formaldehyde (FA) and a low concentration of glutaraldehyde (0.01–0.5%) is a popular fixative for immunoelectron microscopy. Although glutaraldehyde is an excellent fixatives for the preservation of morphology, it severely inactivates immunoreactions of many antigens and heat-induced antigen retrieval (HIAR) is ineffective for glutaraldehyde-fixed specimens. In this chapter we describe a two-step fixation method with FA solution containing Ca^{2+} and Mg^{2+} ions at pH 7.4 and then at pH 8.5 as a universal technique that preserves cellular fine structure. By subsequently subjecting the samples to HIAR by heating ultrathin sections mounted on a grid in 20 mM Tris–HCl (pH 9.0), a high labeling density of immunogold is obtained for many antigens that have seldom been detected using immunoelectron microscopy. To localize antigens in tissue embedded in LR-White resin with good image contrast, the sections are treated after immunogold labeling with a mixture of tannic acid and glutaraldehyde, then with OsO_4 solution and finally double-stained with uranyl acetate and lead citrate.

2. Materials

2.1. Fixation

1. 8% formaldehyde (FA): Prepare FA solution immediately before use by adding 4 g paraformaldehyde (Sigma

Chemicals, St. Louis, MO) to 50 mL of distilled water heated to about 60°C. Add a few drops of 10 N NaOH to clarify the solution and cool before using to prepare fixative 1 and 2

2. Fixative 1: 4% FA, 2.5 mM $CaCl_2$, 1.25 mM $MgCl_2$, and 2.9% glucose in 0.1 M HEPES-NaOH buffer (pH 7.4) (*see* **Note 1**)
3. Fixative 2: 4% FA, 2.5 mM $CaCl_2$, 1.25 mM $MgCl_2$, and 2.1% glucose in 0.1 M HEPES-NaOH buffer (pH 8.5) (*see* **Note 2**)
4. Anesthetic for perfusion fixation: 20% Nembutal (Abbott Laboratories, Abbott Park, IL) in 0.85% NaCl
5. Rocking platform (MildMixer SI-36, Taitec Co., Tokyo, Japan)
6. Double edge razor blades
7. Dissection pan
8. Soft plastic plate
9. Microdissecting scissors
10. Dissecting forceps
11. Disposable plastic syringe (20 mL) with a 25 G needle
12. Plastic disposable centrifugation tube (50 mL) or glass vial (25 mL) with a screw cap
13. Disposable plastic pipette (5 mL)
14. Adhesive tape

2.2. Dehydration and Embedding

1. *N*, *N*-Dimethylformamide (DMF)
2. 50, 70, and 90% aqueous solution of DMF (*see* **Note 3**).
3. LR-White (hard grade, London Resin Company, Berkshire, UK)
4. A 2:1 mixture and a 1:2 mixture of DMF and LR-White
5. Constant temperature oven (TD-500B, Nisshin EM Corp., Tokyo, Japan)
6. Rocking plate (Taitec Co., Tokyo, Japan)
7. 8 mm gelatin capsules (Nisshin EM Corp., Tokyo, Japan)
8. Disposable plastic conical tubes (5 mL)
9. Single edge razor blades
10. Plastic bags

2.3. Sectioning

1. Toluidine blue solution: 0.5% toluidine blue, 1% sodium borate aqueous solution. Store the solution at room temperature

2. 0.5% Neoprene W: Dilute 2% Neoprene W (Nisshin EM Corp., Tokyo, Japan) with toluene
3. 200 mesh nickel grids (Nisshin EM Corp., Tokyo, Japan): Wash grids with acetone and dry on a piece of filter paper. Pick a grid up with a jewelers forceps, dip the grid in 0.5% Neoprene W for few seconds, and then drop the grid on a piece of filter paper. Keep the grids on a piece of filter paper in a glass Petri dish at room temperature (*see* **Note 4**)
4. Light microscope
5. Dissection microscope
6. Hot plate
7. Glass knives and diamond knife (Nisshin EM Corp., Tokyo, Japan)
8. Jewelers forceps (Vigor #5, Nisshin EM Corp., Tokyo, Japan)
9. Glass slides
10. Single edge razor blades: Clean the blades with acetone
11. Grid case (Nisshin EM Corp., Tokyo, Japan)

2.4. Immunostaining

1. HIAR buffer: 20 mM Tris–HCl, pH 9.0
2. TBS: 20 mM Tris–HCl (pH 7.4) containing 0.15 M NaCl
3. Blocking solution: TBS containing 1% bovine serum albumin (BSA, Fraction V, Sigma Chemicals, St. Louis, MO.) and 0.05% sodium azide
4. Dilution solution: Blocking solution containing 0.5% fish gelatin (GE Healthcare Biosciences, Buckinghamshire, UK)
5. Primary antibodies diluted to appropriate concentration with the blocking solution
6. 10 or 15 nm colloidal gold-labeled anti-rabbit and anti-mouse antibodies (GE Healthcare Biosciences, Buckinghamshire, UK): Dilute gold-labeled antibodies 25-fold with the dilution solution and centrifuge at 14,000 rpm for 5 min in a microfuge
7. Teflon staining tubes: Prepare semicircular tubes by cutting a Teflon tube (length, 15–20 mm; inside diameter, 3 mm; outside diameter, 5 mm) and slit along the center of the inner walls
8. Block heater (Dry ThermoUnit DTU-1B; Taitec Co., Tokyo, Japan)

9. Micro centrifuge (Tomy Seiko Co., Tokyo, Japan)
10. Jewelers forceps
11. Eppendorf tubes (1.5 mL)
12. Disposable plastic Petri dish (5.5 cm)
13. Parafilm sheet
14. Wash bottle
15. Blower or air-spray
16. Filter paper

2.5. Electron Staining

1. Post-fixation solution: 2% glutaraldehyde, 0.05% tannic acid (Wako Pure Chemicals, Osaka, Japan) in 0.1 M phosphate buffer (pH 5.5). Prepare immediately before use using 8% distilled glutaraldehyde (Tabb Laboratories, Reading, UK)
2. 1% OsO_4 solution: Dilute 2% OsO_4 (Nisshin EM Corp., Tokyo, Japan) in 0.1 M phosphate buffer (pH 7.4)
3. CO_2-free distilled water: Boil distilled water for 5–10 min just before use
4. 1 N-NaOH: Prepare 1 N NaOH using CO_2-free distilled water immediately before preparing Reynolds lead solution
5. Reynolds lead solution: Add about 30 mL of CO_2-free distilled water to 1.33 g of lead nitrate (Taab Laboratories, Reading, UK) and 1.76 g of sodium citrate in a 50-mL conical centrifugation tube and shake vigorously for 15 min to produce a uniform milky suspension. To this suspension add 8 mL of freshly prepared 1 N NaOH and shake until the suspension clears. Adjust the volume to 50 mL with CO_2-free distilled water. Store the solution in a disposable 20 mL syringe at 4°C. Warm to room temperature and filter through a 0.22 μm filter immediately before use
6. 2% uranyl acetate: Dissolve 1 g of uranyl acetate (Merck KGaA, Darmstadt, Germany) completely in 50 mL distilled water at about 50°C and store the light sensitive solution in a brown bottle at 4°C.
7. Jewelers forceps
8. Wash bottle
9. Blower or air-spray
10. Parafilm sheet

3. Methods

3.1. Fixation (see Note 5)

3.1.1. Perfusion Fixation

1. Inject 0.3 mL of anesthetic (20% Nembutal) intraperitoneally into a mouse. Place the mouse in a dissection pan and fix the limbs and tail with adhesive tape.
2. Remove the rib cage and insert a 25-G needle attached to a 20 mL syringe filled with fixative 1 into the left ventricle of the heart and cut the right atrium with a microdissecting scissors.
3. Perfuse the fixative slowly into the ventricle. If perfusion is working well, the liver and kidney will turn grayish.
4. Place tissues to be sampled in 1–2 mL of fixative 1 on a soft plastic plate and mince them with double edge razor blades into small pieces (1 mm^3) (*see* **Note 5**).

3.1.2. Immersion Fixation

1. Kill a mouse by spinal dislocation or after the anesthesia and remove tissues from the animal as soon as possible. Place tissues to be sampled in 1–2 mL of fixative 1 on a soft plastic plate and mince them with double edge razor blades into small pieces (*see* **Note 6**). Take care to avoid desiccation and mechanical damage by squeezing organs with forceps and scissors.
2. Transfer the small tissue pieces obtained after perfusion fixation or immersion fixation into a 50 mL conical centrifugation tube or a 25 mL vial containing about 20 mL of fixative 1 and incubate for 2 h at room temperature on a rocking platform.
3. Allow tissue pieces to settle to bottom of tube and remove fixative 1 using a pipette.
4. Add about 20 mL of fixative 2 and incubate overnight at room temperature on a rocking platform.

3.2. Dehydration and Embedding

1. Remove fixative 2 and transfer the tissue pieces into a 5 mL conical tube.
2. Dehydrate tissue pieces by incubating for 15 min in 50% DMF, 70% DMF, 90% DMF, and twice for 15 min in 100% DMF at 4°C. Shake the conical tube gently using a rocking plate.
3. Remove the 100% DMF. Incubate the tissue pieces for 30 min with a 2:1 and then a 1:2 DMF and LR-White mixture at 4°C. Shake gently using a rocking plate.
4. Incubate the tissue pieces twice for 1 h with pure LR-White at 4°C. Shake gently using a rocking plate.

5. Place the tissue pieces in fresh LR-White and incubate overnight at 4°C, and then for 2 h at room temperature. Shake gently using a rocking plate.
6. Transfer the tissue pieces into gelatin capsules filled with LR-White, cap and polymerize in a constant temperature oven at 55°C for 24 h (*see* **Note 7**).
7. Cut off the upper portion of the blocks showing incomplete polymerization with a single edge razor blade and store the blocks in a plastic bag at room temperature.

3.3. Sectioning

1. Carefully trim the LR-White blocks with a single edge razor blade under a dissecting microscope forming a pyramid about 2 × 2 mm (*see* **Note 8**).
2. Prepare semithin sections (about 1 μm) with a glass knife, float the sections onto the surface of a drop of distilled water placed on a glass slide, and dry on a hot plate. Stain the sections with the toluidine blue solution for 30 s and then examine under a light microscope (*see* **Note 9**).
3. Trim the blocks for immunoelectron microscopy forming a 0.5 × 0.5 mm pyramid in a region containing cells that are darkly stained with toluidine blue indicative of good fixation.
4. Prepare ultrathin sections (silver-gold color thickness) using a diamond or glass knife and mount the sections on a nickel grid. Blow water off the grids using a blower and store them in a grid case at room temperature.

3.4. Immunostaining

1. Place several grids tissue side up in the slit of a semicircular Teflon tube. All procedures from HIAR to electron staining are carried out using the Teflon staining tube as a grid holder.
2. Fill an Eppendorf tube with HIAR buffer and preheat to 95°C. Place the Teflon tube containing the grids in the Eppendorf tube and incubate for 1 h at 95°C (*see* **Note 10**).
3. Cool the tube to room temperature and wash the grids with TBS. Blot with filter paper to remove TBS. Place the Teflon tube containing the grids on a parafilm sheet and incubate with 150–200 μL of blocking buffer for 30 min at room temperature.
4. Blot with filter paper to remove the blocking solution and incubate the grids with 150–200 μL of primary antibody overnight at 4°C in a 5.5 cm Petri dish. Place a piece of wet filter paper in the Petri dish cover to make a moist chamber.
5. Wash the grids with TBS under a jet using a wash bottle for 2 min and store in an Eppendorf tube containing TBS until the washing of all the grids is finished. Place the Teflon tube

containing the grids on a parafilm sheet and incubate with 150–200 μL of colloidal gold-labeled antibody for 1 h at room temperature.

6. Wash the grids for 3 min with TBS using a wash bottle and then with distilled water. Blow off the water.

3.5. Electron Staining

1. Place the Teflon tube containing the grids on a parafilm sheet and incubate for 5 min with 150–200 μL of the post-fixation solution. Wash with distilled water using a wash bottle and blow off the water (*see* **Note 11**).
2. Incubate the grids with 150–200 μL of 1% OsO_4 solution for 5 min, wash with distilled water, and blow off the water.
3. Incubate the grids with 150–200 μL of 2% uranyl acetate for 5 min, wash with distilled water, and blow off the water.
4. Stain the grids with 150–200 μL of Reynolds lead citrate solution for 1 min, wash with distilled water, and blow off the water.
5. Examine the ultrathin sections with an electron microscope (*see* **Note 12**). Excellent ultrastructural preservation and membrane contrast is obtained in tissue fixed by the perfusion (**Fig. 19.1**) and immersion (**Fig. 19.2**) methods. Furthermore, the strong immunolabeling observed shows that

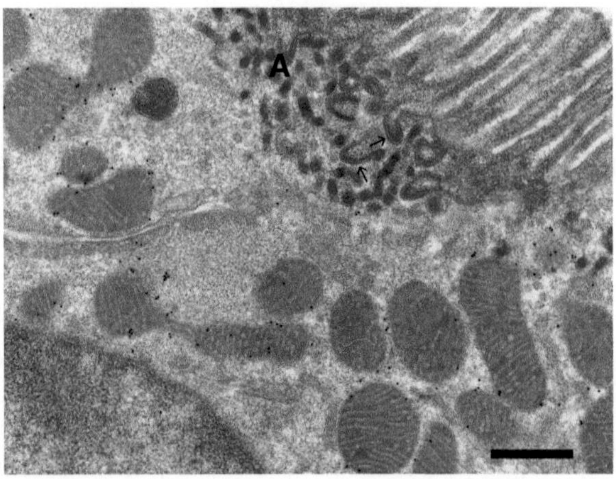

Fig. 19.1. Perfusion-fixed mouse kidney immunogold labeled with a polyclonal antibody to the translocase of the outer mitochondrial membrane (Tom) 20. A mouse was perfusion with fixative 1 and the kidney was removed and minced into small pieces in fixative 1. Tissue pieces were incubated with fixative 1 for 2 h followed by overnight fixation with fixative 2. HIAR was performed by heating the LR-White embedded ultrathin sections in 20 mM Tris–HCl (pH 9.0) at 95°C for 1 h. A strong immunoreaction for Tom 20 is seen along the outer mitochondrial membrane of the proximal tubular cells. Note the high contrast of cellular membranes, well-preserved mitochondria, the numerous clathrin-coated tubules (*arrows*), and the high contrast of the apical canaliculi (**A**). Bar = 0.5 μm.

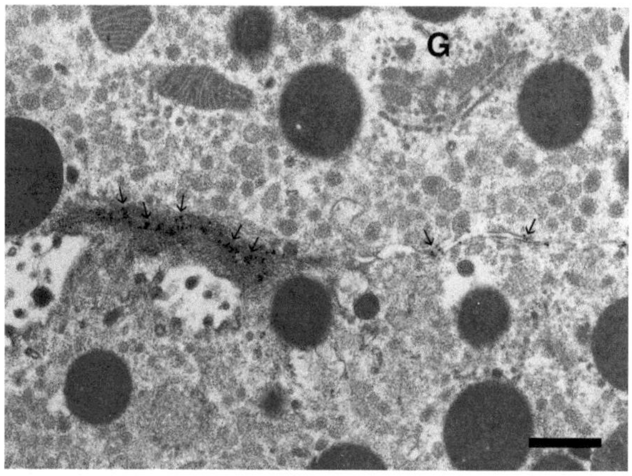

Fig. 19.2. Immersion-fixed mouse pancreas immunogold labeled with a monoclonal antibody to β-catenin. A mouse pancreas was excised, minced into small pieces in fixative 1, and incubated in fixative 1 for 2 h followed by overnight fixation with fixative 2. HIAR was carried out by heating the LR-White embedded ultrathin sections in 20 mM Tris–HCl (pH 9.0) at 95°C for 1.5 h. Strong immunolabeling of β-catenin with the monoclonal antibody (*arrows*) is seen along the cell membrane and the zona adherens. Note the high contrast of cellular membranes, the Golgi complex (*G*), the well-preserved mitochondria, and secretory granules. Bar = 0.5 μm.

HIAR is effective for perfusion and immersion-fixed samples, when they are immunostained with a polyclonal or monoclonal antibody.

4. Notes

1. Two-step fixation using a fixative with a basic pH improves the preservation of cell structures (4); in particular, this protocol is essential for the conservation of the Golgi complexes (4). When FA solution containing Ca^{2+} and Mg^{2+} ions is used as the primary fixative, HEPES buffer provides much better fine cell structure than cacodylate buffer at pH 7.4. Furthermore, deep layers of tissue pieces are fixed well with FA dissolved in HEPES.
2. Osmolarity of the fixation vehicle of fixatives 1 and 2 is about 320 mOsm. Total osmolarity of fixation vehicle is one of the most essential factors for the preservation of fine structures since FA freely diffuses into cells and tissue pieces and is osmotically inactive. A fixation vehicle with about 300–330 mOsm that is in accord with the osmolarity of mammalian body fluid yields excellent cell structures (4). Osmolarities of the glucose-free fixation vehicle for fixative

1 and fixative 2 are about 150 and 200 mOsm, respectively. Osmolarity of 1% glucose is about 58 mOsm.

3. DMF is a better dehydration medium compared to ethanol for the conservation of cellular fine structures in FA-fixed specimens. DMF may reduce abrupt osmotic pressure changes in the tissues and extraction of lipids in membranes.
4. Grids should be treated with a mesh cement, such as Neoprene W (chloroprene rubber) to attach ultrathin sections onto grids tightly, to prevent removal of sections from the grids during immunohistochemical staining, washing, and heavy metal staining and to reduce drift of sections by the electron beam. Since Neoprene W coats only the surface of the grids, antigens localizing to both cutting surfaces of sections are able to bind antibodies. Although grids coated with formvar or collodion film are often used to mount ultrathin sections, these films are labile during heating in HIAR solution.
5. Although immersion fixation is useful for almost all samples including biopsy specimens and cultured cells, perfusion fixation is recommended for the fixation of small animals, such as mice and rats, since it provides rapid and uniform fixation. In particular, perfusion fixation yields excellent morphology of highly vascular tissues such as brain, kidney, and liver.
6. Organs composed of subdivisions, such as the brain, kidney, and adrenal gland, should be entirely removed from animals and immersed in fixative 1 on a plastic plate, and a few millimeter thick slices should be prepared using two sharp-edged blades. A portion of the slices to be examined should be further minced into small pieces (1 mm^3). Thin double edge razor blades (carbon steel) cleaned with acetone should be broken in half, and further broken to make a sharp edge for tissue mincing.
7. Although several kinds of resins have been used as embedding media for the post-embedding method, LR-White is popular. It yields a high labeling intensity and is easy to handle because of its low viscosity and low toxicity. Since oxygen in air inhibits polymerization of LR-White resin, polymerization in a sealed gelatin capsule is recommended. Thermal polymerization of LR-White at 55°C and photopolymerization at 4°C or −20°C show similar morphology preservation and intensity of the immunoreaction (4).
8. Hasty trimming tends to damage the specimen blocks, since the resin is brittle compared to epoxy resins.

9. Examination of semithin sections is important to assess the quality of fixation, to orient the tissue, and to select the constituents to be investigated with the electron microscope. Since marginal portions are fixed well in the immersion-fixed tissue pieces, such portions should be used for immunoelectron microscopy. Cells fixed well are darkly stained with the toluidine blue solution compared to those fixed incompletely. Cutting faces of blocks should be trimmed to a 0.5 mm × 0.5 mm pyramid to prepare ultrathin sections (approximately 100 nm in thickness) without chatter and section compression for examination in the electron microscope.
10. Since HIAR was first introduced by Shi et al. (5) using paraffin sections, heating in various solutions has been examined for antigen retrieval from ultrathin sections embedded in epoxy and acrylic resins (6, 7). We found that 20 mM Tris–HCl (pH 9.0) antigen retrieval solution yields the strongest immunoreaction for specimens prepared by both the pre-embedding and the post-embedding methods (4, 8, 9). Heating for less than 30 min is ineffective for HIAR and yields high immunogold background staining. Heating more than 2 h shows no increase of immunoreaction for most antigens but reduces the contrast of cell organelles. Heating times between 1 and 2 h have been found to be optimal.
11. It is reported that treatment with tannic acid preceding dehydration and embedding is effective for contrasting and preserving cellular fine structure but tannic acid penetrates poorly into tissues and decreases the immunoreaction (10). To avoid these drawbacks of tannic acid treatment, the ultrathin sections are treated with a mixture of tannic acid and glutaraldehyde and then with OsO_4 solution after immunostaining. Cellular membranes; nucleic acids (chromatin, nucleoli, and ribosomes); intracellular filaments (microfilaments, 10 nm filaments, and microtubules); cell organelles such as the endoplasmic reticulum, the Golgi complex, lysosomes, mitochondria; and extracellular components such as collagens are well preserved and seen with high contrast (4).
12. LR-White resin sections are labile in the electron beam compared with epoxy resin sections. To avoid sample destruction by the electron beam, sections should be irradiated evenly at low magnification, 2,000×, then at 3,000× or 4,000×, and then photographed at higher magnifications.

References

1. Roth, J., Bendayan, M., and Orci, L. (1978) Ultrastructural localization of intracellular antigens by the use of protein A-gold complex. *J. Histochem. Cytochem.* **26**, 1074–1081.
2. Slot, J. W. and Geuze, H. J. (1985) A new method of preparing gold probes for multiple-labeling cytochemistry. *Eur. J. Cell Biol.* **38**, 87–93.
3. Fowler, C. B., O'Leary, T. J., and Mason, J. T. (2008) Modeling formalin fixation and histological processing with ribonuclease A: effects of ethanol dehydration on reversal of formaldehyde cross-links. *Lab. Invest.* **88**, 785–791.
4. Yamashita, S., Katsumata, O., and Okada, Y. (2009) Establishment of standardized post-embedding method for immunoelectron microscopy by applying heat-induced antigen retrieval. *J. Electron Microsc.* **58**, 267–279.
5. Shi, S. R., Key, M. E., and Kalra, K. L. (1991) Antigen retrieval in formalin-fixed, paraffin-embedded tissues: an enhancement method for immunohistochemical staining based on microwave oven heating of tissue sections. *J. Histochem. Cytochem.* **39**, 741–748.
6. Brorson, S. H. (2001) Heat-induced antigen retrieval of epoxy sections for electron microscopy. *Histol. Histopathol.* **16**, 923–930.
7. Xiao, J. C., Adam, A., Ruck, P., and Kaiserling, E. (1996) A comparison of methods for heat-mediated antigen retrieval for immunoelectron microscopy: demonstration of cytokeratin No. 18 in normal and neoplastic hepatocytes. *Biotech. Histochem.* **71**, 278–285.
8. Emoto, K., Yamashita, S., and Okada, Y. (2005) Mechanisms of heat-induced antigen retrieval: does pH or ionic strength of the solution play a role for refolding antigens? *J. Histochem. Cytochem.* **53**, 1311–1321.
9. Yamashita, S. (2007) Heat-induced antigen retrieval: mechanisms and application to histochemistry. *Prog. Histochem. Cytochem.* **41**, 141–200.
10. Berryman, M, A., Porter, W. R., Rodewald, R. D., and Hubbard, A. L. (1992) Effects of tannic acid on antigenicity and membrane contrast in ultrastructural immunocytochemistry. *J. Histochem. Cytochem.* **40**, 845–857.

Chapter 20

Double-Label Immunoelectron Microscopy for Studying the Colocalization of Proteins in Cultured Cells

Haruo Hagiwara, Takeo Aoki, Takeshi Suzuki, and Kuniaki Takata

Abstract

Multiple label immunoelectron microscopy localizes and detects multiple antigens in cells and tissues. In double labeling, two kinds of primary antibodies from different animal species are used after being mixed in a single solution. To distinguish the different antigens, secondary antibodies should be labeled with colloidal gold particles of different diameter. Generally, the secondary antibody that is used for detecting the antigen with lower distribution density is labeled with smaller-sized gold particles. In this chapter, double-label immunoelectron microscopy of gelatin-embedded cultured cells using the cryosectioning technique is described.

Key words: Cryosectioning technique, Tokuyasu method, cultured cells, colloidal gold, double immunolabeling.

1. Introduction

Immunoelectron microscopy using antibody probes conjugated with gold particles permits the high-resolution detection and localization of antigens on and within cells. Such detection and localization depends on the antigen-recognition specificity of the primary antibodies, the preservation of the antigenicity of the antigens, and the ability of the antibodies to penetrate the cell so that they are able to bind to the antigens. The diameter of the colloidal gold particles used for immunoelectron microscopy ranges from 2 to 25 nm. Using colloidal gold particles of different sizes, multiple antigens can be detected simultaneously.

S.D. Schwartzbach, T. Osafune (eds.), *Immunoelectron Microscopy*, Methods in Molecular Biology 657,
DOI 10.1007/978-1-60761-783-9_20, © Springer Science+Business Media, LLC 2010

Immunoelectron microscopy using ultrathin frozen sections was developed and improved by Tokuyasu and others (1–5) and has been used for double and triple labeling of antigens in the same specimen (6, 7). This method, which is called the cryosectioning technique (Tokuyasu method), has several advantages for immunolabeling. As the sections are not embedded in resin at the time of immunolabeling, the antibody can more easily access the antigen, resulting in a higher labeling density. Double-label immunoelectron microscopy using ultrathin cryosections is described in this chapter.

2. Materials

2.1. Fixation and Freezing

1. Cultured cells, adherent type
2. Culture medium appropriate to the cell type
3. 100 × 20 mm Falcon tissue culture dish (Becton Dickinson, Franklin Lakes, NJ)
4. 50 mL Falcon polypropylene conical tube (Becton Dickinson, Franklin Lakes, NJ)
5. Cell scraper
6. Phosphate-buffered saline (PBS): To make 1,000 mL PBS, dissolve 8.0 g of NaCl, 0.2 g of KCl, 1.44 g of Na_2HPO_4, and 0.24 g of KH_2PO_4 in 800 mL of distilled water. Adjust the pH to 7.4 with HCl, adjust the volume to 1,000 mL, and store in the refrigerator. A 10× stock solution can be made and stored at room temperature
7. 0.2 M sodium phosphate buffer (PB) (pH 7.4): Prepare by using 0.2 M NaH_2PO_4 to adjust the pH of a 0.2 M Na_2HPO_4 solution to 7.4
8. 6% paraformaldehyde (PFA) stock solution: To make 100 mL solution, mix 6 g of PFA (EM grade, Nacalai Tesque, Kyoto, Japan) powder with 80 mL of distilled water in a Pyrex flask containing a stir bar. Heat to 60°C and add 1 N NaOH dropwise while slowly stirring until the solution clears. Cool down the solution to room temperature, and make it up to 100 mL with distilled water. Fresh solution is recommended, but the solution can be stored for a few weeks in the refrigerator
9. 3% PFA in 0.1 M PB: To make 100 mL solution, mix 50 mL of 6% PFA stock solution and 50 mL of 0.2 M PB
10. 7% (w/v) sucrose in 0.1 M PB
11. 25% glutaraldehyde (GA, EM grade, TAAB, Aldermaston, England)

12. 2% GA in PBS: To make 100 mL solution, mix 8 mL of 25% GA, 50 mL of 0.2 M PB, and 42 mL of distilled water
13. 10% gelatin (Sigma-Aldrich, St Louis, MO) in PBS
14. Razor blades, acetone-washed
15. Liquid nitrogen
16. Specimen carriers (JEOL, Tokyo, Japan)
17. Tissue-Tek OCT compound (Sakura, Tokyo, Japan)
18. 0.5 M sodium phosphate buffer (PB) (pH 7.4): Prepare by using 0.5 M NaH_2PO_4 to adjust the pH of a 0.5 M Na_2HPO_4 solution to 7.4
19. 2.3 M sucrose solution: To make 100 mL sucrose solution, mix 78.7 g of sucrose, 20 mL of 0.5 M PB, and 30 mL of distilled water. After dissolving, adjust the volume to 100 mL with distilled water
20. 0.25 M Na_2CO_3 in 0.1 M PB (pH 7.4): To make 100 mL solution, mix 2.65 g of Na_2CO_3 and 50 mL of 0.2 M PB. After dissolving, adjust the volume to 100 mL with distilled water
21. Polyvinylpyrrolidone (PVP, mean mol wt: 10,000, Sigma-Aldrich, St Louis, MO)
22. Sucrose/PVP mixture: Mix 90 mL of 2.3 M sucrose solution, 2 mL of 0.25 M Na_2CO_3 in 0.1 M PB, and 20 g of PVP

2.2. Cryosectioning

1. 2% methyl cellulose solution: Mix 2 g of methyl cellulose (25 centipoise; Nacalai Tesque, Kyoto, Japan) with cold distilled water to make a 2% solution. Leave the solution in the refrigerator for 3 days and then centrifuge it at $100{,}000\times g$ for 1 h at 4°C. Store the supernatant in the refrigerator. It will remain stable for up to 6 weeks
2. Sucrose/methyl cellulose mixture (pick-up solution): Mix equal volumes of 2.3 M sucrose solution and 2% methyl cellulose solution. Store in the refrigerator
3. Ultramicrotome (Ultracut S, Leica, Vienna, Austria) equipped with an FCS cryosectioning system (Leica, Vienna, Austria)
4. Glass knife for trimming
5. Diamond knife for dry cryosectioning (Diatome, Biel, Switzerland)
6. Nichrome wire (0.26–0.32 in diameter) loop with a diameter of 2 mm
7. 200-mesh nickel grids (Pyser-SGI, Kent, UK) coated with Formvar and carbon

8. A cold 2% gelatin plate: Add 10 mL of 2% gelatin in PBS to a 100 × 20 mm culture dish and leave it to solidify in the refrigerator

2.3. Double Immunogold Labeling

1. 0.02 M glycine (Sigma-Aldrich, St. Louis, MO) in PBS
2. 10% sodium azide: Dissolve 5 g of sodium azide in 50 mL of distilled water. Store at room temperature
3. 3% bovine serum albumin (BSA) fraction V (Roche, Basel, Swiss) in PBS: Add 3 g of BSA to 100 mL of PBS and allow it to dissolve, add 0.2 mL of 10% sodium azide, and store at 4°C
4. 1% BSA in PBS: Add 1 g of BSA to 100 mL of PBS and allow it to dissolve. Add 0.2 mL of 10% sodium azide. Store at 4°C
5. 0.1% BSA in PBS: Add 1 g of BSA to 1,000 mL of PBS and allow it to dissolve. Add 2 mL of 10% sodium azide. Store at 4°C
6. A mixture of primary antibodies of known specificity in 1% BSA in PBS. Each antibody was raised in a different animal species
7. A mixture of colloidal gold-conjugated species-specific secondary antibodies against the primary antibodies used (GE Healthcare, Buckinghamshire, UK). Each secondary antibody should be conjugated to colloidal gold particles of different diameters (*see* **Note 1**)
8. 5% uranyl acetate solution: Dissolve 2.5 g of uranyl acetate (TAAB, Aldermaston, England) in 50 mL of distilled water
9. Methyl cellulose-uranyl acetate solution: Mix 10 mL of 2% methyl cellulose solution with 0.2 mL of 5% uranyl acetate solution
10. Nichrome-wire loops with a diameter of 3.5 mm

3. Methods

3.1. Fixation and Freezing (*see* Note 2)

1. Gently remove the culture medium and rinse the dish once with PBS.
2. Aspirate the PBS, add 3% PFA in 0.1 M PB, gently swirl the mixture, and incubate the cells for 30 min at 4°C or at room temperature.
3. Remove the fixative by aspiration and rinse the cells three times with 7% sucrose in 0.1 M PB.

4. Scrape the cells from the dish with the cell scraper, gently resuspend the cells in 10 mL of PBS, and pour into a 50 mL conical tube.
5. Sediment the cells by centrifugation at 200×*g* for 5 min at 4°C.
6. Decant the supernatant.
7. Resuspend the cells in 10 mL of 10% gelatin in PBS preheated to 37°C, incubate for 1 min at 37°C, and centrifuge at 1,500×*g* for 5 min at room temperature.
8. Allow the gelatin to solidify on ice, thus embedding the cells in the gelatin. Remove the gelatin from the tube by knocking on the tapered bottom of the tube and transfer it to a 100 × 20 mm Falcon tissue culture dish on ice (*see* **Note 3**).
9. The bottom of the gelatin containing the embedded cells appears white. Cut this region containing the cells into about 0.5 × 0.5 × 1.0 mm pieces with a clean razor blade.
10. Incubate the gelatin pieces with a sucrose/PVP mixture for 3–4 h at 4°C, until the specimens become transparent.
11. Mount the specimen (*see* **Note 4**) on a specimen carrier made of aluminum with a flat head coated with OCT compound.
12. Plunge the specimens into liquid nitrogen, freeze them, and store them in liquid nitrogen.

3.2. Cryosectioning (see Note 5*)*

1. Cool the cryosectioning chamber of the ultramicrotome to −80°C.
2. Set up a trimming glass knife.
3. Place the specimen carrier in the specimen holder of the ultramicrotome.
4. Ensure that the front of the specimen is facing the trimming glass knife.
5. Trim the specimen in the shape of a rectangle and cut 0.3-μm semithin sections.
6. Suspend a droplet of pick-up solution on a 2-mm Nichrome-wire loop. Insert it into the cryosectioning chamber, gently touch the semithin sections onto the bottom of the pick-up solution, and bring it out from the chamber.
7. By touching the pick-up solution to a slide glass, mount the semithin sections on its surface.

8. Stain the sections with toluidine blue and examine them with a light microscope to check whether the specimen block is suitable for cutting ultrathin sections.
9. Cool the cryosectioning chamber of the ultramicrotome to −110°C to −120°C.
10. Cut 70-nm ultrathin sections from the specimen block using a cryodiamond knife at a speed of 1–2 mm/s.
11. Suspend a droplet of pick-up solution on a 2-mm Nichrome-wire loop. Insert it into the cryosectioning chamber, gently touch the ultrathin sections on the cryodiamond knife onto the bottom of the pick-up solution, and bring it out from the chamber.
12. By touching the pick-up solution to the 200-mesh nickel grid coated with Formvar and carbon, transfer the ultrathin sections to the grid. The grid can be stored for several days in the freezer at −20°C.

3.3. Double Immunogold Labeling (*see* Note 6)

1. Place the grid samples side down on the cold 2% gelatin plate and leave them for 30 min at 37°C. The grids will float on the surface of the melted gelatin.
2. Transfer the grids onto 0.02 M glycine in PBS (*see* **Note 7**) and then wash them seven times for 1 min by floating them on 0.02 M glycine in PBS.
3. Float the grids on 3% BSA in PBS and then incubate them at room temperature for 15 min to block nonspecific antibody binding.
4. Wash the grids for 1 min by floating them on 0.1% BSA in PBS.
5. Transfer the grids onto 10 μL droplets of a mixture of two different primary antibodies diluted in 1% BSA in PBS (*see* **Note 8**).
6. Incubate the grids at 37°C for 40 min.
7. Wash the grids seven times for 1 min by floating them on 0.1% BSA in PBS.
8. Transfer the grids onto 10 μL droplets of a mixture of two different secondary antibodies conjugated with colloidal gold particles of different diameters diluted in 1% BSA in PBS (*see* **Note 9**).
9. Incubate them at 37°C for 40 min.
10. Wash the grids seven times for 1 min by floating them on 0.1% BSA in PBS.
11. Wash the grids seven times for 1 min by floating them on PBS.

12. Fix the grids for 10 min by floating them on a 5 mL drop of 2% GA in PBS.
13. Wash the grids seven times for 1 min by floating them on distilled water.
14. Make three 3 mL drops of methyl cellulose-uranyl acetate solution in plastic dishes on ice. Float the grids on the first drop for 1 min, on the second drop for 1 min, and on the last drop for 10 min. Shield the grids from light during this process.
15. Pick up the grid with a 3.5-mm Nichrome-wire loop and drain the excess methyl cellulose solution with filter paper.
16. Allow the grids to air-dry in the loop at room temperature (*see* **Note 10**).
17. Remove the grid from the loop and examine the localization of the gold particles with an electron microscope (6). **Figure 20.1** shows the colocalization of caveolin-1 (10-nm colloidal gold, arrows) and Rab5 (5-nm colloidal gold, arrowheads) in intracellular vesicles of human umbilical vein endothelial cells.

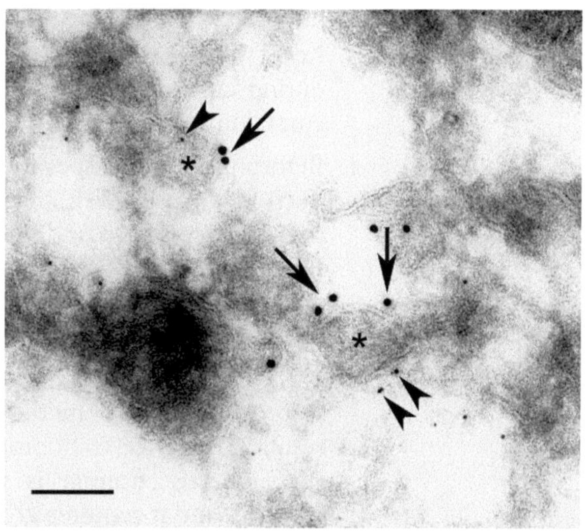

Fig. 20.1. Double-label immunoelectron microscopy using the cryosectioning technique showing colocalization of caveolin-1 and Rab5 in the intracellular vesicles of human umbilical vein endothelial cells (HUVEC). HUVEC were treated with vanadate to induce caveolin phosphorylation and the internalization of caveolae from the cell surface. Caveolin-1 (*arrows*) was immunolocalized using a rabbit anti-caveolin-1 primary antibody and 10-nm colloidal gold particles conjugated to goat anti-rabbit secondary antibody. Rab5 (*arrowheads*) was immunolocalized using a mouse anti-Rab5 primary antibody and 5-nm colloidal gold particles conjugated to goat anti-mouse secondary antibody. Note the colocalization of caveolin-1 and Rab5 in intracellular vesicles (*) suggesting the interaction of the internalized caveolae with early endosomes. (Reproduced from (7) with permission from the publisher.) Bar: 0.1 μm.

4. Notes

1. The diameter of the colloidal gold particles affects the density of the label. When processed under the same conditions, the larger the diameter of gold particles, the lower the number of gold particles labeled in the section. Therefore the use of smaller diameter colloidal gold particles to label antigens with lower distribution density is recommended. A combination of colloidal gold particles of 5 and 10 nm is usually used in order to obtain excellent resolution of two antigens in a section.
2. Specimens are chemically fixed to preserve the ultrastructure of cells and to fix the antigens at their proper sites. The GA-based fixatives that are routinely used for electron microscopic examination usually reduce the antigenicity of antigens. Therefore, formaldehyde-based fixatives, e.g., 2–4% PFA with or without 0.1–0.2% GA, are commonly used. PLP (periodate-lysine-paraformaldehyde) fixative is effective in some cases. After the immunoreaction, the specimen is refixed with glutaraldehyde to preserve the ultrastructure and to immobilize the immunoprobes.
3. Keep the gelatin cool on ice to prevent it from melting during cutting. 10% agarose in PBS can also be used for embedding cultured cells.
4. Before placing the specimen on the carrier, put the specimen in a plastic dish and carefully pick it up in order to remove excess sucrose/PVP mixture around the specimen. Excess sucrose/PVP mixture around the specimen often results in detachment of the specimen from the specimen carrier during cryosectioning.
5. Ultrathin cryosectioning was first used in the 1970s. Improvements in methodology and equipment have resulted in better sectioning and preservation of cell structures. Cryosectioning is the most critical part of this method, and it requires great skill.
6. The antigens on the surface of cryosections are labeled with specific antibodies by the immunoreaction. By using secondary antibodies conjugated with colloidal gold particles of different diameters, it is possible to localize as many as three antigens on a single ultrathin section, although most experiments localize only two antigens.
7. For preincubation, blocking, and washing, the grids should be floated on 0.5–1.0 mL drops of the solution with the specimen side facing the solution. If the grids sink into

the drops, the sections will become soiled. Use 3.5-mm Nichrome-wire loops for transferring the grids.

8. The titer of the primary antibodies ranges from 1 to 50 μg/mL. Prior to immunoelectron microscopy, the optimal titer must be determined by staining the cells with different dilutions of the antibody and checking them with a light microscope.
9. Dilute the secondary antibody solution at the optimal concentration in accordance with the manufacturer's instructions.
10. A thin film of methyl cellulose-uranyl acetate is formed on the surface of the grid after drying. The contrast and brightness of the images depends on the thickness of this film. Gold-colored films are good for 70-nm thick sections.

Acknowledgments

This work was supported, in part, by Grants-in-Aid for Scientific Research from the Ministry of Education, Culture, Sports, Science, & Technology of Japan.

References

1. Tokuyasu, K. T. (1978) A study of positive staining of ultrathin frozen sections. *J. Ultrastruct. Res.* **63**, 287–307.
2. Tokuyasu, K. T. (1986) Application of cryoultramicrotomy to immunocytochemistry. *J. Microsc.* **143**, 139–149.
3. Tokuyasu, K. T. (1989) Use of poly(vinylpyrrolidone) and poly(vinyl alcohol) for cryoultramicrotomy. *Histochem. J.* **21**, 163–171.
4. Liou, W., Geuze, H. J., and Slot, J. W. (1996) Improving structural integrity of cryosections for immunogold labeling. *Histochem. Cell Biol.* **106**, 41–58.
5. Webster, P. and Webster, A. (2007) Cryosectioning fixed and cryoprotected biological material for immunocytochemistry. *Methods Mol. Biol.* **369**, 257–289.
6. Fujimoto, T., Tokuyasu, K. T., and Singer, S. J. (1987) Direct morphological demonstration of the coexistence of vimentin and desmin in the same intermediate filaments of vascular smooth muscle cells. *J. Submicrosc. Cytol.* **19**, 1–9.
7. Aoki, T., Hagiwara, H., Matsuzaki, T., Suzuki, T., and Takata, K. (2007) Internalization of caveolae and their relationship with endosomes in cultured human and mouse endothelial cells. *Anat. Sci. Int.* **82**, 82–97.

Chapter 21

Serial Section Immunoelectron Microscopy of Algal Cells

Tetsuaki Osafune and Steven D. Schwartzbach

Abstract

Electron microscopy when combined with immunogold labeling provides a 2D image of intracellular protein distribution. Cells are however 3D structures. We describe a method of serial section immunogold electron microscopy that allows a 3D cellular image to be reconstructed from a series of electron micrographs. Cells are fixed to preserve cellular ultrastructure and they are embedded in plastic allowing ultrathin sections to be obtained. The ribbon of ultrathin serial sections produced as the microtome sequentially cuts through the sample is labeled with a monospecific antibody to the protein of interest and then with protein-A gold making the antigen–antibody complex visible in the electron microscope. A common field of view from each serial section is photographed in the electron microscope. Using image analysis software, each digitized micrograph is sequentially aligned; immunolabel and cellular structures of interest are traced onto each micrograph; the micrographs are stacked; and the structures of interest are rendered as solid surfaces producing a 3D image of protein distribution within the cell.

Key words: *Euglena*, algae, immunoelectron microscopy, serial sections, 3D image, computer graphics, intracellular localization.

1. Introduction

Electron microscopy provides a way to visualize the highly complex compartmentalized structure of eukaryotic cells. Combining electron microscopy with immunolabeling provides a tool for determining the functions of these compartments. Samples are prepared for immunoelectron microscopy by embedding the biological sample in a plastic resin, using a microtome to cut uniform thin sections from the tissue containing plastic block, immunolabeling the sections, staining the sections with heavy metals, and viewing the immunolabeled cells in the electron microscope. Views of individual sections provide a 2D image of the 3D

S.D. Schwartzbach, T. Osafune (eds.), *Immunoelectron Microscopy*, Methods in Molecular Biology 657,
DOI 10.1007/978-1-60761-783-9_21,

structure under investigation. In many cases this 2D image provides sufficient information regarding cellular function and protein distribution to elucidate structure–function relationships.

A complete picture of cellular structure–function relationships, however, requires a 3D image of cellular structures and protein distribution. When a biological sample embedded in a plastic block is sectioned, the microtome cuts into the block in uniform increments producing a ribbon of uniformly thick ultrathin sections sequentially passing through the biological specimen. Sequentially aligning and stacking the sections produces a 3D model of cellular structures and antigen distribution. Three-dimensional reconstruction has progressed from the rudimentary technique of tracing profiles of serial sections onto transparent plastic sheets scaled to the approximate section thickness, cutting the tracings with a jigsaw and staking the cut sheets producing a 3D model (1) to computerized reconstructions of digitized images (2–4).

Successful reconstruction of a protein's distribution within cellular structures from serial immunolabeled sections is dependent upon obtaining uniform serial sections traversing the entirety of the structure being studied. The embedding resin used is one critical factor affecting section quality and immunolabeling. Resins commonly used for transmission electron microscopy are the epoxy resins such as epon and the hydrophilic resins such as Lowicryl and LR White. Advantages of epoxy resins are that they exhibit excellent microstructure preservation and high-quality sections that are stable in the electron beam are readily obtained. Epoxy resins are polymerized by heating which could compromise antigenicity making them less than ideal for immunomicroscopy applications. Hydrophilic resins can be UV photopolymerized at low temperatures maximizing preservation of antigenicity making them a preferred embedding medium for immunomicroscopy. The hydrophilic resins, however, provide a lower level of ultrastructure preservation; sections are unstable in the electron beam; and most importantly, it is difficult to obtain a uniform ribbon having a large number of serial sections. For immunoelectron microscopy studies of the 3D distribution of high-abundance proteins, epoxy resins are clearly the resin of choice. For studies of less abundant antigens, the final choice of embedding resin requires an evaluation of the tradeoff between increased antigenicity and the difficulty of obtaining a long ribbon of uniform serial sections.

The quality of reconstructed 3D images of cellular structures is only as good as the serial sections used to create the image. Serial sections are obtained from a trimmed trapezoidal block having three straight sides and an angled side for orientation. The top and bottom sides of the block being sectioned must be parallel if one expects to obtain a long straight continuous ribbon of sections. If the sides are not parallel, a curved ribbon with nonuniform sections will be produced and the entire group of sections to be used for 3D reconstruction must be placed on multiple slot grids. Sections can be lost or damaged during segmentation of the curved ribbon into smaller pieces that will fit onto individual slot grids and during transfer of ribbon segments to slot grids making it difficult to obtain a group of sections that traverse the cellular structure under investigation.

The immunogold labeling of serial sections is a simple straightforward procedure. All sections used to reconstruct a 3D image must be labeled simultaneously using the same batch of reagents in order to ensure that gold label is proportional to antigen concentration. Gold particles are treated like any other structural feature during reconstruction of a 3D image from serial sections. Relative to the time and skill needed to prepare serial sections, the additional time necessary to immunolabel serial sections when a suitable antibody is available is trivial compared to the significant information obtained regarding sometimes unexpected protein subcellular localizations. This is evident from our own work using the methods described in this chapter to determine the 3D localization of ribulose-bis-phosphate carboxylase within the chloroplast predominantly to the *Euglena* pyrenoid (3) and the light harvesting chlorophyll a/b binding protein exclusively to the *Euglena* thylakoid membrane and Golgi apparatus (5).

The methods described in this chapter for serial sectioning, immunolabeling, image alignment, and reconstruction of solid surface 3D models of cellular structures are applicable to all biological specimens. The fixation and embedding methods are specific to algae and applicable for most other unicellular organisms. By adapting the fixation and embedding protocol to the specific biological specimen under investigation, the protocols in this chapter can be used to determine the 3D distribution of a specific protein within any cell type.

2. Materials

2.1. Sample Preparation

1. 50 % glutaraldehyde (EM grade, Electron Microscopy Sciences, Hatfield, PA). Store at 0–4°C (*see* **Note 1**)
2. Prepare a 0.1 M potassium phosphate buffer pH 7.2 using 0.1 M KH_2PO_4 to adjust the pH of a 0.1 M solution of K_2HPO_4 to pH 7.2.
3. 2% (w/v) agarose (Bio-Rad, Hercules, CA) prepared in deionized water.
4. 50 , 70, and 90% ethanol.

5. Acetone
6. Specimen rotator (Ted Pella, Redding, CA)

2.2. Embedding Samples in Epon

1. Tri-Corn plastic disposable beakers (Electron Microscopy Sciences, Hatfield, PA)
2. Wooden tongue depressors (Electron Microscopy Sciences, Hatfield, PA)
3. Epon (Embed-812, Electron Microscopy Sciences, Hatfield, PA)
4. DDSA (Dodecenyl Succinic Anhydride, Electron Microscopy Sciences, Hatfield, PA)
5. NMA (Nadic Methyl Anhydride, Electron Microscopy Sciences, Hatfield, PA)
6. DMP-30 [(2,4,6-Tri(dimethylaminomethyl) phenol) Electron Microscopy Sciences, Hatfield, PA].
7. Embedding mix A: In a disposable beaker combine 44 mL Embed-812 with 67 mL DDSA. Separately warm the resin and anhydride to 60°C to reduce viscosity, combine the two components, and mix thoroughly with a wooden tongue depressor (*see* **Note 2**).
8. Embedding mix B: In a disposable beaker combine 67 mL Embed-812 with 56 mL NMA. Separately warm the resin and anhydride to 60°C to reduce viscosity, combine the two components, and mix thoroughly with a wooden tongue depressor (*see* **Note 2**).
9. Embedding resin: In a disposable beaker combine equal volumes of room temperature embedding mix A and embedding mix B and add DMP-30 for a final concentration of 2%. Mix thoroughly using a wooden tongue depressor and degas for 15 min in a vacuum desiccator. Store tightly sealed at room temperature and use that day (*see* **Note 3**).
10. Gelatin capsules size 00 (Electron Microscopy Sciences, Hatfield, PA).

2.3. Preparation of Ultrathin Serial Sections

1. Formvar/Carbon film-coated 2×1 mm nickel slot grids (Electron Microscopy Sciences, Hatfield, PA) (*see* **Note 4**).
2. Butler block trimmer (Electron Microscopy Sciences, Hatfield, PA).
3. Cryotrim 45 diamond trimming tool (Electron Microscopy Sciences, Hatfield, PA)
4. Diamond knife (Electron Microscopy Sciences, Hatfield, PA) (*see* **Note 5**).
5. Eyelash with handle (Ted Pella Inc., Redding, CA)
6. Whatman #1 filter paper (Fisher Scientific, Pittsburgh, PA)

7. Anti-capillary self-closing tweezers (Ted Pella Inc., Redding, CA)
8. Chloroform
9. Q-tip
10. Grid storage box (Ted Pella Inc., Redding, CA)

2.4. Immunogold Labeling of Sections

1. 3% (w/v) hydrogen peroxide: Prepare immediately before use by mixing 600 μL 50% (w/v) hydrogen peroxide (Sigma, St. Louis, MO) with 9.4 mL distilled water.
2. 0.01 M phosphate buffer pH 7.4: Prepare by adjusting the pH of a 0.01 M solution of Na_2HPO_4 to pH 7.4 using 0.01 M KH_2PO_4.
3. PBS (0.01 M phosphate-buffered saline pH 7.4): Prepare by adding 0.85 g NaCl to 100 mL 0.01 M phosphate buffer pH 7.4.
4. PBS–BSA: Add BSA (Sigma, St. Louis, MO) for a final concentration of 1% to PBS.
5. PBS–Tween: Add Tween 20 (Sigma, St. Louis, MO) for a final concentration of 0.05% to PBS.
6. Antibody to protein to be localized diluted to appropriate concentration with PBS–BSA immediately before use.
7. Protein-A gold: Protein-A conjugated to (10-) 15-nm gold particles (EY laboratories, San Mateo, CA) (*see* **Note 6**).
8. 3% uranyl acetate (Electron Microscopy Sciences, Hatfield, PA): Mix 1.5 g uranyl acetate with 50 mL deionized water. Stir overnight in a foil-covered container, add 10 drops glacial acetic acid (Fisher Scientific, Pittsburgh, PA).
9. Parafilm (Fisher Scientific, Pittsburgh, PA).

2.5. 3D Reconstruction from Serial Sections

1. Computer
2. IMOD image reconstruction software (6) freely available from http://bio3d.colorado.edu/imod/ (*see* **Note 7**)

3. Methods

3.1. Sample Preparation

1. Mix approximately 45 mL algal cell culture (*see* **Note 8**) with 0.9 mL 50% glutaraldehyde (*see* **Note 9**) for a final concentration of approximately 1% glutaraldehyde (*see* **Note 10**) and incubate at 4°C for 60 min. Recover the cells by centrifugation for 2 min in a tabletop centrifuge.
2. Resuspend the cell pellet in 10 mL 0.1 M phosphate buffer pH 7.2, incubate 5 min at room temperature on a specimen

rotator, and recover the cells by centrifugation for 2 min in a tabletop centrifuge. Repeat two times (*see* **Note 11**).

3. Resuspend the cell pellet in 1 mL 0.1 M potassium phosphate buffer pH 7.2 and transfer to a microfuge tube. Pellet the cells.
4. Embed the cell pellet in 2% (w/v) agarose by resuspending the cell pellet in 60°C 2% agarose and immediately centrifuge for 30 s to pellet the cells. Remove the tube and place on ice to solidify the agarose.
5. Use a needle to remove the agarose plug from the microfuge tube and cut off the region of the plug containing cells. Cut the agarose into small cubes and transfer them to a 15-mL conical centrifuge tube.
6. Dehydrate the samples by incubating on a specimen rotator for 20 min in 10 mL 50% ethanol and recover the sample by gentle centrifugation. Repeat once.
7. Dehydrate the samples by incubating on a specimen rotator for 20 min in 10 mL 70% ethanol and recover the sample by gentle centrifugation. Repeat once.
8. Dehydrate the samples by incubating on a specimen rotator for 20 min in 10 mL 90% ethanol and recover the sample by gentle centrifugation. Repeat once.
9. Resuspend the sample in 10 mL acetone, incubate on a specimen rotator for 20 min, and recover the sample by gentle centrifugation. Repeat three times.

3.2. Embedding Samples in Epon

1. Resuspend samples in 3 mL of a 1:2 embedding resin: acetone mixture, place uncapped on a specimen rotator in a hood, and incubate 4 h to overnight. Recover the sample by gentle centrifugation and remove the embedding resin using plastic pipettes (*see* **Note 12**).
2. Resuspend the samples in 3 mL of a 2:1 embedding resin:acetone mixture, place uncapped on a specimen rotator in the hood, and incubate 4 h. Recover the sample by gentle centrifugation and remove the embedding resin using plastic pipettes.
3. Resuspend the sample in 3 mL 100% embedding resin and incubate in a vacuum desiccator for 1–2 h. Recover the sample by gentle centrifugation and remove the embedding resin using plastic pipettes.
4. Resuspend the cell sample in a small volume of embedding resin. Fill a gelatin capsule about half full with embedding resin and overlayer with the cell sample. Place the capsule in a centrifuge tube positioning it upright in the tube using tissue paper and centrifuge at full speed in a clinical centrifuge for

10 min to pellet the cells through the embedding resin to the bottom of the capsule.

5. Remove the capsule from the centrifuge tube, top off the capsule with embedding resin, insert a sample identification label into the embedding resin at the top of the capsule, and polymerize by incubation in a 60°C oven for 24 h (*see* **Note 13**).
6. Allow the polymerized block to cool for 24 h. Remove the block from the gelatin capsule by placing in water at 37°C until the capsule dissolves.

3.3. Preparation of Ultrathin Serial Sections

1. Mount the block in a block trimmer and using a razor hand trim the end containing the sample into a four-sided pyramid with walls at a 45° angle and a 0.5–0.75 mm^2 top surface.
2. Identify the region of the hand-trimmed block that will be sectioned (*see* **Note 14**). Determine where to trim the block so as to place the region to be sectioned within a trapezoid approximately 160 μm long, 40 μm wide, and 15 μm high. The top and the bottom of the trapezoid relative to the direction of sectioning must be trimmed perfectly parallel, one side trimmed at a 90° angle and the other angled for orientation.
3. Mount the diamond trimming tool on the microtome. Mount the hand-trimmed block on the microtome so that the section of the block that will form the sides of the trapezoid is perpendicular to the trim tool blade. Rotate the block 10° counter clockwise. Align the right edge of the trimmer blade where the angled left side of the trapezoid will be formed. Cut 15 1-μm sections forming the left-angled side of the trapezoid. Rotate the block 10° clockwise. Move the trimmer blade and position the left edge of the trimmer blade 160 μm to the right of the previous cut. Cut 15 1-μm sections forming the right straight side of the trapezoid. Rotate the block 90° clockwise. Position the right edge of the trimmer blade at the exact position where the bottom of the trapezoid will be formed (*see* **Note 15**). This should be approximately 20 μm from the center of the block region containing the sample. Cut 15 1-μm sections forming the bottom edge of the trapezoid. Move the trimmer blade and position the left edge of the trimmer blade 40 μm to the right of the previous cut. Cut 15 1-μm sections forming the top of the trapezoid. Rotate the block 90° counterclockwise and remove the trim tool.
4. Mount the diamond knife on the microtome. Align the bottom edge of the trapezoid so that it is parallel to the knife edge. Fill the diamond knife trough with distilled water so it is level with the cutting edge but does not form a meniscus.

5. Cut a ribbon of silver ultrathin sections approximately 80- to 90-nm thick. The ribbon should be straight with sections adhering to each other (*see* **Note 16**).
6. Use the eyelash hair to position the ribbon in the center of the knifeboat. Place a slot grid held with a pair of tweezers into the knifeboat at a 45° angle oriented with the slot parallel to the ribbon and with ¼ of the slot above the water surface. Use the eyelash tool to move the ribbon toward the grid. Move the grid under the ribbon and when the ribbon is positioned over the slot, raise the ribbon (*see* **Note 17**).
7. Blot the jaws of the tweezers and the bottom of the grid with Whatman #1 filter paper to absorb all of the liquid. Place the grid sample side up on a dry piece of filter paper in a Petri dish and allow to dry overnight. Store the dried grids in a grid box.

3.4. Immunogold Labeling of Sections

1. Ultrathin sections on grids are floated section side down on a 300-μL drop (*see* **Note 18**) of freshly prepared 3% hydrogen peroxide solution for 10 min (*see* **Note 19**).
2. Wash ultrathin sections on grids four times with PBS by floating grids section side down on a 300-μL drop of PBS for 30 min.
3. Block ultrathin sections on grids by floating grids section side down on a 300-μL drop of PBS–BSA for 30 min.
4. Incubate ultrathin sections on grids with primary antibody by floating grids section side down on a 300-μL drop of primary antibody diluted in PBS–BSA and incubating at 37°C for 20 min (*see* **Note 20**).
5. Wash ultrathin sections on grids twice with PBS–Tween by floating grids section side down on a 300-μL drop of PBS–Tween for 30 min.
6. Incubate ultrathin sections on grids with protein-A gold by floating grids section side down on a 300-μL drop of protein-A gold diluted 1:10 or 1:20 in PBS and incubating for 20 min at room temperature.
7. Wash ultrathin sections on grids twice with PBS–Tween by floating grids section side down on a 300-μL drop of PBS–Tween for 10 min.
8. Wash ultrathin sections on grids twice with deionized water by floating grids section side down on a 300-μL drop of deionized water for 10 min.
9. Stain ultrathin sections on grids with uranyl acetate by floating grids section side down on a 300-μL drop of 3% uranyl acetate for 10 min (*see* **Note 21**).

10. Wash ultrathin sections on grids twice with deionized water by floating grids section side down on a 300-μL drop of deionized water for 10 min.
11. Blot the bottom of the grid with Whatman #1 filter paper and place sample side up on a dry piece of filter paper in a Petri dish and allow to dry overnight.
12. Examine the ribbon in the electron microscope at low magnification to determine the quality of the individual sections and identify a region in the series that can produce high-quality serial images of the structure under investigation (*see* **Note 22**). **Figure 21.1** is an example of a low-magnification image of the middle portion of a serial section ribbon. An arrow indicates a cell whose sectioning starts in section 6 and continues throughout the length of the ribbon. Notice how the diameter of the cell increases from sections 6 to 25 as we section from the outer surface into the interior of the cell.

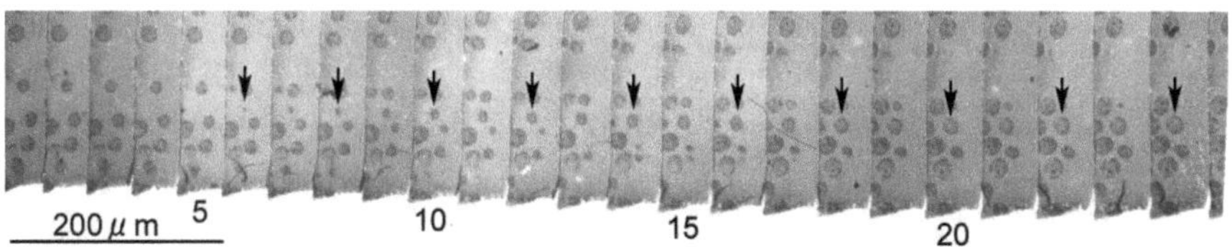

Fig. 21.1. A low-magnification image of the middle portion of a serial section ribbon of *Chlamydomonas reinhardtii* cells. *Chlamydomonas* cells were embedded in Epon, a ribbon of silver ultrathin sections was prepared and examined at low magnification in the electron microscope. The central portion of the ribbon was photographed and the section at the end of the ribbon was arbitrarily defined as section 1. *Arrows* in a section identify a cell that is readily identifiable in each section positioned above a cluster of landmark cells that can be used to orient the identified cells so each section can be viewed at high magnification with the same field of view. The cell identified by the *arrow* is first visible in section 6 and sectioning through the cell continues throughout the length of the ribbon. Notice how the diameter of the cell increases from sections 6 to 25 as we section from the outer surface into the interior of the cell.

13. Identify landmark features in the sections that can be used to orient the region under investigation in the microscope so that each section can be viewed at high magnification with the same field of view. In **Fig. 21.1**, the cell cluster just under the cell marked by an arrow provides convenient field of view orientation points.
14. Examine and photograph each section in the series using the landmark features to orient the sections for the same field of view. **Figure 21.2** shows high-magnification micrographs of selected sections from a 30-section ribbon passing through the proplastid and immunogold labeled with antibody to the small subunit of ribulose-bis-phosphate carboxylase. The propyrenoids are not seen in section 8 Two propyrenoids are clearly seen in section 13 and as the sections pass through the proplastid the propyrenoids

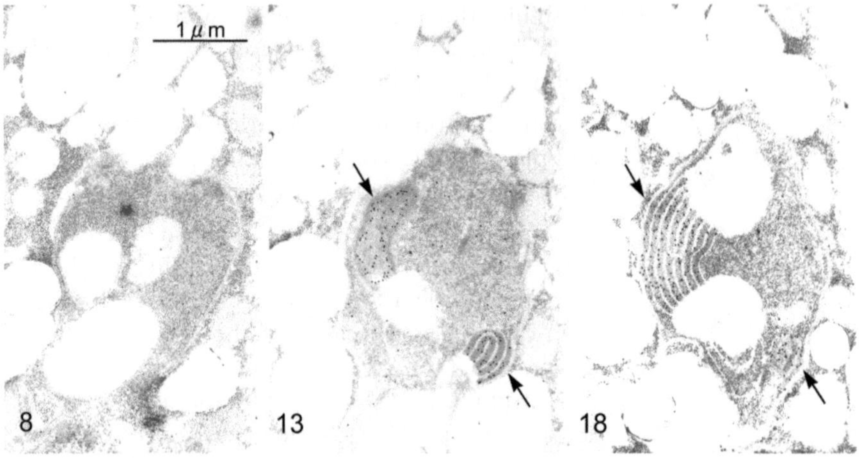

Fig. 21.2. Selected high-magnification serial sections from a 30-section ribbon through the proplastid of dark grown *Euglena* immunolabeled with antibody to the small subunit of *Euglena* ribulose-bis-phosphate carboxylase. The number at the bottom of each figure represents the order of the section in the 30 consecutive section ribbon traversing a proplastid in *Euglena*. Propyrenoids are not seen in section 8. As the sections continue to pass through the proplastid, two propyrenoids (*arrows*) differing in size are seen in section 13 and they constitute a much larger fraction of the total proplastid volume in section 18. Gold particles indicating immunoreaction with RuBisCo antibody can be seen concentrated over each propyrenoid and to a lesser extent in the stroma of the proplastid. (Modified from **Fig. 1**, Osafune et al. (2) with permission.)

are seen to occupy a larger fraction of the proplastid volume in section 18. Notice the immunolabel concentrated over each propyrenoid and to a lesser extent in the stroma.

3.5. 3D Reconstruction from Serial Sections

1. Using a high-resolution scanner, scan the photographic negatives of each section to obtain digital images for 3D reconstruction.
2. Identify a number of reference points visible in adjacent sections and use software to manually or automatically sequentially align each section with the previous section starting in the middle of the ribbon (*see* **Note 23**). For the ribbon shown in **Fig. 21.1**, section 12 would be the starting section and section 13 would be aligned to 12, 14 to 13, 12 to 11, 11 to 10 until all the sections have been aligned.
3. Trace the boundaries of the structure(s) of interest and the position of gold particles onto each digitized image. For the 3D reconstruction of the distribution of ribulose-bis-phosphate carboxylase in the proplastid of dark grown *Euglena* (**Fig. 21.3**), the outline of the prolamellar body, the outline of the propyrenoid, and the position of gold particles were traced onto each micrograph of the 30-section ribbon representative sections of which are seen in **Fig. 21.2**.

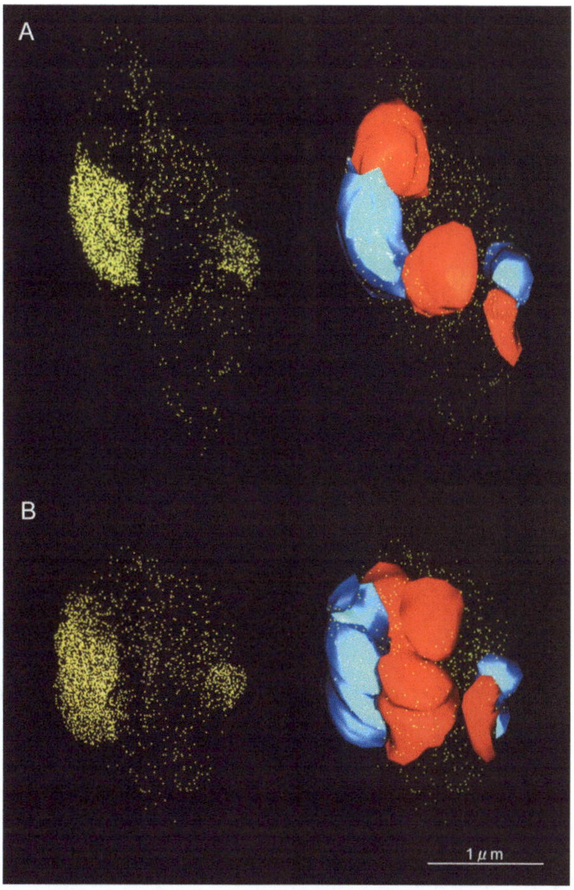

Fig. 21.3. A 3D reconstruction of the proplastid of dark grown *Euglena* viewed from two different angles showing that ribulose-bis-phosphate carboxylase is concentrated in the propyrenoid. Dark grown *Euglena* were fixed and serial section ribbons were stained with antibody to the small subunit of *Euglena* ribulose-bis-phosphate carboxylase. High magnification micrographs from a 30-section ribbon portions of which are seen in **Fig. 21.2** were digitized. Using the IMOD program (6), the micrographs were aligned, the position of the propyrenoid, prolamellar body, and immunogold label was traced onto the aligned micrographs and from the stacked aligned traced micrographs, solid surface renderings of the position of the small subunit of ribulose-bis-phosphate carboxylase, *yellow dots*, the two propyrenoids, *blue*, and the prolamellar body, *red*, were produced. The 3D reconstruction of the distribution of ribulose-bis-phosphate carboxylase is shown on the *left*, while the 3D reconstruction of the propyrenoid and prolamellar body superimposed upon the 3D distribution of ribulose-bis-phosphate carboxylase is shown on the *right*. The images in panel B represent the image in panel A rotated 45° on the *y*-axis. What appear to be two distinct prolamellar bodies surrounding the larger of the two propyrenoids in **A** are clearly seen to be a single interconnected prolamellar body in the rotated image in **B**. Note that the solid surface rendering of the propyrenoids obscures the highly concentrated distribution of ribulose-bis-phosphate carboxylase within the structure. (Modified from **Fig. 3,** Osafune et al. (2) with permission.)

4. Use computer software to stack the individual-aligned images producing a digital representation of the original sample that was embedded in the block.

5. Use computer software to generate a 3D solid surface color image of each of the digitized traced structures. **Figure 21.3** is a 3D reconstruction of the proplastid of dark grown *Euglena* viewed from two different angles with the position of the immunogold-labeled small subunit of ribulose-bis-phosphate carboxylase rendered as yellow dots, the solid 3D surface reconstruction of the propyrenoids rendered in blue, and the solid 3D surface reconstruction of the prolamellar bodies rendered in red. Solid surfaces can be removed to visualize the 3D distribution of immunolabel within an organelle as shown in the left images of **Fig. 21.3**. A comparison of the left panels to the right panels shows that the solid surface rendering of cellular structures, propyrenoid and prolamellar body, can obscure the 3D distribution of immunogold-labeled proteins contained within the structure.
6. Rotate the solid surface 3D image to view the structures from different perspectives. **Fig. 21.3B** was generated by rotating the model in **Fig. 21.3A** by 45° on the Y-axis. What appear to be two distinct prolamellar bodies surrounding the larger of the two propyrenoids in **Fig. 21.3A** is clearly seen to be a single interconnected prolamellar body in the rotated image in **Fig. 21.3B**.

4. Notes

1. Glutaraldehyde and the other chemicals used for fixation and staining are extremely hazardous. Solutions should be prepared and used in a hood. Wear gloves and protective eyeware whenever handling chemicals.
2. Unused embedding mix can be stored for up to 6 months at −20°C but it is preferable to use a freshly prepared mix.
3. Resin waste should be collected and allowed to polymerize before disposal.
4. Nickel grids and gold grids are equally suitable for immunoelectron microscopy with the nickel grids being preferable due to reduced cost. Copper grids are not recommended for immunoelectron microscopy due to the possible reaction of buffer components with the copper producing a precipitate on the surface of the specimen during immunolabeling and possible copper-mediated reduction of antibody–antigen interaction.
5. Diamond knives are available with a 35° and 45° cutting angle. The 45° knife is suitable for most applications. The

35° knife reduces section compression and is especially suited for sectioning samples embedded in hydrophilic resins.

6. By using primary antibodies raised in two different animals and secondary species-specific anti-IgG conjugated with two different sized gold particles for sequential immunolabeling, two antigens can be co-localized on a single section.
7. A number of computer programs are available for using serial sections to produce 3D reconstructions of cellular structures. We have used the program IMOD (6) developed at the Boulder Laboratory for 3D Electron Microscopy of Cells and freely available from http://bio3d.colorado.edu/imod/ for 3D reconstructions of *Chlamydomonas* mitochondria (4), the distribution of the light harvesting chlorophyll a/b binding protein in *Euglena* (5), and the distribution of the small subunit of ribulose-bis-phosphate carboxylase (3) in *Euglena*. Two additional freely available programs are ImageJ (7) available for download from http://rsbweb.nih.gov/ij/ and Reconstruct (8) available for download from http://www.bu.edu/neural/Reconstruct.html or http://synapse-web.org/tools/index.stm. Additional commercially available software packages are listed in (8). Although the actual procedure for 3D image reconstruction is software specific, a common analysis procedure is used regardless of the actual software.
8. The minimum number of cells that will provide sufficient material for sectioning is 4×10^7 cells. Using less cells produces a small cell pellet that after embedding is difficult to position in the microtome for sectioning. Starting with a larger number of cells facilitates the sectioning process.
9. Glutaraldehyde fixation can be performed directly in any medium the cells are grown in. Isolated organelles are fixed directly in the final purification buffer.
10. The amount of glutaraldehyde used for fixation represents a compromise between preserving cellular microstructure and retaining antigenicity. Better microstructure preservation can be obtained at higher glutaraldehyde concentrations but this may result in a loss of antigenicity resulting in a decline in colloidal gold-labeling density.
11. Cells can be stored in 0.1 M potassium phosphate buffer pH 7.2 for up to 1 week at 0–4°C.
12. It is important to add enough acetone resin mixture to the tube so that the specimen containing agarose pieces are not exposed to air when the acetone evaporates from the mixture.

13. Epoxy resins are easy to use and provide excellent preservation of microstructure after sectioning, and long ribbons of serial sections are easily obtained. The necessity to heat the resin for polymerization does, however, lead to a loss of antigenicity making their use most suitable for localization of high-abundance proteins or high titer antibodies. Hydrophilic resins such as Lowicryl K4M and LR White which can be polymerized by ultraviolet light at low temperature are better at preserving antigenicity. They are more widely used for immunoelectron microscopy than epoxy resins although they exhibit a lower level of ultrastructure preservation and the sections are more easily damaged by the electron beam. It is difficult to obtain long ribbons of uniform serial sections from specimens embedded with hydrophilic resins making the epoxy resins the preferred resins for serial reconstruction projects. It has been reported (9) that spraying the surface of a hydrophilic resin block with salon quality hair spray and allowing the block to dry overnight allows continuous ribbons of uniform sections to be obtained.
14. If a visual inspection does not identify the region of the block containing the sample, thick sections can be cut and examined in the light microscope to identify the region of the block containing the sample.
15. The top and bottom sides of the block relative to the direction of sectioning must be trimmed perfectly parallel to produce a straight ribbon of joined sections. If they are not parallel, the ribbon will be curved, the sections poorly joined, and there is a strong possibility the ribbon will break up into smaller segments.
16. A curved ribbon cannot be mounted on a single slot grid but must be divided and mounted on several slot grids. The ribbon can be separated into shorter segments using a wetted eyelash hair by gently pushing the ribbon at the junction of two sections.
17. The most convenient way to pickup and transfer grids is to use a self-closing anti-capillary tweezers. Extra care must be taken to remove all moisture when other types of instruments are used for transfer.
18. A piece of parafilm should be placed in a Petri dish containing a piece of buffer-saturated Whatman #1 filter paper. Place drops of washing solution side by side on the parafilm sheet and move the grid sequentially from drop to drop incubating with the lid closed. Incubation in a water-saturated environment prevents the drops from evaporating. Depending on the scarcity of the reagent and time of incubation, drop size can be varied from 100 to 1,000 μL.

19. Incubation in hydrogen peroxide increases immunoreactivity by etching the resin surface making the embedded antigen more accessible to the antibody. Concentrations of 0.3–3% (w/v) hydrogen peroxide can be used depending on the extent of antigen accessibility required. Samples embedded in hydrophilic resins such as Lowicryl K4M and LR White do not need to be pretreated with hydrogen peroxide solution and should be incubated directly in PBS.
20. The optimal antibody dilution and incubation conditions must be determined empirically for each antibody. Typical dilutions are 50- to 1,000-fold for a 20-min incubation with antibody at 37°C. Incubation times as long as 24 h at 4°C can be used (10). The lower the abundance of the antigen, the higher the antibody concentration and/or incubation time needed. When excessive background labeling is seen, increasing the amount of BSA in the blocking solution, lowering the antibody concentration, and/or shortening the incubation time may decrease non-specific labeling. Control experiments with pre-immune serum should be routinely performed to verify the specificity of the immunoreaction.
21. Care must be taken to withdraw the solution from the top of the bottle to avoid depositing precipitated crystals on the grid.
22. Low-magnification viewing of the sections exposes the entire series and grid coating to low electron beam energy which produces more stable images with less image drift when viewed at high magnification.
23. Differences in the position of the section within the microscope viewing screen and deformation of sections due to compression result in the photographs of each section from a ribbon being unaligned with other sections. Section alignment involves identifying a number of reference points seen in adjacent sections and then using the software to align the sections. Since the misalignment is nonlinear, it is best to choose reference points throughout the section rather than focusing solely on an area near the object of interest within the section.

Acknowledgments

This work was supported by National Science Foundation Grant MCB-0080345 and MCB-0196420 to S.D.S and Grant in Aid for Scientific Research No. 20570062 from the Ministry of Education, Sciences, Sports and Culture Japan to T.O.

References

1. Osafune, T. (1973) Three-dimensional structures of giant mitochondria, dictyosomes and "concentric lamellar bodies" formed during the cell cycle of *Euglena gracilis* (Z) in synchronous culture. *J. Electron Microsc.* **22**, 51–61.
2. Osafune, T., Ehara, T., Sumida, S., Hase, E., and Schiff, J. A. (1990) Light-independent processes in the formation of thylakoids and pyrenoid in proplastids of dark-grown cells of *Euglena gracilis. J. Electron Microsc.* **39**, 245–253.
3. Osafune, T., Sumida, S., Ehara, T., and Hase, E. (1989) Three-dimensional distribution of ribulose-1,5-bisphosphate carboxylase/oxygenase in chloroplasts of actively photosynthesizing cell of *Euglena gracilis. J. Electron Microsc.* **38**, 399–402.
4. Ehara, T., Osafune, T., and Hase, E. (1995) Behavior of mitochondria in synchronized cells of Chlamydomonas reinhardtii (Chlorophyta). *J. Cell Sci.* **108 (Pt 2)**, 499–507.
5. Schwartzbach, S. D., Osafune, T., and Löffelhardt, W. (1998) Protein import into cyanelles and complex chloroplasts. *Plant Mol. Biol.* **38**, 247–263.
6. Kremer, J. R., Mastronarde, D. N., and McIntosh, J. R. (1996) Computer visualization of three-dimensional image data using IMOD. *J. Struct. Biol.* **116**, 71–76.
7. Abramoff, M. D., Magelhaes, P. J., and Ram, S. J. (2004) Image processing with image. *Biophotonics Int.* **11**, 36–42.
8. Fiala, J. C. (2005) Reconstruct: a free editor for serial section microscopy. *J Microsc. 218*, 52–61.
9. Harris, K. M., Perry, E., Bourne, J., Feinberg, M., Ostroff, L., and Hurlburt, J. (2006) Uniform serial sectioning for transmission electron microscopy. *J. Neurosci.* **26**, 12101–12103.
10. Osafune, T., Ehara, T., Hase, E., and Schiff, J. A. (1993) Stage-dependent association of nuclear and chloroplast DNA molecules through bridges transiently formed between the two organelles in synchronized cells of *Euglena gracilis. J. Electron Microsc.* **42**, 197–201.

Chapter 22

Freeze-Etch Electron Tomography for the Plasma Membrane Interface

Nobuhiro Morone

Abstract

To visualize the basal or apical cytoplasmic surface just beneath the plasma membrane, we developed two different methods ("unroof" and "rip-off"). The immunoreplica technique for "unroof" and "rip-off" sample preparation that will be presented in this chapter can determine the distributions of actin, actin-binding proteins, transmembrane proteins, and membrane lipids at the interface of the plasma membrane. We have currently developed freeze-etch electron tomography, which could visualize the 3D molecular architecture of membrane-associated structures (membrane skeleton, clathrin-coated pits, and caveolae) on the cytoplasmic surface of the plasma membrane with 0.85-nm-thick consecutive sections made ≈100 nm from the cytoplasmic surface using rapidly frozen, deeply etched, platinum-replicated plasma membranes. The membrane skeletons that are closely apposed to the plasma membrane interface are suggested to form the boundaries of the membrane compartments responsible for the temporary confinement of membrane molecules.

Key words: Freeze-etch tomography, immunoreplica, actin, membrane skeleton, clathrin-coated pits, caveolae.

1. Introduction

The membrane-associated undercoat structure on the cytoplasmic surface of the plasma membrane plays a crucial role in a variety of cell functions that are related to cell shape, transport, endocytosis, exocytosis, and a variety of signaling pathways. Along with clathrin-coated pits (1), caveolae (2), and lipid rafts (3), the membrane-associated undercoats contain the membrane skeleton [*MSK*, (4–6)], as the portion of the cytoskeleton at the interface with the plasma membrane is often called.

S.D. Schwartzbach, T. Osafune (eds.), *Immunoelectron Microscopy*, Methods in Molecular Biology 657,
DOI 10.1007/978-1-60761-783-9_22,

The MSK is expected to differ from the conventional bulk cytoskeleton in terms of its structure and protein composition. It has been observed and analyzed particularly in the thick cortical actin layers in immune cells and in the spectrin–actin network in red blood cells. Recently, the 3D molecular architecture of the MSK, including the clathrin-coated pits and the caveolae, was reconstructed by electron tomography using rapidly frozen, deeply etched, platinum-replicated plasma membranes (6). Freeze-etch electron tomography is a new tool for structural analysis of plasma membranes and has two appropriate methods for visualizing the apical or basal cytoplasmic surface of the cell membrane during the initial sample preparation. These two visualization methods which are known as "unroofing" (7, 8) and "rip-off" techniques (2, 6) will be presented in this chapter. The structural analysis of the plasma membrane by the freeze-etch tomography method presented in this chapter is reasonably well correlated with the ultrahigh-speed single-particle tracking of membrane lipids and proteins (6, 9). This correlation microscopy supported the novel concept of the actin-based membrane skeleton underlying the plasma membrane structure (6, 10).

2. Materials

2.1. Cell Culture on Glass Coverslips

1. Normal rat kidney fibroblasts (NRK; ATCC, Manassas, VA)
2. HAM-F12 medium (SIGMA, St. Louis, MO) supplemented with 10% fetal bovine serum (FBS; Invitrogen/GIBCO, Grand Island, NY)
3. Solution of 0.05% trypsin and 1 mM EDTA (Invitrogen/GIBCO, Grand Island, NY)
4. 3 or 5 mm, standard #1 or #00 small glass coverslips (Matsunami Glass, Tokyo, Japan)
5. Cleaning solution (Fisher Scientific, Fair Lawn, NJ) for washing glass coverslips
6. Tissue culture dish treated by vacuum gas plasma (Becton Dickinson/Falcon, Franklin Lakes, NJ)
7. CO_2 incubator (Sanyo, Tokyo, Japan)

2.2. Unroofing: Membrane Preparation for Basal Cytoplasmic Surface

1. Mammalian "Ringer's" solution plus: 5 mM HEPES-NaOH pH 7.4, 3 mM NaH_2PO_4, 155 mM NaCl, 3 mM KCl, 2 mM $CaCl_2$, 1 mM $MgCl_2$, 10 mM glucose. Store at 4°C and unless otherwise indicated warm to room temperature (RT) prior to use (7, 8).

2. Mammalian "Ringer's" solution minus: 5 mM HEPES-NaOH pH 7.4, 3 mM NaH_2PO_4, 155 mM NaCl, 3 mM KCl, 3 mM EGTA, 5 mM $MgCl_2$, 10 mM glucose. Store at 4°C and unless otherwise indicated warm to RT prior to use.
3. Medium molecular weight poly-L-lysine: Prepare a 5 mg/mL 10× stock solution in Milli-Q water (Millipore Corporate, Billerica MA) using ≈40–70 kDa poly-L-lysine (SIGMA, St. Louis, MO). Prepare a 1× working solution by diluting one part of the 10× solution with nine parts of Ringer's minus solution.
4. Buffer A: Prepare a 2× stock solution of a 60 mM HEPES-KOH pH 7.4, 140 mM KCl, 10 mM $MgCl_2$, 6 mM EGTA. Store at 4°C and unless otherwise indicated warm to RT prior to use. Prepare a 1× working solution by diluting one part stock solution with one part of Milli-Q water.
5. 2.5% (w/w) Glutaraldehyde (GA) in buffer A: Dilute one part of a 25% (w/w) GA stock solution (distilled EM grade, TAAB, UK) with nine parts buffer A at RT prior to the start of each experiment.
6. Probe-type low power (≈1 watt) ultra sonic device (TOMY-SEIKO, Tokyo, Japan).
7. High reflectance mirror (Sigma Koki, Tokyo, Japan) (*see* **Note 1**).
8. Olympus SZ61 Light microscope (Olympus, Tokyo, Japan) with back/front LED illumination (*see* **Note 2**).
9. #7 Forceps (Dumont, Switzerland) to handle small coverslips.

2.3. Rip-Off: Membrane Preparation for Apical Cytoplasmic Surface

1. 35 mm cell culture dish (BD-Falcon, Franklin Lakes, NJ) (6).
2. 1% Alcian blue: Prepare a 1% 8GX Alcian blue (Waken, Osaka, Japan) solution in distilled water at RT.
3. High molecular weight poly-L-lysine: Prepare a 5 mg/mL solution in 0.1 M KCl using ≈300 KDa poly-L-lysine (SIGMA, St. Louis, MO).
4. Buffer B: Prepare a 2× stock of 20 mM PIPES-KOH pH 7.4, 200 mM KCl, 10 mM $MgCl_2$, 6 mM of EGTA. Store at 4°C and unless otherwise indicated warm to RT prior to use. Prepare working solution by dilution of one part stock solution with one part of Milli-Q water.
5. 2.5%GA (TAAB, distilled EM grade) in buffer B.

2.4. Immunostaining for Replica

1. Buffer C: Prepare a 2× stock solution of 200 mM NaCl, 60 mM HEPES-NaOH pH 6.8, 4 mM $CaCl_2$. Store at 4°C and unless otherwise indicated warm to RT prior to use (6).

2. Quenching solution: 10 mM glycine, 10 mM NH_4Cl in buffer C at RT.
3. Blocking solution: Prepare a 2% neutral BSA (SIGMA, St. Louis, MO) solution in buffer C.
4. 8% (w/v) paraformaldehyde (PFA; Nacalai Tesque, Kyoto, Japan): Prepare an 8% (w/v) PFA, 1 M NaOH stock solution in Milli-Q water by heating with constant stirring to 50–60°C. Prepare a working 4% PFA solution by diluting one volume 8% PFA with one volume buffer C.
5. Tris-buffered saline: 20 mM Tris-HCL pH 8.0, 225 mM NaCl.
6. Primary antibody diluted in 0.1% BSA in buffer C.
7. 5 nm (or 10 nm) colloidal gold conjugated anti-IgG secondary antibody (EY laboratories, San Mateo, CA): Prepare working solution by dilution of one volume antibody with 19 volumes of 0.1% BSA in 20 mM Tris-HCl pH 8.
8. Phosphate-buffered saline (PBS): Prepare a 25× stock of 203 mM Na_2HPO_4, 37 mM KH_2PO_4, 3.4 M NaCl, 67 mM KCl adjusted to pH 7.4 with HCl if necessary. Prepare working solution by dilution of one part stock solution with 24 parts of Milli-Q water.
9. 2.5% GA: Dilute one volume of a 25% (w/w) GA stock solution (distilled EM grade, TAAB, UK) with nine parts phosphate-buffered saline minus to post-fix the immunocolloidal gold with the antigen.

2.5. Rapid-Freeze, Deep-Etch Replica

1. Filtered Milli-Q water: Filter Milli-Q water first with a 0.45 and then a 0.22-μm disk filter (Millipore, Ireland) (6–8).
2. Eiko (Tokyo, Japan) rapid-freezing device to freeze and rapidly bring samples into contact with pure copper metal block cooled in liquid helium.
3. 20-mL glass storage vials (NICHIDEN-RIKA Glass Tokyo, Japan) to store frozen samples in liquid nitrogen tank (*see* **Note 3**).
4. Olympus SZ61 light microscope (Olympus, Tokyo, Japan) to check the water content of the sample surface before freezing.
5. HITACHI-FR7000S (Hitachi, Tokyo, Japan) or BAL-TEC BAF-060 (Bal-Tec, Liechtenstein) freeze-etching systems.
6. Collodion in isopentyl acetate (Nishin EM, Tokyo, Japan).
7. Hydrofluoric acid (WAKEN, Osaka, Japan).
8. Photo-Flo 200 (Kodak, Rochester, NY).
9. Thin bar 200 mesh copper grid (Gilder, Lincolnshire, UK) coated with polyvinyl formvar (Nisshin EM, Tokyo, Japan).

2.6. Freeze-Etch Tomography

1. FEI Tecnai "sphera" 200 KV and "spirit" 100-KV electron microscopes (Eindhoven, Netherlands) for tomography (6).
2. CCD cameras: GATAN GIF camera with 1 K × 1 K pixels (16 bit; Abingdon, UK) and GATAN Ultrascan camera with 2 K × 2 K pixels (32 bit) CCD cameras.
3. IMOD 3D reconstruction software (11) freely available from http://bio3d.colorado.edu/imod/ operating on a Linux system.
4. AMIRA-DEV 3D rendering software package (Mercury Computer Systems, San Diego, CA) on a Linux system.

3. Methods

3.1. Cell Culture on Glass Coverslips

1. Wash small glass coverslips with the cleaning solution and distilled water.
2. Culture NRK cells on coverslips for unroofing or in 35-mm plastic tissue culture dishes for rip-off, in HAM-F12 medium supplemented with 10% FBS under a 5% CO_2 atmosphere at 37°C.

3.2. Unroofing: Membrane Preparation for Basal Cytoplasmic Surface

1. Grow the cells on glass coverslips to ~90% confluence, usually 2–3 days after inoculation (7, 8).
2. Wash cells by incubation with mammalian "Ringer's" solution plus at 37°C for 10 min.
3. Treat cells with medium molecular weight 1× poly-L-lysine in mammalian "Ringer's" solution minus for 10 s.
4. Rinse cells with buffer B diluted to 1:3 with distilled water for 5 s (three times).
5. Unroof the apical cell membranes by exposure to a 1-s ultrasonic burst by carefully focusing the top of the probe sonicator on the membrane surface as shown in **Fig. 22.1**.
6. Fix the specimen attached to the coverslip after unroofing with 2.5% GA in buffer A for 30 min at RT.
7. Wash the specimen three times with buffer A for 15 min.

3.3. Rip-Off: Membrane Preparation for Apical Cytoplasmic Surface

1. Prepare Alcian blue (or high molecular weight poly-L-lysine)-coated coverslips (No. 1, 5 × 5 mm) by immersing them for 10 min at RT in their solutions, washing them with distilled water, and then drying them in air (6).
2. Maintain the cells in the medium supplemented with 10% FBS in 35-mm plastic dishes under a 5% CO_2 atmosphere at 37°C.

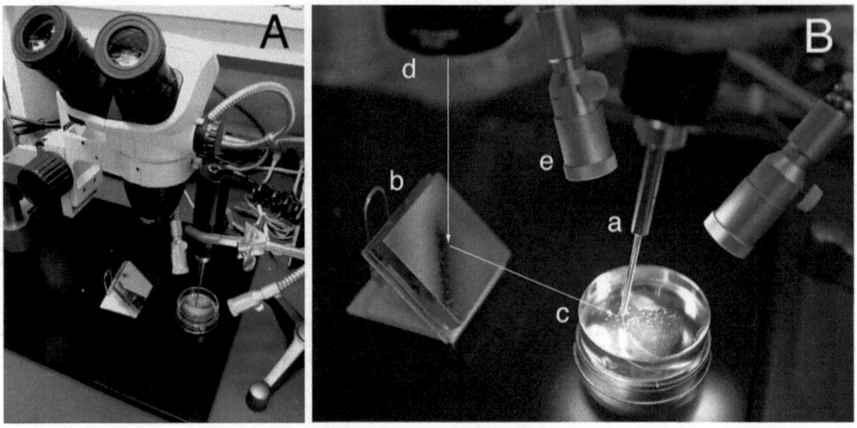

Fig. 22.1. Schematic photograph of some accessories with stereomicroscope and ultrasonicator in the "unroofing" procedure. (**A**) Overall view. (**B**) Close-up view. (*a*) The small thin-probe of the ultrasonic device. (*b*) The mirror with high reflectance. (*c*) A 35-mm diameter plastic dish with cells growing on a coverslip. (*d*) An optional lens of a light microscope. (*e*) Light sources of high illumination.

3. Grow the cells to ~70% confluence, usually in 2–3 days after inoculation.
4. Wash the cells three times with ice-cold buffer B (which mimics the environment in the cytoplasm somewhat, but is slightly hypotonic) and then expose the cells for 15–30 s to ice-cold 70% buffer B.
5. After the buffer in the culture dish is drained, remove the remaining excess water using filter paper, but prevent it from drying up.
6. Put the plastic dish on an ice-cold aluminum plate.
7. Place coverslips coated with positively charged Alcian blue or poly-L-lysine on top of the upper cell surface and incubate on ice for 5–15 min. During this period, good contact between the cell surface and the coverslip should be developed.
8. Slowly add ice-cold 2.5% GA in buffer B into the space between the culture dish and the coverslip.
9. To expose the cytoplasmic surface of the upper cell membrane, remove the upper cell membrane from the rest of the cell adhering to the coverslip placed on top of the cell layer by floating the coverslip gently off from the cells using the surface tension of the buffer.
10. Fix the membrane by incubating the coverslips in fresh, ice-cold 2.5% GA in buffer B for 30 min.
11. Wash the coverslips three times for 15 min with buffer B.

3.4. Immunostaining for Replica (see Note 4)

1. Fix cell membranes on coverslip in 1% PFA, 0.25% GA for immunoreplica in buffer C for 30 min at RT (6).
2. Wash the specimen with buffer C three times for 15 min at RT.
3. Quench the specimen for 15 min with 10 mM glycine, 50 mM NH_4Cl_2 at RT.
4. Wash the coverslips three times with buffer C for 15 min at RT.
5. Block the specimen for 30 min with 2% BSA in buffer C at RT.
6. Wash the specimen three times with buffer C for 15 min at RT.
7. Incubate the specimen with the primary antibody (10 μg/mL) at RT for 1 h in buffer C containing 0.1% BSA.
8. Wash twice with buffer C containing 0.1% BSA for 15 min.
9. Wash the specimen for 15 min with 25 mM Tris-HCl pH 8.0.
10. Incubate the specimen on the coverslips for 1 h with the colloidal gold conjugated secondary antibody at RT in Tris-buffered saline containing 0.1% BSA.
11. Wash the specimen twice with Tris-buffered saline.
12. Wash the specimen three times for 15 min with PBS.
13. Fix the labeled specimen for 5 min in 2.5% GA in PBS at RT.
14. Wash the specimen three times for 15 min in PBS.

3.5. Rapid-Freeze, Deep-Etch Replica

1. Wash the specimen with distilled water for 1 min before rapid freezing (6–8).
2. Attach the specimen holder to the plunger tip of the rapid-freezing device.
3. Place the coverslip with the specimen on a small piece (around 7 mm in diameter) of rabbit lung on the sample holder and remove the excess water using filter paper. The piece of rabbit lung functions as a cushion during impact of the specimen against the copper block of the rapid-freezing device.
4. Slam down (free fall) the specimen onto a polished, pure copper block, which has been precooled in liquid helium.
5. Put the rapidly frozen specimens into liquid nitrogen.
6. Transfer the frozen specimen into the freeze-etching chamber kept at −140°C or below.

7. Shave off the excess ice covering the cytoplasmic surface of the membrane with a prechilled knife, using a microtome placed in the chamber.
8. Raise the specimen temperature to –90°C for 10–30 min (*see* **Note 5**).
9. Rotary-shadow the etched specimen surfaces with platinum at an angle of 22° from the surface and then with carbon from the top (*see* **Note 6**).
10. Apply a drop of 0.05% collodion to the replicated specimen immediately after the frozen replicas melt to fortify them.
11. Remove the platinum–carbon replica from the glass coverslip by incubating at RT for 5 min in a 35-mm plastic dish containing 1% hydrofluoric acid in distilled water and a drop of Photo-Flo 200 solution to prevent fragmentation (6).
12. Remove the hydrofluoric acid being careful not to damage the replicas and wash the replicas three times with distilled water containing a drop of Photo-Flo 200 solution to prevent fragmentation.
13. Mount the replica floating at the water surface onto the 200 mesh formvar-coated copper grids by inserting the grid under the water surface and lifting the replica onto the grid.

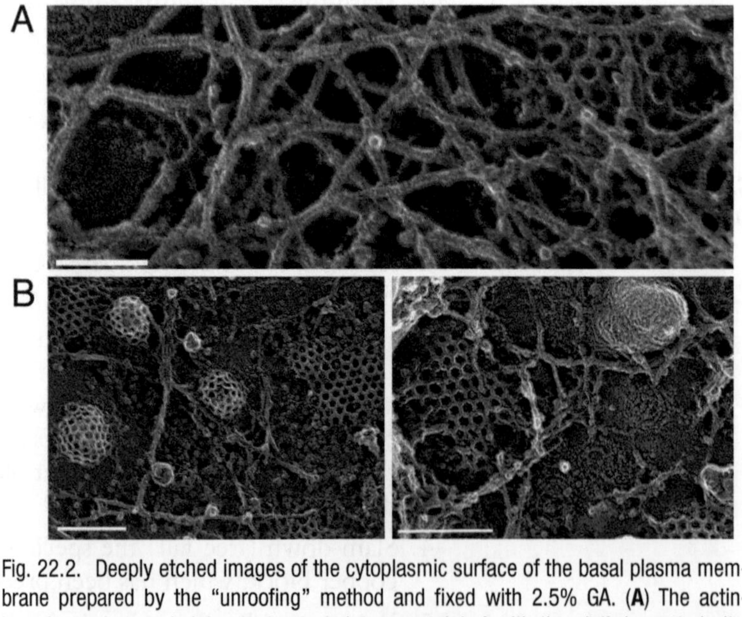

Fig. 22.2. Deeply etched images of the cytoplasmic surface of the basal plasma membrane prepared by the "unroofing" method and fixed with 2.5% GA. (**A**) The actin-based membrane skeleton and cytoskeleton associated with the clathrin-coated pits. Scale bar = 50 nm. (**B**) The membrane undercoats consist of the clathrin-coated pits, caveolae, and actin filaments, and represent a variety of morphologies. Scale bar = 100 nm.

14. Observe the specimen at magnifications of 10–70 K with an FEI Tecnai "spirit" 120-KV electron microscope. **Figure 22.2 A, B** show typical electron micrographs produced by the unroofing method showing actin-based membrane skeletons, clathrin-coated pits, and caveolae. **Figure 22.3A–C** represent examples of immunoreplica micrographs produced by the rip-off method, in which the white small particles are the nanogold particles conjugated to the secondary antibody that binds the primary antibody to phosphatidylinositol biphosphate (**Fig. 22.3A**), spectrin (**Fig. 22.3B**), and the EGF receptor (**Fig. 22.3C**) localized to the cytoplasmic surface of the plasma membrane.

3.6. Freeze-Etch Tomography

1. Place the sample in the FEI Tecnai "*Sphera*" TEM equipped with a CCD camera (6, 11).

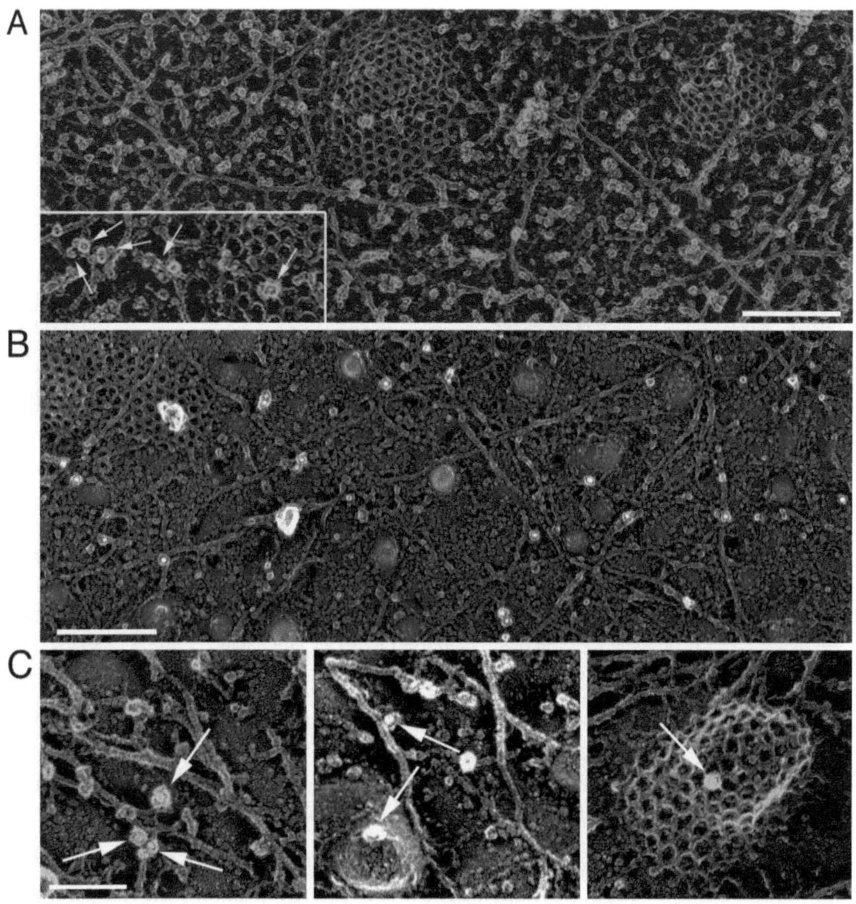

Fig. 22.3. Immunoreplica images deeply etched on the cytoplasmic surface of the plasma membrane by "ripping off" the apical plasma membrane. The white small particles represent the secondary antibody–colloidal gold conjugates bound to primary antibody to (**A**) phosphatidylinositol biphosphate (PIP2), (**B**) spectrin, and (**C**) EGF receptor beside the membrane skeleton on the clathrin-coated pits or the caveolae. Scale bar = 100 nm.

2. Record a series of automatically tilted images from a single field of view of the replica at tilt angles of every 1.0° in the range between ±70° for a total of 141 images. An example of a series of tilted images with an original pixel size on the micrograph of 0.85 nm is shown in **Fig. 22.4A**.

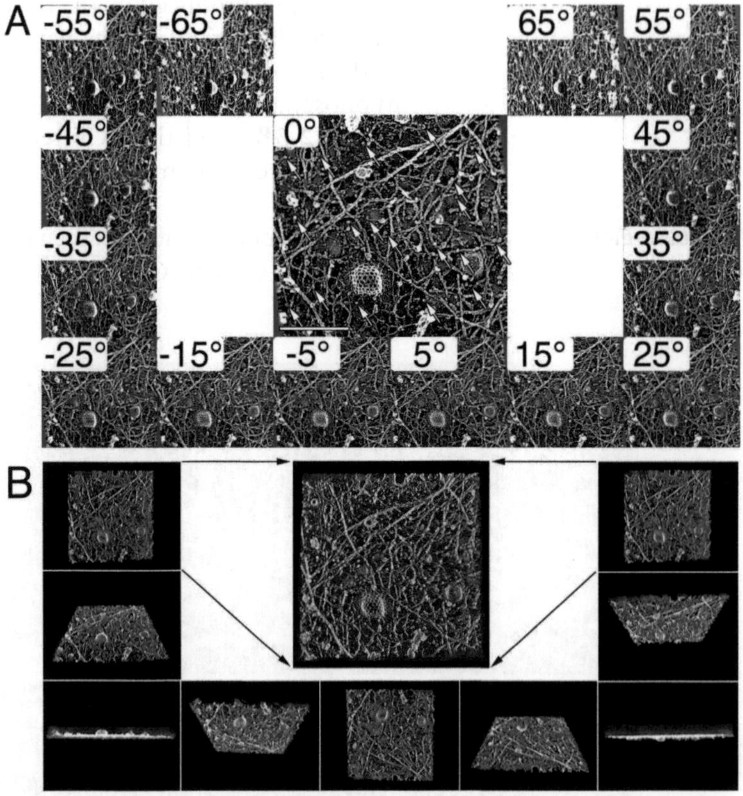

Fig. 22.4. Freeze-etch tomography at the interface with the plasma membrane. (**A**) The tilted images at 5° interval in the range between ±65°were observed for a single field of view in which some points (*white arrows*) were used to correct the original image to show concentricity against the same tilt axis using the IMOD software. Image width = 677 nm. From these tilted images, a series of sliced images could be calculated. (**B**) A series of sliced images were calculated from the corrected tilted images and 3D images were reconstructed from these sliced images.

3. Use the IMOD program to correct for the position of the specimen against the tilt axis and the long undulations of the membrane, using more than 20 points (*see* white arrows at 0° in **Fig. 22.4A**) (*see* **Note 7**).
4. Use the IMOD program to obtain a series of the 121 image sections at every 0.85 nm by a calculation based on the set of 141 tilt images at tilt angles of every 1.0° in the range between ±70°.
5. Reconstruct the 3D images at the interface structure of the plasma membrane with 0.85-nm-thick, consecutive *se*ctions

made $\approx$100 nm from the cytoplasmic surface, and perform a rendering (displaying 3D images in different ways) by using the AMIRA-DEV software (6). A whole reconstructed image is shown in **Fig. 22.4B**. The portion of the cytoskeleton within around 10 nm from the cytoplasmic surface of the cell membrane could be called the membrane skeleton. The distribution of the mesh size of the membrane skeleton, as determined by the freeze-etch tomography, agrees well that of the compartment size determined from high-speed single-particle tracking of phospholipid diffusion (6, 9, 10).

4. Notes

1. A high-reflectance mirror makes it easy to observe how the cells are grown on the surface of the small coverslips.
2. The LED-transmitted illumination base of the Olympus stereomicroscope system SZ61 provides excellent observation of the platinum–carbon replica at the air–water interface.
3. The glass vial with melamine resin cap is resistant to liquid nitrogen.
4. The coverslips with the attached cytoplasmic surface of basal plasma membrane prepared by unroofing and the coverslips with the attached apical cytoplasmic surface of the plasma membrane prepared by rip-off are processed in an identical manner.
5. The sublimation of ice from the membrane surface is practically called deep etching.
6. The molecules as well as the gold probes localized on the cytoplasmic surface of the cell membrane should be immobilized by the deposited platinum.
7. Since 3D reconstruction needs a series of high-precision tilted images, the original images were corrected to show the concentricity against a tilting axis by using the IMOD software.

Acknowledgments

The author would like to thank Drs. Takeshi Baba and Shinichi Ohno at Yamanashi University Medical School for their help in preparing large platinum replicas, Dr. Jiro Usukura at Nagoya

University, Dr. John Heuser of Washington University at St. Louis, and Dr. Shigeki Yuasa at National Center of Neurology and Psychiatry for their helpful discussions, and Dr. Akihiro Kusumi at Kyoto University for his kind direction in guiding and initiating us in the field of plasma membrane research. This research was supported in part by Health Labor Sciences Research Grant Nano-001, JST Sentan, and KAKENHI Kiban to N. Morone.

References

1. Heuser, J., and Kirchhausen, T. (1985) Deep-etch views of clathrin assemblies. *J. Ultrastruct. Res.* **92**, 1–27.
2. Rothberg, K. G., Heuser, J. E., Donzell, W. C., Ying, Y. S., Glenney, J. R., and Anderson, R. G. (1992) Caveolin, a protein component of caveolae membrane coats. *Cell* **68**, 673–682.
3. Simons, K., and Ikonen, E. (1997) Functional rafts in cell membranes. *Nature* **387**, 569–572.
4. Heuser, J. E., and Kirschner, M. W. (1980) Filament organization revealed in platinum replicas of freeze-dried cytoskeletons. *J. Cell Biol.* **86**, 212–234.
5. Hirokawa, N., and Heuser, J. E. (1981) Quick-freeze, deep-etch visualization of the cytoskeleton beneath surface differentiations of intestinal epithelial cells. *J. Cell Biol.* **91**, 399–409.
6. Morone, N., Fujiwara, T., Murase, K., Kasai, R. S., Ike, H., Yuasa, S., Usukura, J., and Kusumi, A. (2006) Three-dimensional reconstruction of the membrane skeleton at the plasma membrane interface by electron tomography. *J. Cell Biol.* **174**, 851–862.
7. Enomoto, A., Murakami, H., Asai,N., Morone, N., Watanabe, T., Kawai. K., Murakumo, Y., Usukura, J., Kaibuchi, K., and Takahashi, M. (2005) Akt/PKB regulates actin organization and cell motility via Girdin/APE. *Dev. Cell* **9**, 389–402.
8. Heuser, J. E., and Anderson, R. G. (1989) Hypertonic media inhibit receptor-mediated endocytosis by blocking clathrin-coated pit formation. *J. Cell Biol.* **108**, 389–400.
9. Fujiwara, T., Ritchie, K., Murakoshi, H., Jacobson, K., and Kusumi, A. (2002) Phospholipids undergo hop diffusion in compartmentalized cell membrane. *J. Cell Biol.* **157**, 1071–1081.
10. Kusumi, A., Umemura, Y., Morone, N., and Fujiwara, T. (2008) Paradigm shift of the molecular dynamics concept in the cell membrane: High-speed single-molecule tracking revealed the partitioning of the cell membrane. In *Anomalous Transport: Foundations and Applications*, Klages, R., Radons, G., and Sokolov, I. M., eds. Wiley-VCH, Weinheim, 545–574.
11. Kremer, J. R., Mastronarde, D. N., and McIntosh, J. R. (1996) Computer visualization of three-dimensional image data using IMOD. *J. Struct. Biol.* **116**, 71–76.

Chapter 23

Localization of rDNA at Nucleolar Structural Components by Immunoelectron Microscopy

Seiichi Sato and Yasushi Sato

Abstract

Fluorescence in situ hybridization (FISH) shows that DNA encoding ribosomal RNA cistron (rDNA) is localized to small speckles scattered in the nucleolus and the nucleolus-associated chromatin (NAC). This technique cannot however precisely locate rDNA in the nucleolar ultrastructural components such as the fibrillar center (FC), dense fibrillar component (DFC), and granular component (GC). In situ hybridization at the electron microscopic level is suitable for localization of rDNA at the ultrastructural level. We have tried to determine the precise localization of rDNA in the nucleolus of a higher plant by electron microscopic (EM) in situ hybridization using biotin-labeled 18S rDNA and demonstrated that it is exclusively localized in the fibrillar centers (FCs) and the nucleolus-associated chromatin (NAC). A secondary antibody coupled to the smallest (5 nm) colloidal gold particles was used in this technique to increase the label. Another important factor to increase the label was pretreatment with proteinase. Convincing results are obtained when the samples are pretreated with 1 μg/mL proteinase K for 45 min at 37°C before immunogold labeling.

Key words: *Allium cepa*, immunoelectron microscopy, in situ hybridization, nucleolus, rDNA.

1. Introduction

The nucleolus, a ribosome-producing apparatus, is the most obvious structure in the nucleus and ubiquitous to all eukaryotic cells. Electron microscopy shows that it is constructed from three major structural components: a fibrillar center (FC), a dense fibrillar component (DFC), and a granular component (GC) (1). The development of immunoelectron microscopy has allowed some nucleolus-specific substances such as fibrillarin and the upstream ribosomal binding factor (UBF) to be localized in the nucleolus

S.D. Schwartzbach, T. Osafune (eds.), *Immunoelectron Microscopy*, Methods in Molecular Biology 657,
DOI 10.1007/978-1-60761-783-9_23, © Springer Science+Business Media, LLC 2010

at the electron microscopic level (2, 3). The ribosomal DNA (rDNA) was first localized at the ultrastructural level of the nucleolus by application of in situ hybridization using rDNA probes prior to immunoelectron microscopy. Some report that the DFC is the site where actively transcribing rDNA is localized (4) but others report that rDNA resides in the FCs and its transcription takes place on their surface (5). These controversial conclusions might be caused by a weak signal.

This chapter presents a method for precise localization of rDNA in the nucleolus of a higher plant by in situ hybridization followed by immunoelectron microscopy (EM in situ hybridization) (6, 7). The rDNA probe is a 1,782-bp 18S rDNA, which is amplified by PCR and biotin labeled by nick translation. The ultrathin sections on grids are heated to denature DNA and then incubated overnight in a hybridization mixture containing single-stranded biotinylated 18S rDNA probe. The localization of rDNA is examined by immunoelectron microscopy using murine anti-biotin monoclonal antibody and then goat anti-mouse IgG antibody coupled to colloidal gold.

The success of this technique is dependent upon increasing the density of colloidal gold on a targeted structure with as little ultrastructural damage as possible. The size of colloidal gold is also important for labeling density. Quantitative analysis using protein A-gold probes indicates that labeling density is inversely proportional to the size of the gold particles (8). This suggests that small gold particles are more effective for detection of low-density substances. Removal of proteins from DNA molecules by digestion with proteinase is another important factor for obtaining high labeling density. As protease treatment inevitably sacrifices ultrastructure, intensive protease treatment should be avoided. Satisfactory results are obtained by the treatment with 1 μg/mL proteinase K at 37°C for 45 min. Unfortunately, this technique has not been successfully applied to other genes at the present time.

2. Materials

2.1. Sample Embedding

1. 66.7 mM Sϕrensen's phosphate buffer (pH 7.0): Mix seven volumes of 66.7 mM Na_2HPO_4 and three volumes of 66.7 mM KH_2PO_4.
2. Freshly prepared 4% (w/v) paraformaldehyde (EM grade, TAAB, Aldermaston, UK) in 66.7 mM Sϕrensen's phosphate buffer (*see* **Note 1**).
3. 15%, 30%, 50%, 75%, 95%, and 99.5% ethanol (99.5% ethanol is commercially obtained).

4. Premixed LR White resin (London Resin, Reading, UK).
5. Saran Wrap (Asahikasei, Tokyo, Japan) cut into small pieces about 10 mm^2.
6. Number 3 gelatin capsules (Electron Microscopy Sciences, Hatfield, PA).

2.2. Preparation of 18S rDNA Probe (see Note 2)

1. DNA extractor WP kit (Wako Pure Chem., Osaka, Japan). Store at 4°C.
2. Primers for PCR: Forward primer, 5′-GTAGTCATATG-CTTGTCTCA-3′ and reverse primer, 5′-TCCGCAGG-TTCACCTACGGA-3′. Dilute an aliquot of the original solution of primers to 10 μM with sterile distilled water for making up PCR reactions. Store the 10 μM stocks and remaining original solution of primers at −20°C (*see* **Note 3**).
3. DNase- and RNase-free 10× PCR buffer (Sigma, St Louis, MO). Store at −20°C.
4. 25 mM $MgCl_2$ in sterile distilled water. Store at 4°C.
5. dNTPs: Solution containing 2 mM final concentration of each dNTP (Takara Bio, Ohtsu, Japan). Store at −20°C.
6. Takara Ex Taq polymerase (Takara Bio). Store at −20°C.
7. Thermocycler.
8. BioNick^{TM} Labeling System (Invitrogen, Carlsbad, CA). Store at −20°C.
9. TE buffer: 10 mM Tris-HCl, pH 8.0, 1 mM EDTA. Store at room temperature.
10. 20× SSC stock solution: 3 M sodium chloride, 0.3 M trisodium citrate. Use 0.1 volume 20× SSC stock to prepare solutions containing a final concentration of 2× SSC.
11. Hybridization mixture: 50% (v/v) formamide, 10% (w/v) dextran sulfate, 2× SSC. Store at 4°C.
12. Microfuge.

2.3. Sectioning Procedures

1. 200 mesh nickel grids (Electron Microscopy Sciences) freshly etched for 15 s at 3 mA with Ion Coater (Eiko Engineering Co., Hitachinaka, Japan). (*see* **Note 4**).
2. Glass knifes made by a MESSER (Sukay Lab. Inc., Tokyo, Japan) and equipped with a trough.
3. Non-magnetic titanium tweezers (Grobet, Carlstadt, NJ) (*see* **Note 5**).
4. Aron alpha instant adhesive (Toagosei, Tokyo, Japan).
5. A surgical knife with an exchangeable blade (Futaba, Tokyo, Japan).

6. #2 Filter paper (Toyo Roshi Kaisya, Tokyo, Japan).
7. Reichert Gridbox (Leica, Wetzlar, Germany).
8. Sterile bi-distilled water.

2.4. Electron Microscopic In Situ Hybridization

1. PBS: 136 mM NaCl, 6 mM Na_2HPO_4, 1.5 mM KH_2PO_4.
2. 1 μg/mL proteinase K (Sigma, St. Louis) in PBS (*see* **Note 6**).
3. 0.2 mg/mL RNase A (Sigma, St. Louis) in PBS (*see* **Note 7**).
4. 20× SSC stock solution: 3 M sodium chloride, 0.3 M trisodium citrate. Use 0.1 volume 20× SSC stock to prepare solutions containing a final concentration of 2× SSC.
5. 70% (v/v) formamide in 2× SSC.
6. 50% (v/v) formamide in 2× SSC.
7. 70, 90, and 99.5% Ethanol.
8. Washing medium: 0.1% Tween 20 in PBS.

2.5. Immunogold Labeling

1. PBS: 136 mM NaCl, 6 mM Na_2HPO_4, 1.5 mM KH_2PO_4.
2. Blocking solution: 1% BSA in PBS.
3. Washing medium: 0.1% Tween 20 in PBS.
4. Murine anti-biotin antibody (Sanbio, Uden, the Netherlands): Dilute 1:100 in blocking solution.
5. Goat anti-mouse IgG (Fc) coupled to 5-nm colloidal gold (Amersham Pharmacia Biotech, Little Chalfont, UK): Dilute 1:20 in blocking solution.
6. Uranyl acetate solution: Prepare by adding 1.25 g uranyl acetate to 25 mL 50% methanol.

3. Methods

3.1. Sample Embedding

1. Excise actively growing root tips of *Allium cepa* about 2-mm long and place them in freshly prepared 4% paraformaldehyde in 66.7 mM Sϕrensen phosphate buffer. Incubate for 2 h at 4°C.
2. Rinse the samples three times by incubating for 1 h at room temperature in the same buffer.
3. Dehydrate the samples (*see* **Note 8**) by incubating them in 15% ethanol for 60 min, 30% ethanol for 60 min, 50% ethanol for 60 min, 75% ethanol for 60 min, 95% ethanol for 60 min, and 99.5% ethanol twice for 60 min (*see* **Note 9**).

4. Infiltrate with LR White resin by incubating for 60 min in LR White and 99.5% ethanol (1:1) followed by incubating twice in pure LR White resin for 60 min.
5. Place the samples in gelatin capsules, fill to the brim with resin, seal the top with a piece of Saran Wrap to avoid contact with oxygen, and then cap the capsule (*see* **Note 10**).
6. Polymerize the samples by incubating overnight at 60–65°C (*see* **Note 11**).

3.2. Preparation of 18S rDNA Probe

1. Freeze the root tips in liquid N_2 and grind them into a powder using a mortar and pestle. Transfer the powder to an Eppendorf 1.5-mL microtube and isolate DNA using the DNA extraction WP kit according to the manufacturer's instructions.
2. Prepare a PCR reaction mixture in a PCR tube by mixing 10 μL 10× PCR buffer, forward primer for a final concentration of 0.5 μM, reverse primer for a final concentration of 0.5 μM, total genomic DNA for a final concentration of 0.5 μM, $MgCl_2$ for a final concentration of 1.5 mM, dNTPs for a final concentration of 0.2 mM, 0.5 μL Taq polymerase (5 U/μL), and sterile distilled water for a 100 μL reaction volume.
3. Set the PCR microtube containing the reaction mixture in a Thermocycler and amplify 18S rDNA using the following thermal profile: initial denaturation at 94°C, 4 min, 25 cycles of 30 s at 94°C, 30 s at 48°C, 60 s at 72°C followed by a final 4 min elongation at 72°C.
4. Label the amplified 18S rDNA with biotin using the BioNick^{TM} Labeling System according to the manufacturer's instruction.
5. Add 2–2.5 volumes of ice cold 99.8% ethanol, gently mix, and incubate at −20°C for 1–2 h to precipitate DNA.
6. Recover precipitated DNA by centrifugation for 15 min at 12,500×*g* in a microfuge at 4°C and discard the supernatant.
7. Dissolve the DNA pellet in PBS and repeat the ethanol precipitation to separate the unincorporated nucleotides from the labeled 18S rDNA probe.
8. Suspend the 18S rDNA probe in the hybridization mixture at a final concentration of 2.5 ng/μL. Store at −20°C until use (*see* **Note 12**).

3.3. Sectioning Procedures

1. Strip the gelatin capsules from the hardened resin block and cut the block into half-sized fragments with a razor blade. Glue the fragments containing root tips to another resin block using Aron alpha instant adhesive (*see* **Note 13**).

2. Set the block on a trimming stand and trim it under a binocular microscope so that the entire block face is about 0.25–0.5 mm^2 and contains resin-embedded tissue (*see* **Note 14**).
3. Mount the trimmed block and a glass knife on an ultramicrotome so that the block surface is parallel to the edge of the glass knife.
4. Move the knife holder so that the block surface is within about 1 mm of the edge of the glass knife by looking through the ultramicrotome microscope.
5. Pour bi-distilled water into the glass knife trough forming a flat surface level with the knife edge.
6. Automatically cut gold sections about 90–150-nm thick.
7. Hold the nickel grid with a pair of non-magnetic tweezers and push it down on the sections floating in the trough to collect the sections on the grid (*see* **Note 15**).
8. Remove water from the tweezers and grids with a piece of filter paper and place the grid sample side up on filter paper in a Petri dish (*see* **Note 16**).

3.4. Electron Microscopic In Situ Hybridization (see Note 17)

1. Place a drop of 1 μg/L proteinase K in PBS on Parafilm in a Petri dish and submerge the grids in the drop for 45 min at 37°C (*see* **Note 18**).
2. Wash the grids three times by dipping them in drops of PBS.
3. Submerge the grids sample side down in a drop of 0.2 mg/mL RNase in PBS for 60 min at 37°C.
4. Wash the grids three times by dipping them into drops of PBS and then once into a drop of water.
5. Submerge the grids in 70% formamide in 2× SSC and incubate them at 80°C for 5 min to denature DNA molecules and immediately cool them by dipping into ice water.
6. Pass the grids through 70%, 90%, and then absolute ethanol followed by air-drying.
7. Heat the hybridization mixture containing 2.5 ng/mL 18S rDNA probe in a microfuge tube at 90°C for 4 min and immediately cool it down by submerging the tubes in ice water. Submerge the grids in the hybridization mixture and incubate them overnight at 37°C.
8. Wash the grids twice by dipping them in 50% formamide in 2× SSC at 37°C for 15 min and then three times in washing medium for 10 min (*see* **Note 19**).

3.5. Immunogold Labeling

1. Block sections on grids for 30 min at room temperature by submerging them in a drop of blocking solution.

2. Immediately transfer the grids onto a drop of primary antibody in blocking solution and incubate overnight at 15°C (*see* **Note 20**).
3. Wash sections on grids three times by dipping them into a drop of washing medium.
4. Incubate sections on grids for 60 min at 37°C with secondary antibody by submerging them in a drop of goat anti-mouse IgG (Fc) in blocking solution.
5. Wash sections on grids three times by dipping them into drops of washing medium and then three times by dipping them into drops of bi-distilled water.
6. Dry sections on grids by blotting them with filter paper and then placing on filter paper.
7. Stain sections on grids in the dark for 30 min at room temperature with uranyl acetate by placing them section side down on a drop of aqueous uranyl acetate.

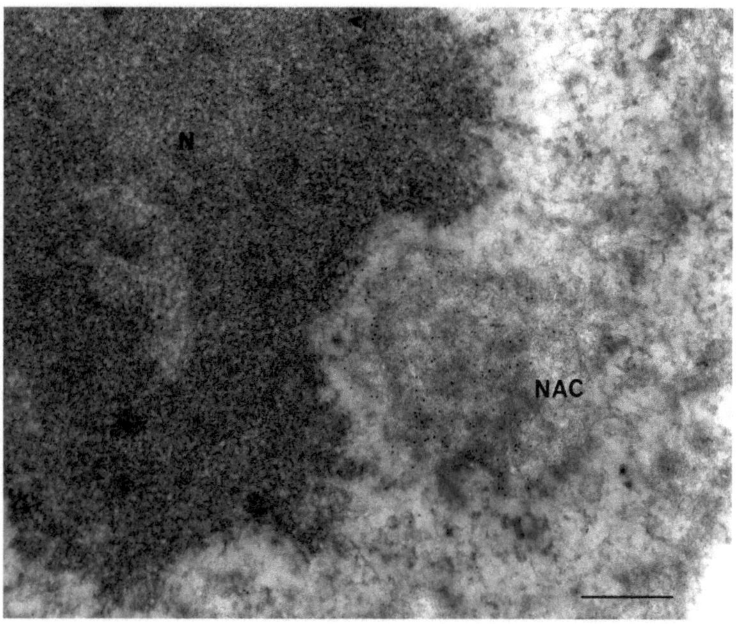

Fig. 23.1. Electron microscopic in situ hybridization using 18S rDNA probe showing strong labeling of the nucleolus-associated chromatin (*NAC*) in a root tip cell of *Allium cepa*. This technique consists of two major procedures: DNA annealing between in situ DNA in the section and a biotin-labeled rDNA probe (in situ hybridization) followed by immunogold labeling of the rDNA probe. The section was pretreated with 1 μg/mL proteinase K for 45 min at 37°C and heated prior to DNA annealing causing the NAC to become somewhat dispersed. Gold particles are specifically distributed on the NAC with only a few gold particles on the nucleolus (*N*). Bar: 0.2 μm. Reproduced from (6) with permission from Oxford University Press.

8. Wash sections on grids three times with bi-distilled water by placing them section side down on a drop of bi-distilled water for 10 min.
9. Dry sections on grids by blotting them with filter paper and then placing on filter paper.
10. Examine sections with an electron microscope. **Figure 23.1** shows strong label on the nucleolus-associated chromatin (NAC) in a root tip cell of *A. cepa* and **Fig. 23.2** shows strong label on the fibrillar centers (FCs).

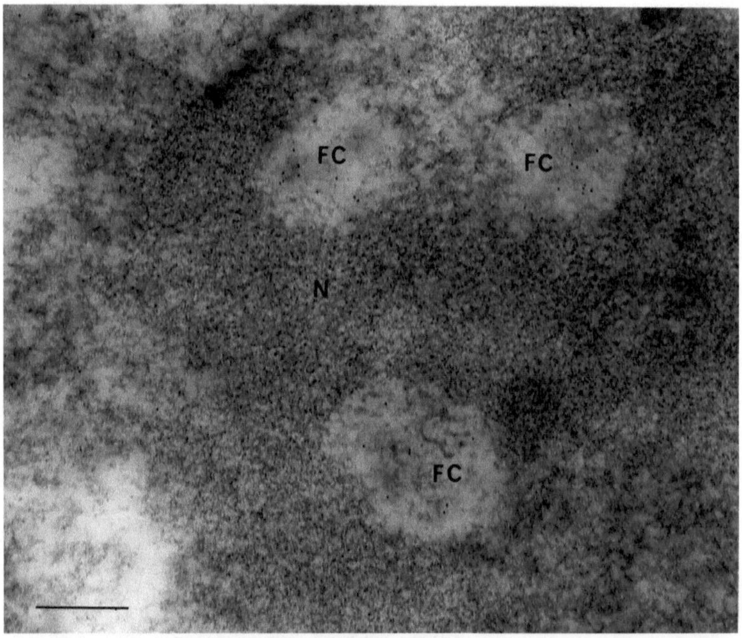

Fig. 23.2. Electron microscopic in situ hybridization using 18S rDNA probe showing strong labeling of the fibrillar centers (*FCs*) of the nucleolus (*N*) in a root tip cell of *Allium cepa*. The section was prepared as described in **Fig. 23.1**. Gold particles are mainly seen on the FCs, suggesting that the FCs are the site where rDNA is located. Bar: 0.2 μm.

4. Notes

1. Paraformaldehyde should be handled in a fume cupboard and waste paraformaldehyde should go into a waste bottle because it is hazardous.
2. Disposable rubber gloves should be used when solutions are prepared and distilled water should be sterilized by autoclaving to avoid contamination with bacteria and DNase.

3. Primers were designed for the 18S rDNA nucleotide sequence highly preserved among plant species.
4. Naked grids (without support such as formvar and collodion film) are used because the sections are resistant enough to be observed under the electron microscope. An alternative way to etch the grids is to submerge them in 1 N NaOH for 10 min followed by rinsing in distilled water.
5. Non-magnetic titanium tweezers are used because stainless tweezers tend to become magnetized, attracting the nickel grids.
6. Proteinase K should be pretreated at 37°C for 1 h to digest DNase before use.
7. RNase A should be boiled for 10 min to inactivate DNase before use.
8. Do not use acetone for dehydration because it prevents proper polymerization of LR White resin.
9. Absolutely dry ethanol is not required because LR White resin is somewhat water miscible.
10. The resin is not properly polymerized in the presence of oxygen. The resin dissolves BEEM capsules made of polypropylene.
11. The resin can be polymerized to be moderately hard without an accelerator by incubating overnight at 60–65°C. The toxicity of this resin is not fully understood at the present time. You should polymerize the rest of resin and the resulting plastic block can be disposed of as normal laboratory trash.
12. Nick-translated biotinylated rDNA probe is reported to be very stable and usable for at least 1 year when stored at −20°C.
13. The sample block in gelatin capsules is too thin to fit into the block holder. Thus, it needs to be glued to another thick block which can fit into the block holder.
14. LR White resin is hydrophilic and readily absorbs water when the specimen is sectioned. Thus, you should be careful that the area of the trimmed block face does not exceed 0.25–0.5 mm^2.
15. If the grids are not etched with an ion coater, they are hydrophobic and hence the sections do not adhere to them.
16. The sections on a grid can be kept in a sealed case containing silica gel at −20°C for at least several days.
17. The sections on a grid are fully submerged in all solutions to detect signals on both sides of them in the following steps.

18. High concentrations of proteinase K will markedly damage ultrastructure in a short time. We recommend a treatment of less than 1 h at a very low concentration.
19. To avoid nonspecific labeling, do not dry the grids from this step to **Section 3.5**, Step 9.
20. It is empirically found that a long incubation at low temperature results in stable and strong label compared to a short incubation at 37°C.

References

1. Jordan, E. G. (1984) Nucleolar nomenclature. *J. Cell Sci.* **67**, 217–220.
2. Puvion-Dutilleul, F. and Christensen, M. E. (1993) Alterations of fibrillarin distribution and nucleolar ultrastructure induced by adenovirus infection. *Eur. J. Cell Biol.* **61**, 168–176.
3. Roussel, P., Andre, C., Masson, C., Geraud, G., and Hernandez-Verdun, D. (1993) Localization of the RNA polymerase I transcription factor hUBF during the cell cycle. *J. Cell Sci.* **104**, 327–337.
4. Scheer, U., Messner, K., Hazan, R., Raska, I., Hansmann, P., Falk, H., Spiess, E., and Franke, W. W. (1987) High sensitivity immunolocalization of double and single-stranded DNA by a monoclonal antibody. *Eur. J. Cell Biol.* **43**, 358–371.
5. Thiry, M. and Thiry-Blaise, L. (1989) In situ hybridization at the electron microscope level: an improved method for precise localization of ribosomal DNA and RNA. *Eur. J. Cell Biol.* **50**, 235–243.
6. Yano, H. and Sato, S. (2002) Combination of electron microscopic in situ hybridization and anti-DNA antibody labeling reveals a peculiar arrangement of ribosomal DNA in the fibrillar centers of the plant cell nucleolus. *J. Electron Microsc.* **51**, 231–239.
7. Sato, S., Yano, H., Makimoto, Y., Kaneta, T., and Sato, Y. (2005) Nucleolonema as a fundamental substructure of the nucleolus. *J. Plant Res.* **118**, 71–81.
8. Yokota, S. (1988) Effect of particle size on labeling density for catalase in protein A-gold immunocytochemistry. *J. Histochem. Cytochem.* **36**, 107–109.

Chapter 24

Immunogold Labelling for Scanning Electron Microscopy

Martin W. Goldberg and Jindriska Fiserova

Abstract

Scanning electron microscopes are useful biological tools that can be used to image the surface of whole organisms, tissues, cells, cellular components and macromolecules. Processes and structures that exist at surfaces can be imaged in pseudo or real 3D at magnifications of anything from about ×10 to ×1,000,000. Therefore a whole multicellular organism, such as a fly, or a single protein embedded in one of its cell membranes can be visualised. In order to identify that protein at high resolution, or to see and quantify its distribution at lower magnifications, samples can be labelled with antibodies. Any surface that can be exposed can potentially be studied in this way. Presented here is a generic method for immunogold labelling for scanning electron microscopy, using two examples of specimens: isolated nuclear envelopes and the cytoskeleton of mammalian culture cells. Various parameters for sample preparation, fixation, immunogold labelling, drying, metal coating and imaging are discussed so that the best immunogold scanning electron microscopy results can be obtained from different types of specimens.

Key words: Scanning electron microscopy, immunogold, labelling, *Xenopus*, nuclear, envelope.

1. Introduction

Scanning electron microscopes (SEMs) have traditionally been used at relatively low resolutions to bridge the gap between light and transmission electron microscopy (TEM). A common perception is that SEM provides depth of focus and 3D information to complement inherently 2D light microscopy and TEM. However, information is seen as limited to the surface and resolution perceived as low. With the advent of 3D methods in light microscopy and TEM (confocal microscopy and EM tomography) biologists' interest in SEM has waned. Meanwhile, driven by the semiconductor industry, SEMs have been developed that rival TEMs in

S.D. Schwartzbach, T. Osafune (eds.), *Immunoelectron Microscopy*, Methods in Molecular Biology 657,
DOI 10.1007/978-1-60761-783-9_24, © Springer Science+Business Media, LLC 2010

resolution, with in-lens field emission SEMs achieving better than 0.5 nm resolution.

The main challenge remaining for high-resolution biological SEM is to expose the surface of interest in as close to its native morphology as possible. The next challenge is to identify the observed structures. Sometimes morphology gives this information, for example microvilli protruding from the cell surface. If the membrane is removed with detergent, a mass of cytoskeletal filaments will be seen whose dimensions give clues as to their identity but not proof. However, many features such as particles on the membrane surface cannot be identified morphologically. Then, immunolabelling provides the tool to identify the structures.

Confocal microscopy is limited to looking at a handful of cellular components at once, because each component has to be individually fluorescently labelled. SEM, on the other hand, reveals many cellular components morphologically, and then individual components can be identified by immunogold labelling. TEM also gives morphological identification of cellular structures and the ability to label with antibodies but is limited generally to thin sections.

SEM sample preparation is often straightforward, but not always. However, once the method to expose and preserve the surface of interest has been determined (1), it is usually straightforward to localise proteins and other biological molecules with colloidal gold tagged probes. Labelling is in principle similar to immunofluorescence, therefore methods can be adapted accordingly. One of the main challenges is finding suitable antibodies that are able to efficiently recognise aldehyde-fixed epitopes. Careful controls, such as using competing peptides or using cells where the antigen has been knocked out by siRNA, are important to ensure specificity. Knowledge of where the antigen is likely to be located can help identify spurious results. Quantification is often important to ensure that labelling is above background, as labelling levels are often quite low. One of the most important factors, however, is having a well-characterised antibody. This should label strongly and specifically by immunofluorescence after paraformaldehyde fixation (fixation methods such as methonal/acetone, commonly used in immunofluorescence, which destroy ultrastructure are inappropriate). It should also label a single band on a Western blot and be inhibited by competing antigen (2).

For SEM, the surface of interest must be exposed. If the outside of the plasma membrane of mammalian culture cells is of interest, this is easy: simply wash off the medium and replace it with a suitable fixative. If the surface is buried deep in the cell or integrated into a tissue, there are a variety of options to expose the buried surface. Some cellular components can be uncovered by fracturing techniques such as freeze fracture or dry fracturing (1), but these methods usually preclude immunolabelling because the sample is either frozen or fixed and dried, respectively. It is possible, however, to permeabilise cells with detergents, fix lightly with paraformaldehyde to preserve antigens and immunogold label before strongly fixing, drying and then fracturing. This approach, however, is only relevant to detergent-resistant structures.

For many intracellular components it may be necessary to isolate them. In this way membranes can be preserved but morphological changes are possible. For instance, nuclear envelopes from *Xenopus laevis* oocytes are easily isolated and labelled and methods for isolating nuclei from yeast, suitable for SEM and labelling (3) have been developed. Unlike purification of organelles for biochemical analysis, however, it is unnecessary, and in fact usually undesirable, to aim for highly purified preparations because damage is likely to occur. Macromolecular complexes such as chromatin, microtubules, viruses or any protein complex can be isolated, fixed, immunolabelled and prepared for SEM.

In contrast to SEM,samples for TEM are often embedded in resin to enable sectioning. This also supports structures and protects them from the electron beam. SEM samples on the other hand cannot be embedded, and therefore must be robustly fixed in order to limit damage and artefacts. Strong fixation, however, is not compatible with antibodies. For immuno-SEM it is therefore necessary to first lightly fix the sample with freshly prepared paraformaldehyde, then immunolabel, followed by a strong fixation with glutaraldehyde. It is often useful to add agents such as tannic acid to help preserve protein filaments in the second fixation (4, 5). To preserve membranes it is essential to post-fix with osmium tetroxide. Fixation is followed by dehydration through a series of increasing concentrations of ethanol. Samples are then critical point dried. Biological samples cannot be air dried for SEM because the surface tension of the evaporating water crushes the sample, flattening whole cells and obliterating fine details.

Biological samples are not electrically conductive. This is a problem for SEM because large numbers of electrons enter the sample during imaging. Because SEM is generally used to look at bulky samples (whole cells, organelles, tissues), the electrons build up causing sample instability and bursts of emitted electrons, making imaging difficult. Fixation with heavy metals such as osmium reduces electron build up, but usually it is necessary to coat the sample with a thin metal film (6) making the sample conductive so that the electrons hitting it flow to earth. Coating also increases the number of secondary electrons emitted from the sample surface (i.e. increases the signal).

For high-resolution SEM and immuno-SEM, gold sputter coating, often used for SEM, is inappropriate. Although gold has a high atomic number and is therefore a very efficient emitter of secondary electrons, as a noble metal it is highly mobile when it hits the surface of the sample and aggregates into large particles which obscure details of the sample. The other problem with gold is that it is the same material as the gold attached to the secondary antibody. Therefore there is no contrast between the coating and the gold marker. Although better, platinum is also unsuitable for similar reasons. Sputtered chromium has become the metal of choice. The grain size is too small to resolve and hence provides a coating that does not obscure detail. Importantly, the atomic number of chromium is significantly different from gold so that the metal-coated sample surface can be easily distinguished from the gold particles using a backscatter detector (which detects primary electrons reflected from dense objects such as colloidal gold particles). Thin (1–2 nm) chromium coats result in little backscatter signal, whereas the gold particles appear as bright white dots in the backscatter image. Therefore it is necessary to acquire two images of the sample in order to resolve morphological features and the location of the immunogold label. The secondary electron image gives high-resolution information about the sample surface, morphological information. The backscatter image shows the position of the immunogold particles. It is then necessary to superimpose the two images to show the position of the gold particles in relation to the structure.

This chapter presents two examples of sample preparation for immuno-SEM: isolation of nuclear envelopes from *Xenopus* oocytes and exposure of the cytoskeleton in cultured cells. However, the subsequent procedures for immunolabelling, sample processing, critical point drying, chromium coating, imaging and image processing can be carried out on any organelle/cell structure/cell surface that can be suitably exposed as long as it is unobscured and undamaged. In using these protocols it is essential to consider how chemical and physical conditions used to prepare samples (such as buffers, centrifugation) could affect the ultrastructure.

2. Materials

2.1. Preparing Samples for Immuno-SEM

2.1.1. Spreading Xenopus Oocyte Nuclear Envelopes for Immuno-SEM

1. *X. laevis* ovary
2. Stereo microscope with oblique illumination
3. Silicon mounts (Agar Scientific Ltd, Stansted, UK)
4. Dumont tweezers 5/45 (Agar Scientific Ltd, Stansted, UK)
5. Diamond scribe for labelling silicon mounts (Agar Scientific Ltd, Stansted, UK)

6. Drawn out glass Pasteur pipettes (*see* **Note 1**)
7. Dissecting scissors (Agar Scientific Ltd, Stansted, UK)
8. Diamond sharpening stone (Agar Scientific Ltd, Stansted, UK)
9. Dissecting needles (sharpened, *see* **Note 2**)
10. Fine glass needles, mounted on a handle (*see* **Note 3**)
11. Cocktail sticks (Plastico, Mitcham, UK)
12. 35 mm Petri dishes (Jencons-Pls, East Grinstead, UK)
13. Bunsen burner (Jencons-Pls, East Grinstead, UK)
14. Whatman 85 mm diameter Filter paper (Fisher Scientific, Loughborough, UK)
15. Acetone
16. 5:1 buffer: 17 mM NaCl, 83 mM KCl, 10 mM Hepes-NaOH pH 7.4. Autoclaved can be stored at room temperature up to a month
17. Immuno-Fix: 4–6% EM grade paraformaldehyde (Agar Scientific Ltd, Stansted, UK) in 0.1 M Hepes-NaOH pH 7.4 (*see* **Note 4**). Make fresh as required

2.1.2. Adherent Culture Cell Cytoskeleton

1. Silicon mounts
2. Fine tweezers (Agar Scientific Ltd, Stansted, UK)
3. 35 mm Petri dishes (Jencons-Pls, East Grinstead, UK)
4. Sticky tape such as Sellotape® or Scotch Tape (3 M)
5. Cells of interest growing in culture flask
6. 10x PBS: Dissolve the following reagents in 750 mL of water: 80 g NaCl, 2.0 g KCl, 14. 4 g Na_2HPO_4,2.4 g KH_2PO_4. Adjust the volume to 1 L with distilled H_2O and sterilise by autoclaving. Store at 4°C. Make 1x fresh as required. Adjust pH to 7.4 with NaOH, if necessary
7. Extraction wash buffer: 10 mM Pipes-NaOH pH 6.8, 100 mM NaCl, 2 mM $MgCl_2$, 1 mM EGTA, 300 mM Sucrose. Store aliquots at −20°C for up to a year
8. Extraction buffer: 10 mM Pipes-NaOH pH 6.8, 100 mM NaCl, 2 mM $MgCl_2$, 1 mM EGTA, 300 mM sucrose, 0.5% Triton X-100. Store aliquots at −20°C for up to a year
9. Immuno-Fix: 4–6% EM grade paraformaldehyde (Agar Scientific Ltd, Stansted, UK) in 0.1 M Hepes-NaOH, pH 7.4 (*see* **Note 4**). Make fresh as required

2.2. Immunolabelling

1. Immuno-Fix: 4–6% paraformaldehyde in 0.1 M Hepes-NaOH pH 7.4 (*see* **Note 4**)
2. PBS: *see* **Section 2.1.2**

3. 100 mM glycine in PBS. Make fresh as required
4. Pre-blocking solution: 1% Bovine serum albumin (BSA) (Sigma-Aldrich, St. Louis, MO) in PBS (*see* **Note 5**). Make fresh as required
5. Whatman 85 mm diameter Filter paper (Fisher Scientific, Loughborough, UK)
6. 35 mm Petri dishes (Jencons-Pls, East Grinstead, UK)
7. "Wet chamber": Place a wet filter paper in the lid of a 9 cm Petri dish and invert so the lid is now the base and the base is used as the lid. A dry slide or parafilm is placed on the filter paper
8. Parafilm (Jenkons-Pls, East Grinstead, UK)
9. Primary antibody diluted in PBS (*see* **Note 6**)
10. 5 or 10 nm gold secondary antibody (Agar Scientific Ltd, Stansted, UK) in PBS (*see* **Notes 6, 7**)

2.3. Processing Samples for SEM

1. SEM fix: 2% EM grade glutaraldehyde (Agar Scientific Ltd, Stansted, UK), 0.2% tannic acid (Sigma-Aldrich, St. Louis, MO), 0.1 M Hepes-NaOH pH 7.4 (*see* **Note 8**). Make fresh as required.
2. 0.1 M sodium cacodylate-NaOH pH 7.4 (*see* **Note 9**). Store at room temperature up to a month.
3. 0.1% OsO_4 (*see* **Note 10**) in 0.1 M sodium cacodylate-NaOH pH 7.4 or 1% OsO_4 if tannic acid is omitted from SEM fix (*see* **Note 8**). Store at room temperature up to a month.
4. 100% ethanol dried and stored with Molecular Sieve (Sigma, St. Louis, MO). Use to prepare 50, 70 and 95% ethanol.

2.4. Critical Point Drying

1. Bal-Tec CPD030 (now Leica Microsystems GmbH, Wetzlar, Germany) (other makes are also suitable) with exit gas flow meter and connected to a high purity CO_2 cylinder with liquid withdrawal
2. 100% ethanol dried and stored with Molecular Sieve (Sigma, St. Louis, MO)
3. CPD carrier (Leica Microsystems GmbH, Wetzlar, Germany)

2.5. Chromium Coating

1. Cressington 308UHR coating unit with sputter head, chromium target, swinging shutter, "cryo-pump" and film thickness monitor (Cressington Scientific Instruments Ltd, Watford, UK)
2. High purity research grade argon gas (BOC, Guildford, UK)

3. Liquid nitrogen
4. Clean glass slide (VWR International, Lutterworth, UK)

2.6. Imaging in SEM

Scanning electron microscope, preferably with field emission source (*see* **Note 7**), and with secondary and backscatter detectors (e.g. S-5200 in-lens feSEM, Hitachi High-Tech, Tokyo)

2.7. Image Processing

Adobe Photoshop

3. Methods

3.1. Preparing Samples for Immuno-SEM

3.1.1. Spreading Xenopus Oocyte Nuclear Envelopes for Immuno-SEM

1. Scratch numbers on silicon mounts using diamond scribe
2. Remove mounts from backing and place in glass dish containing acetone
3. Pick up mounts with fine tweezers and dry with tissue to clean
4. Fill four 35 mm Petri dishes with 5:1 buffer
5. Place two dishes on the stereo microscope stage and two next to it
6. Remove a pea sized piece of ovary and place it in one of the Petri dishes on the bench containing 5:1 buffer
7. After a few seconds, transfer to the other dish on the bench containing 5:1 buffer to rinse
8. Transfer to one of the dishes on the microscope stage
9. Place a cleaned numbered mount in the other dish on the microscope stage
10. Using a fine angled (Dumont©5/45) tweezers, grip a piece of connective tissue close to the oocyte
11. Using a sharpened dissecting needle, pierce the dark hemisphere containing the nucleus. Ideally the knobbly transparent nucleus should emerge from the hole (*see* **Note 11**)
12. Fill a drawn out Pasteur pipette with buffer from the dish
13. Place the end of the pipette under the surface of the buffer and expel a small amount of buffer
14. Very carefully pick up the nucleus with the pipette so that it only rises a millimetre or two into the pipette (*see* **Note 12**)
15. Lift the pipette containing the nucleus out of the dish and move the second dish containing the mount under the microscope field of view

16. Dip the end of the pipette into the dish directly over the mount and let the nucleus fall out and float down onto the mount. Do not expel it as it will be lost (*see* **Notes 12** and **13**)
17. When the nucleus settles onto the mount it should stick (*see* **Note 14**)
18. Once the nucleus has stuck, take a fine glass needle, pierce it at the base and cut across the bottom. You should see a small bluish glob of nuclear contents float out, which can be removed by a gentle stream of buffer from the pipette
19. Transfer the mount into Immuno-Fix

3.1.2. Adherent Culture Cell Cytoskeleton

1. Number silicon mounts with diamond scribe, clean with acetone, dip in 70% ethanol and sterilise in a flame
2. Place mounts and a coverslip in cell culture dishes and allow cells to grow to about 70% confluence (*see* **Note 15**)
3. Wash cells briefly with PBS
4. Wash cells briefly with Extraction Wash Buffer by transferring the mounts into a Petri dish filled with Extraction Wash Buffer
5. Transfer mounts to Extraction Buffer and incubate at room temperature for 5 min
6. Transfer mounts to Immuno-Fix

3.2. Immunolabelling (see Note 16)

1. Incubate mounts in Immuno-Fix for 2 h at room temperature or overnight at 4°C
2. Transfer the mounts to PBS for 5 min
3. Quench unreacted aldehyde by placing the mounts in 100 mM glycine in PBS for 5 min
4. Block non-specific binding by placing the mounts in 1% BSA in PBS for 20–60 min (*see* **Note 5**)
5. Remove the mounts from the dish and place on dry filter paper to dry the back of the mount (*see* **Note 17**)
6. Place the mounts on a dry slide or Parafilm in a wet chamber
7. Pipette 10 μL primary antibody onto the mount and incubate 30–120 min at room temperature or overnight at at 4°C (*see* **Notes 18** and **19**)
8. Wash three times for 5 min then once for 15 min in PBS
9. Place the mount on filter paper to dry the back of the mount (*see* **Note 17**)
10. Pipette 10 μL secondary-gold antibody onto the mount and incubate 30–60 min at room temperature (*see* **Note 18**)

11. Wash three times for 5 min and then once for 15 min in PBS
12. Place in SEM fix for 10 min at room temperature or overnight at 4°C

3.3. Processing Samples for SEM

1. Fill two Petri dishes with sodium cacodylate and one with 0.1% osmium tetroxide in sodium cacodylate (if tannic acid was used in the SEM fix) or 1% osmium tetroxide in sodium cacodylate (if tannic acid was omitted from the SEM fix)
2. Transfer the mounts from SEM fix into the first dish containing sodium cacodylate for 1 min
3. Transfer the mounts into the second dish containing sodium cacodylate for 1 min
4. Transfer the mounts into the dish with osmium tetroxide for 10 min (*see* **Note 20**)
5. Set out six Petri dishes. Fill one with distilled water and the remainder with 50, 70, 95, 100% ethanol, respectively
6. Transfer the mounts into each of these dishes for 2 min (*see* **Note 21**)

3.4. Critical Point Drying

1. Transfer the mounts to a CPD carrier under 100% dry ethanol
2. Fill CPD chamber with 100% ethanol and place CPD carrier in chamber
3. Close lid and start cooling
4. Exchange ethanol for liquid CO_2 with at least 10 changes until all the ethanol is replaced
5. Leave to stand for 30 min (*see* **Note 22**)
6. Perform 10 additional exchanges
7. Warm chamber to 40°C
8. Release CO_2 gas slowly over about 10 min
9. At this point samples such as cells grown on silicon mounts (*see* **Section 3.1.2**) can be "dry-fractured". Simply, place a piece of sticky tape over half of the silicon mount, gently smooth it down with very light pressure and remove. In this way half of the sample will be intact and in the other half cells will be fractured to differing degrees

3.5. Chromium Coating

1. Place silicon mounts with the sample on a clean glass slide in the vacuum chamber of the coating unit and pump to $\sim 10^{-6}$mBar
2. Open isolation valve on the top plate of the cryo-pump
3. Allow vacuum to recover to $\sim 10^{-6}$mBar

4. Fill cryo-pump with liquid nitrogen
5. Wait for a vacuum of at least 5×10^{-7} mBar
6. Ensure shutter is in place between sample and sputter head
7. Introduce argon atmosphere to a pressure of around 10^{-3} mBar
8. Sputter for 30 s onto shutter. The plasma should be a sky-blue color
9. Open shutter until 1–1.5 nm chromium is deposited on the sample (*see* **Note 23**)
10. Turn off the current
11. Close valve on the cryo-pump
12. Vent the vacuum chamber
13. Remove the glass slide with specimens and place on a sheet of white paper. The colour of the metal coat on the glass slide should be grey (*see* **Note 24**)
14. Image samples in SEM as soon as possible (*see* **Note 25**)

3.6. Imaging

1. Insert the mount containing the specimen into SEM
2. Select 10 kV accelerating voltage (*see* **Note 26**)
3. Set emission current as high as possible (*see* **Note 27**)
4. Use a relatively large spot size (*see* **Note 28**)
5. Use the secondary electron detector to acquire a high-resolution surface image, the secondary image. This detects low energy electrons ejected from the sample surface (known as secondary electrons), giving an image of the sample surface (**Fig. 24.1A**). However, there is little contrast to distinguish objects of different materials in the secondary image (**Fig. 24.1A**). Therefore colloidal gold particles coated with antibody molecules will look much like any other globular particle of the same size. In **Fig. 24.1C** the black arrow indicates a colloidal gold particle and the white arrow points to a similar structure that is not. Sometimes a bright circle is observed within the antibody-gold particle (**Fig. 24.1C**, white arrowhead and see inset) but this cannot always be unequivocally identified with the secondary electron detector
6. Use the backscatter detector to acquire the backscatter image (*see* **Note 29**). The backscatter detector detects high energy primary electrons that are reflected from the surface of dense particles such as gold causing it to appear in the image as a bright dot on a dark background (**Fig. 24.1D**). Some dense biological structures such as nuclear pore complexes do backscatter and can be seen with the backscatter

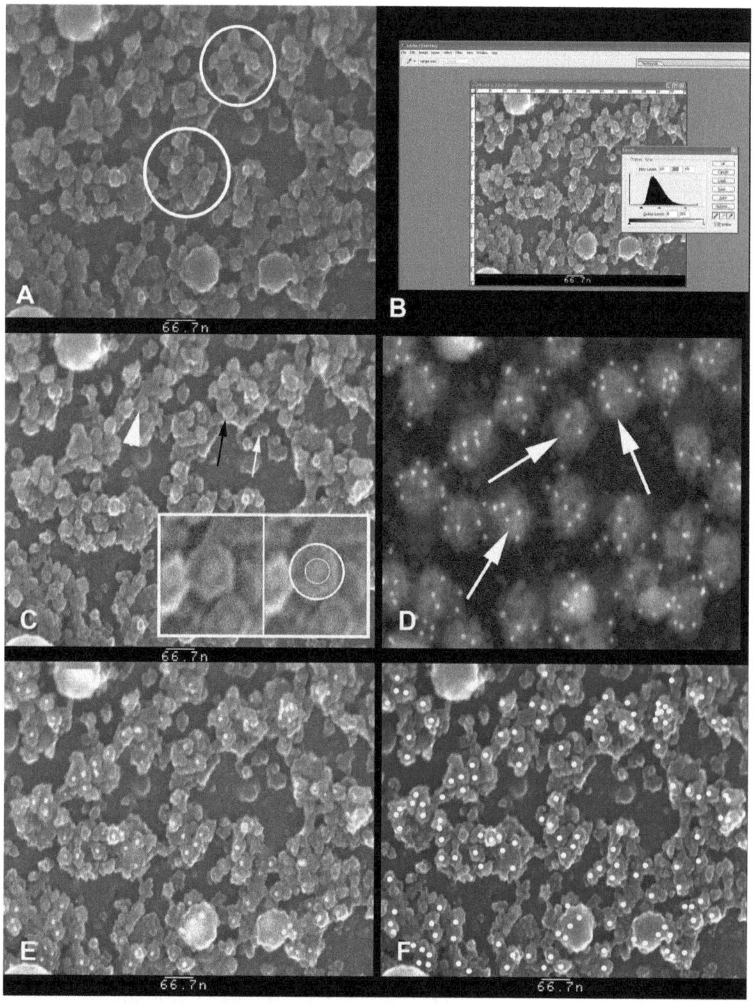

Fig. 24.1. (**A**) Field emission scanning electron micrograph of the cytoplasmic face of an isolated *X. laevis* nuclear envelope showing nuclear pore complexes two of which are circled. (**B**) Adobe Photoshop screen shot showing optimisation of contrast using levels. (**C**) Secondary electron image with optimised contrast. A 10 nm diameter colloidal gold particle coated with antibody molecules is indicated with the black arrow and a similar size particle that is not an immunogold particle is indicated with the white arrow. The white arrowhead indicates an immunogold particle where the extra secondary electron signal generated by the gold particle can be seen as a brighter white centre. The left inset image shows the same particle at a higher magnification and the right inset image is the same image with the boundary of the gold particle in the center and the antibody coat indicated by the rings. (**D**) A backscatter electron image of the same area as shown in **C**. The bright white dots can be unequivocally identified as gold particles, whereas the fuzzy larger white objects (*white arrows*) are the signal generated from the nuclear pore complexes. The membranes, which are thin, generate very few backscatter electrons. (**E**) is a combined secondary/backscatter image and in (**F**) the position of the gold particles has been marked before deleting the backscatter image as described in the text.

detector as dim noisy structures, but they are clearly distinguishable from colloidal gold particles (**Fig. 24.1D**)

3.7. Locating Gold Labels to Specific Structures

1. Open the secondary image file (**Fig. 24.1A**) in Photoshop
2. Open the backscatter image file (**Fig. 24.1D**) in Photoshop
3. If necessary, convert both images to greyscale as follows: click Image, then Mode, then Greyscale
4. Optimise the image contrast. For the secondary image, click on Image, Adjust, Levels, then move the top and bottom sliders to select only the grey level values that are being used (*see* **Fig. 24.1B**) and move the middle slider to give the best brightness. The result is shown in **Fig. 24.1C** (*see* **Note 30**)
5. For the backscatter image (**Fig. 24.1D**) the contrast needs to be maximised rather than optimised. Go to Levels, leave the left slider at the maximum position, move the right one to the bottom end of the peak and then slide the middle one towards the right in order the reduce low-level background signal (*see* **Note 31**)
6. Select the backscatter image and press Ctrl-A, to select the whole image
7. Press Ctrl-C, to copy it to the clipboard
8. Click on the secondary image (**Fig. 24.1C**)
9. Press Ctrl-V to paste backscatter image onto the secondary image as a new layer. At this stage only the backscatter image is seen. Click Window on the tool bar and select the Layers floating window
10. With the backscatter layer selected, select Lighten from the pull down menu. This will make the black background transparent so that white dots (the gold particles) will be seen on the secondary image (*see* **Fig. 24.1E** and **Note 32**)
11. If the backscatter image of the gold particles is difficult to see in the overlaid image (this is often the case in low magnification images or where the background backscatter signal is high). Paste the backscatter image as a second layer onto the secondary image, as in Steps 8 and 9
12. Open a third layer on top of the backscatter image
13. Convert the image to RGB (Image, then Mode, then RGB colour)
14. Select the Brush Tool and set the brush size to roughly the same diameter as the gold particles
15. In the third layer, click on the position of each gold particle superimposing a colored dot on top of each particle

16. Delete the second layer (the backscatter image) leaving colored dots showing the position of each gold particle superimposed on the high-resolution secondary image (**Fig. 24.1F**) (*see* **Note 33**)

4. Notes

1. To pick up and clean oocyte nuclei, the bore of the pipette must be a little larger than the diameter of the nucleus. To make a suitable pipette, take a normal Pasteur pipette and heat the shaft in a Bunsen flame just behind the tapered region while turning slowly. When it begins to go soft quickly pull both ends away from each other to approximately arms length. ALLOW TO COOL then carefully score all around the circumference of the pulled out region at a position where it has an external diameter of approximately 1 mm with the diamond scribe and break.
2. Use a diamond sharpening stone to make the needle as sharp as possible.
3. Fine glass needles can be prepared in a similar way to the drawn out pipettes: simply heat the pipette until it melts completely, then pull the two ends apart very quickly making a long very fine length of glass fibre. Break the fibre into 2 cm long pieces and mount on a cocktail stick with sticky tape.
4. Paraformaldehyde can be dissolved using a water bath heated to 60°C. Use caution when working with paraformaldehyde because the powder and vapours are toxic and easily inhaled. A fume hood should be used as well as protective clothing and gloves.
5. For some antibodies serum or fish skin gelatine (Sigma-Aldrich, St. Louis) are more appropriate blocking agents. It is usually necessary to test different blocking agents to determine the one that most efficiently reduces background staining while minimally inhibiting specific staining.
6. Sometimes it is appropriate to add a blocking agent such as BSA (*see* **Note 5**) to the antibody solution, but in some cases this can reduce the labelling level without increasing the signal:noise ratio. This can only be determined empirically.
7. SEMs with a field emission electron source will easily resolve 10 nm gold particles, if imaging conditions are right

(*see* below). 5 nm gold is also possible, but imaging must be perfectly set up. For SEMs with a tungsten electron source larger gold particle may be necessary (15–20 nm), although labelling sensitivity is inversely related to gold particle size and larger particles will also obscure more of the sample.

8. Tannic acid is added to increase the preservation of cytoskeletal filaments. It is also an osmium mordant (increases osmium binding). Therefore a lower concentration of osmium is used after tannic acid fixation (0.1%). If tannic acid is omitted (i.e. if you are just looking at a membrane surface and don't need to optimally preserve filaments) then a higher concentration of osmium is necessary (1%).
9. Caution when working with sodium cacodylate. May be fatal if swallowed. Known carcinogen in humans. Harmful if inhaled, may be harmful by skin contact. Long-term exposure may lead to kidney and liver damage. Eye and skin irritant. Use gloves, safety glasses, good ventilation. Handle as a carcinogen.
10. Osmium tetroxide is highly toxic and volatile and should be kept and used only in a fume hood and only be disposed of with great care. We place used OsO_4 in a bottle containing corn oil while awaiting professional disposal.
11. Sometimes yellow yolk will come out first. You may have to put a tiny amount of pressure onto the oocyte to encourage the nucleus to come out by pressing **very** gently with the side of the needle. If you press too hard, you might see a small puff of clear liquid coming out of the hole as the nucleus bursts.
12. If the nucleus comes into contact with a liquid/air interface it will lyse and be lost. Therefore it is important to not have air bubbles trapped in the pipette.
13. Nuclei are quite sticky and may stick to the bottom of dishes if you let them settle. Therefore if the nucleus misses the mount you should ensure that it is kept moving and not allowed to settle.
14. You can check that the nucleus has stuck to the mount by gently blowing a stream of buffer over it using the pipette. However, be sure not to expel a bubble over the nucleus as this will destroy it.
15. Silicon mounts are opaque so cell density cannot be monitored on the mount. However, the chemical and physical properties of the silicon surface are similar to glass, so a glass coverslip is placed next to the mount to monitor density. The coverslip can then be used for simultaneous light microscopy.

16. The same procedure is used for most samples but you will have to determine the best conditions for blocking and antibody concentrations based on knowledge, experience (i.e. from previous immune-fluorescence) and trial and error.
17. The back of the mount must be completely dry otherwise the antibody solution will not form a droplet on the mount and runs off the sides, resulting in drying of the sample and hence destruction of ultrastructure.
18. Generally antibodies are used at a higher concentration than in immunofluorescence (10- or even 100-fold greater). However, it is necessary to determine the optimum concentration empirically.
19. Sometimes 1% BSA is included in antibody mixture (i.e. the antibody is diluted in blocking buffer). This is useful to reduce non-specific binding, but for some antibodies it may eliminate all binding and has to be omitted.
20. 10 minutes is sufficient for thin samples such as nuclear envelopes or single cells, but longer times (30–60 min) are required for bulkier samples such as pieces of tissue.
21. Ensure mounts are transferred between liquids quickly to ensure there is never a risk of air drying.
22. This allows any ethanol trapped in the sample time to diffuse out.
23. Thicker films of chromium may obscure gold particles.
24. If there is any hint of brown it suggests that the coating is contaminated with chromium oxide. Reasons for this include insufficient initial vacuum prior to introducing argon, vacuum leaks, leaks in the argon pipes, air in the argon pipes after changing the cylinder, insufficient pre-sputtering onto the shutter. If the coater has not been used for some time, chromium oxide can build up on the target which will then require longer pre-sputtering on the shutter. This can happen even when the unit is kept under vacuum. It is a good practice to keep a chromium coating unit pumping continuously to keep the chamber and vacuum meticulously clean.
25. Chromium films on samples oxidise rapidly and samples should be examined ideally on the day they are coated. There is a noticeable loss of signal over a few hours and a significant loss over a few days. The deterioration of the coat can be slowed by storing samples in a vacuum.
26. 10 kV has been found to be optimal in our Hitachi S5200 SEM (7) but 30 kV was better in the Topcon DS130F (8), therefore it is necessary to experiment with different

voltages and other conditions to optimise for the sample, gold size and specific microscope. The best advice is to try different accelerating voltages and see which gives the most stable, high-resolution image compatible with detection of gold particles remembering that high beam currents and a large spot size (larger beam diameter) increase the electron dose and hence the backscatter signal.

27. If beam damage is observed the current may have to be reduced. Some analytical SEMs (such as the Hitachi SU70) are capable of very high beam currents which would be inappropriate.
28. The spot size will be a compromise between a loss of resolution and the ability to detect the gold because high resolution is achieved by a small spot size, but this results in a lower electron dose and hence lower backscatter electron emission.
29. If the two images have to be acquired consecutively then there is a possibility of drift occurring between acquiring the two images. In this case, the two images will have to be carefully aligned using objects, such as the immunogold particles, that can be recognised in both images.
30. This spreads the grey level values over the entire 256 level range and gives an optimum contrast.
31. It may also be useful to filter this image using the Median filter (Filters, then Noise, then Median, select a value of one and click OK). The median filter degrades the image, so it is important to check that only noise is removed and not any gold particles. This is particularly a problem at low magnifications when the gold size is closer to single pixel size.
32. It is best if the secondary and backscatter images are acquired simultaneously so that exactly the same area is capture. If the two images were obtained consecutively it may be necessary to align them, if there was any sample drift between acquiring the two images. Select the backscatter image and use the Move tool to line up any features that can be seen in both images (such as the gold particles).
33. The disadvantage of this procedure is that certain information other than the lateral position of gold particles in the backscatter image is lost. In particular, you may observe in the actual backscatter image, that different gold particles have different intensities and apparent sizes, despite being of a uniform actual size. This is because particles that are right at the surface will backscatter more intensely than those that are beneath the surface. It is therefore possible to determine if an antibody is labelling at or below the surface

depending on the intensity of the backscatter signal. This can be confirmed by determining whether or not there is an appropriate sized particle at the position of the backscatter dot. If there is not, and the signal is relatively low, it means the label is below the surface.

Acknowledgements

This work was funded by the Biotechnology and Biological Sciences Research Council, UK grant number BB/E015735/1.

References

1. Allen, T. D., Rutherford, S. A., Murray, S., Gardiner, F., Kiseleva, E., Goldberg, M.W., and Drummond, S. P. (2007) Visualization of the nucleus and nuclear envelope in situ by SEM in tissue culture cells. *Nat. Protoc.* **2**, 1180–1184.
2. Harlow, E. and Lane, D. (1988) Antibodies: A Laboratory Manual. Cold Spring Harbor Laboratory Press, Cold Spring Harbor, New York, pp. 359–552.
3. Kiseleva, E., Allen, T. D., Rutherford, S., Bucci, M., Wente, S. R., and Goldberg, M. W. (2004) Yeast nuclear pore complexes have a cytoplasmic ring and internal filaments. *J. Struct. Biol.* **145**, 272–288.
4. Ris, H. (1997) High-resolution field-emission scanning electron microscopy of nuclear pore complex. *Scanning* **19**, 368–375.
5. Goldberg, M. W. and Allen, T. D. (1992) High resolution scanning electron microscopy of the nuclear envelope: demonstration of a new, regular, fibrous lattice attached to the baskets of the nucleoplasmic face of the nuclear pores. *J. Cell Biol.* **119**, 1429–1440.
6. Apkarian, R. P. (1994) Analysis of high quality monatomic chromium films used in biological high resolution scanning electron microscopy. *Scanning Microsc.* **8**, 289–299.
7. Goldberg, M. W., Huttenlauch, I., Hutchison, C. J., and Stick, R. (2008) Filaments made from A- and B-type lamins differ in structure and organization. *J. Cell Sci.* **121**, 215–225.
8. Rutherford, S. A., Goldberg, M. W., and Allen, T. D. (1997) Three-dimensional visualisation of the route of protein import: the role of nuclear pore complex structures. *Exp. Cell Res.* **232**, 146–160

Chapter 25

Horseradish Peroxidase as a Reporter Gene and as a Cell-Organelle-Specific Marker in Correlative Light-Electron Microscopy

Thomas Schikorski

Abstract

A modern electron microscopic approach to the investigation of the structural organization of proteins and subcellular structures demands the use of molecular genetic techniques. The successful implementation of genetic techniques is closely tied to a reporter gene such as the green fluorescent protein (GFP). Although GFP has been widely used for light microscopy, it has many limitations for use in electron microscopy. In the search for a reporter gene for electron microscopy, interest in the use of horseradish peroxidase (HRP) DNA has recently increased, and several studies already have proven the feasibility of HRP expression in mammalian cells. Here, we describe a protocol that uses a HRP chimera to label the endoplasmic reticulum of HEK cells.

Key words: Endoplasmic reticulum, Golgi, electron microscope.

1. Introduction

The introduction of green fluorescent protein (GFP) as a fluorescent marker for gene transfer began a revolution in cell biology (1–3). With GFP as a reporter gene, it became possible not only to identify transfected cells but also to observe structural changes during live cell imaging sessions or after fixation. This widespread revolution in cell biology, however, did not spill over into fine structural research, since GFP cannot be directly detected by electron microscopy (EM).

Horseradish peroxidase (HRP) protein has been widely used in neural tract tracing and immunocytochemistry and is an established and excellent marker for both light microscopy (LM) and

S.D. Schwartzbach, T. Osafune (eds.), *Immunoelectron Microscopy*, Methods in Molecular Biology 657,
DOI 10.1007/978-1-60761-783-9_25,

EM. Since HRP remains an active enzyme when conjugated to other proteins (like antibodies and wheat germ agglutinin) it suggests itself as a substitute for GFP that also can be used in EM. As such HRP has a great potential as reporter gene for correlative LM and EM. Indeed several studies have reported the successful expression of HRP in *Drosophila* (4, 5) and in various mammalian cell lines (6, 7). Several HRP chimeras have been constructed each tailored to specific questions. The signal peptide of the human growth hormone (hGH) has been fused to HRP to target HRP toward the Golgi complex and from there via the *trans*-Golgi network into the secretory pathway from which it was secreted (6). The addition of p-selectin to the hGH–HRP chimera directed the HRP protein to large and small secretory vesicles in PC12 cells (7). When an endoplasmic reticulum (ER) retention signal was added to hGH–HRP (instead of p-selectin), the protein was retrieved from the Golgi complex and accumulated inside the ER (6). This construct was an excellent organelle-specific marker for EM. It also is a wonderful reporter gene since the HRP reaction product is confined inside the ER and does not occlude fine structure yet the labeled ER is easily identifiable in electron micrographs. Placing this HRP chimera under the control of a cell-type-specific promoter will label specific cell types in the LM and EM. By using a neuron-specific promoter, we were able to exclusively label neurons and all their processes in culture (8).

There are seemingly unlimited possibilities for the future use of HRP expression. Simply co-expressing HRP with a gene of interest will allow the investigation of protein assemblies. Furthermore, fusing HRP with one of the many organelle-specific targeting sequences will allow the specific labeling of organelles in cells and tissue. Placing such a construct under the control of a cell-type-specific promoter will label a particular organelle in only a specific cell type. These examples only briefly capture the potential of HRP cDNA as a reporter gene for EM, and many more elegant ways to use HRP cDNA are being created. This technique is already an attractive and exciting alternative to immunocytochemistry and, in theory, HRP cDNA may become as powerful as GFP when studying protein function, protein assemblies, and subcellular structures and in neural tract tracing. A limitation on the use of HRP expression as a marker in electron microscopy is the fact that HRP is a plant protein that requires glycosylation to become a functional enzyme. Glycosylation of HRP can be initiated in mammalian cells by targeting the protein toward the Golgi complex.

This protocol describes the use of HRP as a ER marker. Targeting of the construct used in this protocol (8) is achieved by placing the secretory signal sequence of the human growth hormone upstream of the HRP DNA. Use of the signal peptide targets HRP protein toward the Golgi complex but also leads to secretion of the protein via the secretory pathway into the extracellular space. The secretion of HRP is prevented by adding an endoplasmic reticulum (ER) retention signal downstream of HRP. The retention signal, which consists of the four-amino acid sequence KDEL, ensures the retrieval of glycosylated HRP from the Golgi complex and the retention of the functional enzyme inside the ER. Finally, expression of HRP is rendered neuron-specific by placing HRP under the control of the promoter of the synaptic vesicle protein, synapsin (9). The entire construct, including the secretory signal sequence, the retention signal, the HRP gene, and the synapsin promoter, was cloned into the pDC511 vector. Cells and neurons that are transfected with this "ssHRPcKDEL" construct accumulate HRP in the smooth and rough ER of the entire cell. After HRP histochemistry is performed, the ER of transfected cells is densely labeled and the cell body can be recognized using the light and electron microscope (EM). This protocol describes how to transfect tissue culture cells with the HRP construct described above, visualize HRP histochemically, embed these specimens for EM analysis, and how to identify the same group of cells for correlative light and electron microscopy.

2. Materials

2.1. Tissue Culture

2.1.1. Preparation of Coverslips

1. Biosafety cabinet with UV light source (Baker, Sanford, ME).
2. CO_2 incubator (Sanyo, Wood Dale, IL).
3. Sterile 2, 5, and 10 mL disposable pipettes with matching pipette controller (BD Bioscience, San Jose, CA).
4. Pipettors with matching tips (10, 200, and 1,000 μL) (Gilson, Middleton, WI).
5. Filter units with 0.2 μm low protein binding membrane and 250 mL capacity (Nalgene, Rochester, NY).
6. Complete growth medium: 174 mL Dulbecco's Modified Eagle Medium (DMEM, Invitrogen, Carlsbad, CA), 20 mL fetal calf serum (FCS, Invitrogen, Carlsbad, CA), 2 mL penicillin/streptomycin solution (5,000 units/5 mg/mL, Invitrogen, Carlsbad, CA). Combine the reagents in a 250 mL filter unit and sterile filter. Place the medium into a 5% CO_2 incubator at 37°C with the lid slightly unscrewed to allow for CO_2 equilibration.
7. Round 12-mm coverslips (Bellco, Vineland, NJ, *see* **Note 1**).

8. Coating solution: just before use mix 25 μL of poly-D-lysine (1 mg/mL frozen stock solution, Sigma-Aldrich, St. Louis, MO) with 475 μL of sterilize demineralized water. Add 500 μL of rat tail collagen (1 mg/mL frozen stock solution, Sigma-Aldrich, St. Louis, MO).
9. Sterile 24-well plate (BD Bioscience, San Jose, CA).

2.1.2. Propagating Tissue Culture

1. HEK (human embryonic kidney) cell line (American Type Culture Collection (ATCC), Manassas, VA).
2. Sterile 75 cm^2 flask with a vented cap (BD Bioscience, San Jose, CA).
3. 5 and 10 mL sterile disposable pipettes (BD Bioscience, San Jose, CA).
4. Complete growth medium (*see* step 6 of **Section 2.1.1**).
5. Clinical centrifuge with a rotor for 15 mL conical centrifuge tubes (Cole-Parmer, Vernon Hills, IL).
6. 15 mL sterile conical centrifuge tubes (BD Bioscience, San Jose, CA).
7. Trypsin solution: 0.05% trypsin, 0.02% EDTA (Invitrogen, Carlsbad, CA).
8. Trypan Blue (Invitrogen, Carlsbad, CA).
9. Hemocytometer (Fisher Scientific, Pittsburgh, PA).

2.2. Transfection of Cultures

1. Plasmid DNA: ssHRPcmycKDEL DNA in a pDC511 vector ((8), *see* **Note 2**).
2. Lipofectamine 2000 (Invitrogen, Carlsbad, CA).
3. DMEM (Invitrogen, Carlsbad, CA).
4. Sterile microcentrifuge tubes.
5. Complete growth medium (*see* step 6 of **Section 2.1.1**).

2.3. HRP Histochemistry

1. Fixative: 4% paraformaldehyde, 5% glutaraldehyde (Electron Microscopy Sciences (EMS), Hatfield, PA, *see* **Note 3**), 0.2 mM calcium chloride, 80 mM sodium cacodylate-HCL pH 7.4 (10).
2. 24-well culture plate (BD Bioscience, San Jose, CA).
3. Cacodylate buffer: 90 mM sodium cacodylate–HCL, pH 7.4, 0.2 mM $CaCl_2$.
4. Tris-buffered saline (TBS): 25 mM Tris–NaOH, pH 7.6, 150 mM NaCl.
5. 0.1 mg/mL diaminobenzidine (DAB, Dako, Carpinteria, CA) in Tris-buffered saline (*see* **Note 4**).
6. Hydrogen peroxide (H_2O_2): 1% stock solution in water.

2.4. EM Processing of Tissue Cultures

1. 4% osmium tetraoxide (OsO_4, Electron Microscopic Sciences (EMS), Hatfield, PA (*see* **Note 5**), 80 mM cacodylate–HCl, pH 7.4, 1% potassium ferrocyanide.
2. Cacodylate buffer: 90 mM cacodylate–HCl, pH 7.4, 0.2 mM $CaCl_2$.
3. 2% uranyl acetate in water (EMS, Hatfield, PA).
4. Ethanol, 100%.
5. Ascending ethanol series 50, 70, and 90% prepared from 100% ethanol.
6. Propylene oxide (EMS, Hatfield, PA).
7. 8 mL snap-cap glass vials (EMS, Hatfield, PA).
8. Resin mixture for embedding. Prepare separately (a) Epon resin: 20 g EMbed-812, 11 g DDSA, 9 g NMA, 0.8 mL DMP-30 and (b) Spurr resin: 10 g ERL, 6 g DER, 26 g NSA, 0.4 mL DMAE (all from EMS, Hatfield, PA). When both resins are completely mixed, combine Epon and Spurr resin batches and mix thoroughly.
9. Aclar film (SPI supplies, West Chester, PA).
10. BEEM capsules, size 00 (EMS, Hatfield, PA).
11. Concentrated hydrofluoric acid.

2.5. Sectioning

1. Diamond knife (EMS, Hatfield, PA).
2. Ultramicrotome (Leica, Bannockburn, IL).
3. Double-edge stainless steel razor blades (EMS, Hatfield, PA).
4. Copper slot grids (2×1) Formvar coated (EMS, Hatfield, PA).
5. Forceps for grid handling.
6. Grid storage box (EMS, Hatfield, PA).
7. 2% aqueous uranyl acetate.
8. Hanaichi lead solution (*see* **Note 6**): Calcined lead citrate 0.20 g, lead nitrate 0.15 g, lead acetate 0.15 g, sodium citrate 1.00 g, distilled water 41.00 mL. Prepare calcined lead citrate by heating crystal lead citrate for several hours in a melting pot (200–300°C) until the color changes to a light brownish yellow. Overheated lead citrate with a dark brownish or black color can't be used. After cool-down, place the calcined lead citrate and other reagents in a 50 mL volumetric flask and mix well to produce a yellowish milky solution. Add 9.0 mL of 1 N NaOH to the solution and mix well until the solution becomes clear with a light yellowish color.

Transfer the solution to an amber glass with a screw cap bottle for storage.

9. Parafilm.

3. Methods

3.1. Tissue Culture

3.1.1. Preparation of Coverslips

1. The following steps are carried out inside a biosafety cabinet (*see* **Note 7**).
2. The day before propagating the HEK cell line (Step 3.1.2), prepare a 24-well plate by placing one coverslip into each well.
3. Place ~2 μL of coating solution in the center of each coverslip, then distribute evenly with a sterile blunt-ended plastic or wooden stick.
4. Dry the 24-well plate in a biosafety cabinet for 20 min.
5. Add ~500 μL of sterile demineralized water to each well to wash the coverslip. Remove water after about 1 min.
6. Sterilize the 24-well plate with UV light in a biosafety cabinet for 20 min (lid removed).
7. Add a few drops of complete growth medium so that the coverslips are just covered.
8. Place the 24-well plate into the incubator for use the next day.

3.1.2. Propagating Tissue Culture

1. Remove a 75 cm^2 flask with HEK cells grown almost to confluency and place into the biosafety cabinet.
2. Remove and discard the culture medium.
3. Briefly rinse the cell layer with trypsin solution to remove all traces of serum that contains trypsin inhibitor.
4. Add 2.5 mL of trypsin solution to the flask and observe cells under an inverted microscope until the cell layer is dispersed (usually within 5–15 min). To avoid clumping do not agitate the cells by hitting or shaking the flask while waiting for the cells to detach. Cells that are difficult to detach may be placed at 37°C to facilitate dispersal.
5. Add 2.5 mL of complete growth medium (to make a total of 5 mL together with the trypsin solution) and aspirate cells by gently pipetting.
6. Transfer the suspension to a 15-mL conical centrifuge tube and gently pellet cells in a centrifuge at ~10×*g* for ~1 min.

7. Discard the supernatant and resuspend cells in 5 mL of complete growth medium.
8. Mix 50 μL of the cell suspension with 50 μL Trypan Blue (*see* **Note 8**). Count live cells, cells not stained by Trypan blue, with a hemocytometer, and plate 15,000 cells in 500 μL complete growth medium into each well that contains a coverslip. Place the 24-well culture plate into the incubator at 37°C.
9. Add 300,000–500,000 cells of the remaining cell suspension to a new 75 cm^2 culture flask and 10 mL complete growth medium. Place into the incubator at 37°C.
10. When cells reach confluency (typically this takes 3 days) the steps in **Section 3.1.2** should be repeated. When no experiments are planned the steps involving plating cells on coverslips are omitted.

3.2. Transfection of Tissue Cultures

1. The day after plating cells into the 24-well plate, mix 0.8 μg of plasmid DNA with 50 μL of DMEM for each well to be transfected.
2. Mix 1 μL of lipofectamine 2000 with 50 μL of DMEM and incubate for 5 min.
3. Add the 50 μL of the DNA mixture and mix gently. Incubate at 37°C for 20–40 min.
4. In the meantime, remove the complete growth medium and add 300 μL of serum free DMEM.
5. Add 100 μL DNA/lipofectamine mixture to each well. Mix gently by rocking the plate back and forth. Incubate at 37°C for 2–6 h.
6. Aspirate DNA/lipofectamine mixture and replace with complete growth medium. Incubate at 37°C.

3.3. Fixation and HRP Histochemistry

1. Two days after transfection, remove the coverslips with transfected HEK cells and place them in Karnovsky fixative in a new 24-well plate. Gently agitate and fix for 20 min.
2. Once the cells are fixed all of the following steps are performed outside the biosafety cabinet on a standard lab bench.
3. Wash three times in TBS.
4. Pre-incubate each well in 350 μL of 0.1 mg/mL DAB in TBS and place on a shaker for 5 min.
5. Mix DAB with 0.005% H_2O_2 (final concentration) and use this solution to replace the DAB solution in the culture wells. Develop for 10 min to 2 h (*see* **Note 9**).

6. Wash thoroughly with TBS and examine in the light microscope. Photograph a region of interest at various magnifications (×10–×60). These micrographs are used to analyze the LM data and later, to identify the same region of interest in the EM. **Figure 25.1A** shows a confluent HEK cell monolayer with numerous HRP-labeled cells. A region of interest is enclosed in a trapezoid and a cell which will subsequently be analyzed in the EM is identified by an arrowhead.
7. Store the specimens in the culture wells in Karnovsky fixative until cells are processed for EM.

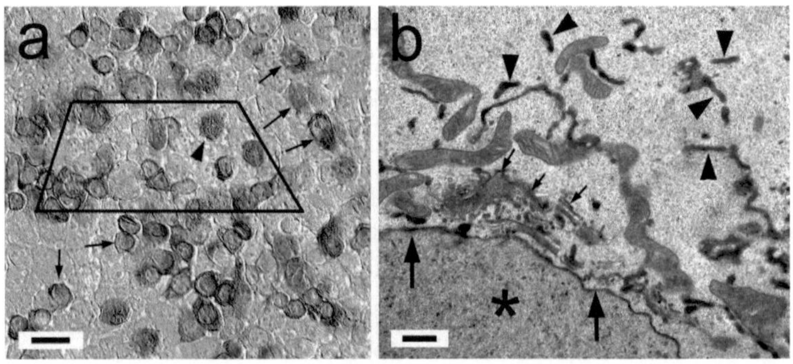

Fig. 25.1. HEK cells transfected with plasmid ssHRPcmycKDEL and labeled with HRP. (**A**) A light micrograph of HEK cells transfected with horseradish peroxidase (*HRP*). *Arrows* point to examples of transfected cells that are surrounded by a confluent background of untransfected cells. An *arrowhead* points to a cell that will subsequently be analyzed in the electron microscope. A trapezoid identifies the region of the light micrograph containing the cells whose "Gestalt" and position to one another are used to identify the cell to be analyzed (*arrowhead*) at high magnification in the EM. (**B**) The fine structure of a cell body of a HEK cell that was transfected with and expressed HRP in the endoplasmic reticulum (*ER*). HRP reaction product can be found in several organelles. The *large arrows* point to the HRP-labeled nuclear envelope which is continuous with the ER. The *arrowheads* mark HRP-labeled ER and the *small arrows* point to the HRP-labeled *cis*-side of a Golgi complex. Bar equals 15 μm in a, 500 nm in (**B**).

3.4. EM Processing of Tissue Cultures

1. Wash coverslips in the culture wells three times in cacodylate buffer. During the following steps, the coverslips remain in the original culture wells until they are transferred to a glass vial in step 10 of **Section 3.4** (*see* **Note 10**).
2. Postfix in 500 μL of 1% OsO_4, 1.5% potassium ferrocyanide, 80 mM cacodylate–HCl, pH 7.4. Place on a shaker inside a fume hood and incubate at room temperature for 20–60 min. Longer postfixation times may result in higher contrast of the final EM graphs.
3. Wash once in cacodylate buffer and then wash thoroughly with demineralized water.

4. For block contrasting stain cells in 400 μL of 2% uranyl acetate for 1 h.
5. Wash thoroughly with demineralized water.
6. Dehydrate in an ascending series of 50, 70, 90, and 100% ethanol by successively changing solutions in the culture wells. Incubate each ethanol step for 5–10 min. Repeat the 100% ethanol step three times to ensure complete dehydration.
7. In the meantime, prepare 8 mL snap-cap glass vials with ~2 mL of propylene oxide and place a pencil written paper label into each vial.
8. Transfer the coverslips from the 24-well plate to the glass vials containing propylene oxide. Ensure that each coverslip is on top of the paper label with the cells facing up and fully covered by the propylene oxide. Place on a rotator for 10 min.
9. Replace the propylene oxide with a mixture of propylene oxide and resin mixture (1+1). Place on a rotator for 10–20 min.
10. For resin infiltration, replace the propylene oxide/resin mixture with pure resin mixture. Place on a rotator for 15 min. Exchange pure resin two more times to ensure complete removal of propylene oxide. Extend the last infiltration step to 60 min.
11. Place a piece of Aclar film on a microscope slide so that the film is held in place by a small amount of resin between the slide and the film. Place a coverslip next to its label on top of the Aclar film, cells facing up. Under a stereo microscope confirm that the cells are facing up!
12. Place a couple of resin drops on the cells and cover the cells with a second piece of Aclar film avoiding trapping air bubbles. Complete the sandwich with a second glass slide and place it inside an oven. The used glass vials can be used as weights for even, flat embedding.
13. Polymerize at 60°C for 16–48 h.
14. Remove the top slide and the top piece of Aclar film and discard. Under a microscope examine the coverslip on the bottom slide. With the help of the micrographs taken previously (*see* **Section 3.3**, step 6), identify the group of transfected HEK cells to be analyzed in the EM.
15. Surround the area containing the group of cells to be analyzed in the EM (**Fig. 25.1A**, trapezoid) with pencil marks. Place a BEEM capsule filled with resin upside-down on top of the marked region and polymerize at 60°C for another 16–24 h.

16. Break the BEEM capsule off of the glass slide. This leaves the coverslip and the cells attached to the BEEM capsule.
17. Clean the glass with a razor blade to remove any resin and place the block face into a small amount of concentrated hydrofluoric acid for 20 min or until the glass is completely dissolved. *Wash thoroughly!* (*see* **Note 11**). The monolayer of cells is now exposed and can be trimmed and sectioned.

3.5. Sectioning

1. Re-examine the pencil marked area and identify the region of interest under a stereo microscope.
2. Trim the block until only a trapezoid remains that contains the group of transfected and untransfected HEK cells (**Fig. 25.1A**) which were identified by LM and which will be analyzed by EM. For the final perpendicular cut use a new blade and cut the base and top of the trapezoid perpendicular to its surface.
3. Fasten the block into the ultramicrotome and align the block face EXACTLY parallel to the diamond knife edge (*see* **Note 12**).
4. Cut thin sections (50–60 nm) right from the start. Although they may not be complete, the very first sections already contain cells.
5. Collect short ribbons of sections on Formvar-coated slot grids and let dry.
6. Stain with uranyl acetate for 10 min by placing the grids (section down) onto a 50 μL droplet of 2% aqueous uranyl acetate.
7. Wash thoroughly with distilled water.
8. Stain with lead for 3 min by placing the grids on top of a 50 μL droplet of Hanaichi lead solution.
9. Wash thoroughly with distilled water and let dry.
10. Examine the serial sections in the electron microscope.
11. Locate the cells and structures of interest indentified by LM in the EM. At low magnification compare the "Gestalt" and position of the cells to one another with that of the LM micrographs taken previously (*see* **Section 3.3**, step 6). For example, in the trapezoid indicated in **Fig. 25.1A** the upper and lower left corners are occupied by transfected cells and the right corners are occupied by untransfected cells. The clusters of transfected or untransfected cells in a corner of the section in the EM are used for orientation. Once the upper right corner has been matched with the trapezoid in the LM micrograph the first transfected cell toward the left is the transfected cell that is pointed

at by an arrow head in **Fig. 25.1A**. Once a cell is identified take high-magnification EM micrographs for analysis. **Figure 25.1B** shows the fine structure of the transfected HEK cell indicated by the arrow head in **Fig. 25.1A**. The ER is filled with electron-dense DAB polymer.

4. Notes

1. Tissue culture-treated plastic provides excellent conditions for cell culture. However, glass coverslips provide superior optical characteristics for high-resolution light microscopy and are resistant to organic solvents during EM processing. To ensure optimal cell growth on glass coverslips, selection of glass sold specifically for cell culture is recommended.
2. This plasmid can be obtained from the author. Typically, the plasmid is transfected into *Escherichia coli* and multiplied in a growth medium. Plasmid DNA is extracted by a miniprep. For a detail protocol *see* (11).
3. EM grade 16% paraformaldehyde solution and 25% glutaraldehyde solution can be purchased in 10 mL ampoules. This allows for safe handling and ensures the highest quality of the reagent.
4. DAB is very stable when dissolved in TBS and can be frozen (at −20°C) for several weeks. The slightly higher pH increases stability and reduces background staining. High-quality DAB from a trusted vendor should be used. In the past, we have used 10 mg tablets from Dako. Unfortunately, Dako discontinued this product in the middle of 2009, and at present (spring 2009), we cannot recommend an alternative vendor. Sigma-Aldrich DAB should not be used.
5. Osmium can conveniently be bought in 2 mL ampoules that ensure optimal quality and safety in the laboratory.
6. A stable lead solution (12) is based on calcined lead citrate. The solution is stable for storage and prevents the formation of lead precipitates during the staining procedure. It also provides high-contrast on-section staining. This lead solution is stable and free from precipitates when kept at room temperature or in the refrigerator for 1 year.

7. When preparing and working with cell cultures general sterile working practices have to be employed to avoid infection of the cultures with fungi or bacteria. Sterile working practices include the use of a biosafety cabinet, sterile tools (e.g., the sterilization of the dissecting tools with 70% ethanol), sterile supplies (like culture wells and pipettes), and sterile solutions. Many solutions can be bought sterile but it is a good working habit to sterile filter solutions before use.
8. Trypan blue is a stain for live cells. Dead cells take up dye very heavily and thus should not be included when counting live cell density.
9. The sensitivity of the HRP reaction can be increased by increasing the DAB concentration up to 5 mg/mL and also by lowering the pH to 5.5. However, increasing the sensitivity will also increase the background that is visible as nonspecific staining across all cells in the light microscope.
10. Handling the coverslips in a 24-well culture dish is easy and convenient. Lifting the coverslips off the bottom will ensure that trapped solution beneath the coverslip will mix properly when changing solutions. Four hundred microliters of solution is needed to fully immerge a coverslip within a well of a 24-well plate.
11. Vapors or even residual amounts of hydrofluoric acid can etch the surface of optical lenses over time which will destroy the lenses. It is therefore recommended to store treated blocks away from optical instruments.
12. It is very important that the block face is properly aligned with the diamond knife. This can be achieved by observing the backlight that passes between the knife and the block face. This light band must have the same width from left to right. Also the distance of the knife must be equal at the top and bottom of the block. This can be achieved by advancing the block at its bottom edge toward the knife until the light band just disappears. Retract the block in 10 equal steps and move the top of the block to the knife edge. Advance the block toward the knife while counting the steps. When less than 10 steps are needed before the light band just disappears the top part of the block face needs to be tilted away from the knife. When the light band is still visible after 10 steps, the top of the block face needs to be tilted toward the knife. Repeat these steps until the distance to the knife is equal at the bottom and at the top of the block face. During this procedure great caution must be taken not to damage the diamond knife by moving the block into the knife edge.

Acknowledgements

This work was supported by the NIH grants U54 NS039408 and R21 NS0263208.

References

1. Chalfie, M., Tu, Y., Euskirchen, G., Ward, W. W., and Prasher, D. C. (1994) Green fluorescent protein as a marker for gene expression. *Science* **263**, 802–804.
2. Heim, R., Cubitt, A. B., and Tsien, R. Y. (1995) Improved green fluorescence. *Nature* **373**, 663–664.
3. van Roessel, P. and Brand, H. (2002) Imaging into the future: visualizing gene expression and protein interactions with fluorescent proteins. *Nat. Cell Biol. Rev.* **4**, E15–E20.
4. Dubois, L., Lecourtois, M. M., Alexandre, C., Hirst, E., and Vincent, J. P. (2001) Regulated endocytic routing modulates *wingless* signaling in Drosophila embryos. *Cell* **105**, 613–624.
5. Larsen, C. W., Hirst, E., Alexandre C., and Vincent, J. P. (2003) Segment boundary formation in Drosophila embryos. *Development* **130**, 5625–5635.
6. Connolly, C. N., Futter, C. E., Gibson, A., Hopkins, C. R., and Cutler, D. F. (1994) Transport in and out of the Golgi complex studied by transfecting cells with cDNAs encoding horseradish peroxidase. *J. Cell Biol.* **127**, 641–652.
7. Norcott, J. P., Solari, R., and Cutler, D. F. (1996) Targeting p-selectin to two regulated secretory organelles in PC12 cells. *J. Cell Biol.* **134**, 1229–1240.
8. Schikorski, T., Young, S. M., and Hu, Y. (2007) Horseradish peroxidase cDNA as a marker for electron microscopy in neurons. *J. Neurosci. Methods* **165**, 210–215.
9. Kugler, S., Meyn, L., Holzmuller, H., Gerhardt, E., Isenmann, S., Schulz, J. B., et al. (2001) Neuron-specific expression of therapeutic proteins: evaluation of different cellular promoters in recombinant adnoviral vectors. *Mol. Cell. Neurosci.* **17**, 78–96.
10. Karnovsky, M. J. (1965) A formaldehyde-glutaraldehyde fixative of high osmolarity for use in electron microscopy. *J. Cell Biol.* **27**, 137A.
11. Sambrook, J. and Russell, D. W. (2001) *Molecular Cloning. A Laboratory Manual.* CSHL Press, Cold Spring Harbor, New York.
12. Hanaichi, T., Sato, T., Iwamato, T., Malavasi-Yamashiro, J., Hoshino, M., and Mizuno, M. (1986) A stable lead by modification of Sato's method. *J. Electron Microsc.* **35**, 304–306.

Chapter 26

Monitoring Rapid Endocytosis in the Electron Microscope via Photoconversion of Vesicles Fluorescently Labeled with FM1-43

Thomas Schikorski

Abstract

Cells communicate via endo- and exocytosis with their environment and neighboring cells. At synapses of the nervous system, fast exocytosis is coupled to fast endocytosis, which forms the basis for neurotransmitter release. The introduction of the unique fluorescent FM dyes allowed the monitoring of fast synaptic vesicle exo-endocytic cycling during live imaging sessions and after photoconversion of FM dyes into an electron-dense diaminobenzidine polymer synaptic vesicle cycling can be studied in the electron microscope. This protocol describes FM dye labeling of synaptic vesicles of cultured hippocampal neurons and photoconversion of the fluorescent synaptic vesicles for analysis in the electron microscope (EM).

Key words: Membrane retrieval, activity-dependent labeling of synaptic vesicles, hippocampal cell culture.

1. Introduction

Fluorescent dyes have been a major tool in cell biology and the neurosciences for many decades. Particularly in the neurosciences, fluorescent dyes simplified tract-tracing techniques and allowed the intracellular filling and subsequent imaging of, e.g., Lucifer Yellow-filled neurons. Hence, these dyes became indispensable tools for the discovery of neuronal development and morphology and for tracing the connectivity of neuronal circuits. The discovery that many fluorescent dyes can be photoconverted into an electron-dense diaminobenzidine (DAB) polymer (1) made it possible to study the fine structure of fluorescently labeled neurons at the electron microscope level.

S.D. Schwartzbach, T. Osafune (eds.), *Immunoelectron Microscopy*, Methods in Molecular Biology 657,
DOI 10.1007/978-1-60761-783-9_26, © Springer Science+Business Media, LLC 2010

The FM dye family (FM2-10, FM1-43, FM4-64, and others) is a group of fluorescent dyes with unique characteristics (2). The molecules consist of a hydrophilic, positively charged head group that makes the dye water soluble and a lipophilic tail that facilitates the incorporation of the dye into membranes. In addition, the hydrophilic head prevents the dye from flipping over to the opposite leaflet of the plasma membrane, and as a consequence, the dye labels only the exposed leaflet of lipid bilayers. Consequently, FM dyes are excellent markers for membrane traffic because they become trapped in endocytic membrane compartments via endocytosis and cannot escape. Although several other dyes can be used to follow endocytosis (such as Lucifer Yellow or horseradish peroxidase), the fast kinetics with which FM dyes incorporate into membranes enable visualization of fast membrane turnover, which is not possible with conventional dyes.

Synaptic vesicles (SV) fuse with the plasma membrane during neurotransmitter release and are rapidly retrieved from the plasma membrane. Because neurotransmitter release and membrane retrieval are activity-dependent, SV can be labeled with FM1-43 (and other FM dyes) in an activity-dependent manner. This technique was first applied to the neuromuscular junction (3) and has been widely used in neuronal cell culture (4) and also in brain slices (5) to directly measure presynaptic function.

Like many other fluorescent dyes, FM dyes (especially FM1-43 and FM4-64) can be photoconverted (6–8) and thus allow the visualization of endocytic vesicles as electron-dense membrane compartments in the electron microscope. This technique can be used for all known endocytic pathways including caveoli, bulk endocytosis, and clathrin-mediated pathways as well as the very fast, synapse-specific pathway called "kiss and run" (4). The specific activation of an individual endocytic pathway can be used to label the corresponding endocytic compartments in electron micrographs. Varying the time of dye exposure and/or the time to fixation after activation allows tracking of endocytic membrane compartments over time.

The following protocol describes how to prepare primary neuronal cell cultures, induce synaptic activity and SV cycling in the presence of FM1-43, photoconvert the fluorescent dye, and process the specimens for EM.

2. Materials

2.1. Hippocampal Cell Culture

2.1.1. Preparation of Coverslips

1. Biosafety cabinet with UV light source (Baker, Sanford, ME).
2. Sterile 2, 5, and 10 mL disposable pipettes with matching pipette controller (BD Bioscience, San Jose, CA).

3. Pipettors with matching tips (10, 200, and 1,000 μL) (Gilson, Middleton, WI).
4. Filter units with 0.2 μm low protein binding membrane and 250 mL capacity (Nalgene, Rochester, NY).
5. 10% FCS medium (*see* **Note 1**): 172 mL Minimum Essential Medium (MEM, Invitrogen, Carlsbad, CA), 20 mL fetal calf serum (FCS, Invitrogen, Carlsbad, CA), 4 mL 1 M dextrose (Fisher, Pittsburgh, PA), 2 mL penicillin/streptomycin solution (5,000 units/5 mg/mL, Invitrogen, Carlsbad, CA), 2 mL N2 supplement (100x, Invitrogen, Carlsbad, CA). Combine the reagents in a 250 mL filter unit and sterile filter. Place the medium into a 5% CO_2 incubator at 37°C with the lid slightly unscrewed to allow for CO_2 equilibration.
6. Round 12-mm coverslips (Bellco, Vineland, NJ, *see* **Note 2**).
7. Concentrated nitric acid.
8. Concentrated hydrochloric acid.
9. Rotator (LabScientific, Livingston, NJ).
10. 95% denatured ethanol.
11. Coating solution: just before use, mix 25 μL of poly-D-lysine (1 mg/mL frozen stock solution, Sigma-Aldrich, St. Louis, MO) with 475 μL of sterilized demineralized water. Add 500 μL of rat-tail collagen (1 mg/mL frozen stock solution, Sigma-Aldrich, St. Louis, MO).
12. Sterile 24-well plate (BD Bioscience, San Jose, CA).

2.1.2. Cell Culture Preparation

1. CO_2 incubator (Sanyo, Wood Dale, IL).
2. Water bath (LabScientific, Livingston, NJ).
3. Sterile 2, 5, and 10 mL disposable pipettes with matching pipette controller (BD Bioscience, San Jose, CA).
4. Pipettors with matching tips (10, 200, and 1,000 μL) (Gilson, Middleton, WI).
5. Clinical centrifuge with rotor for 15 mL conical centrifuge tubes (Cole-Parmer, Vernon Hills, IL).
6. Filter units with 0.2 μm low protein binding membrane and 250 mL capacity (Nalgene, Rochester, NY).
7. Neurobasal medium (*see* **Note 1**): 174 mL Neurobasal A medium (Invitrogen, Carlsbad, CA), 2 mL penicillin/streptomycin solution (Invitrogen, Carlsbad, CA, 5,000 units/5 mg/mL), 0.5 mL GlutaMax-1 (100x, Invitrogen, Carlsbad, CA), 4 mL B27 supplement (50x, Invitrogen, Carlsbad, CA). Combine the reagents in a 250 mL filter unit and sterile filter. Warm the medium and

equilibrate the pH in a 5% CO_2 incubator with the lid slightly unscrewed at 37°C.

8. Sterile syringe filters with 0.2 μm low protein binding membrane (BD Bioscience, San Jose, CA).
9. Sterile 10 mL syringes (BD Bioscience, San Jose, CA).
10. Sterile 15 and 50 mL conical centrifuge tubes (Biologix Research Company, Lenexa, KS).
11. Enzyme solution: 5 mL Neurobasal medium, 100 μL EDTA (50 mM sterile filtered stock, Sigma-Aldrich, St. Louis, MO), 100 units papain (Sigma-Aldrich, St. Louis, MO), and 2 mg cysteine (Sigma-Aldrich, St. Louis, MO). Sterile filter the solution into a 15 mL centrifuge tube by using a syringe filter with a 10 mL syringe. For activation of the enzyme, incubate the enzyme solution at 37°C for 45 min.
12. Newborn rat pups (1–3 days old).
13. Dissecting solution: place 50 mL ice-cold Hanks Balanced Salt Solution (HBSS, Invitrogen, Carlsbad, CA) into a 50 mL centrifuge tube. Add 50 μL HEPES free acid (1 M sterile stock) which adjusts the pH to 7.2.
14. Dissecting tools: one pair of medium and one pair of fine scissors, small scalpel, small spatula, two very fine forceps.
15. 70% ethanol prepared from 95% denatured ethanol.
16. Sterile petri dishes 100 and 60 mm in diameter (BD Bioscience, San Jose, CA).
17. Triturating solution: add 100 μL of sterile 50 mM EDTA to 5 mL of Neurobasal medium.
18. 0.05% trypsin–EDTA solution (Invitrogen, Carlsbad, CA, store at −20°C).
19. Trypan Blue Stain (Invitrogen, Carlsbad, CA).
20. Hemocytometer (Fisher Scientific, Pittsburgh, PA).
21. 75 cm^2 flask with a plug-seal cap (BD Bioscience, San Jose, CA).
22. FUDR stock: 5 mg of 5-fluoro-2-deoxyuridine and 12.5 mg uridine dissolved in 25 mL of Neurobasal A medium. Sterile filter using a syringe filter and a 30 mL syringe. Store 300 μL aliquots in sterile microcentrifuge tubes at −20°C.

2.1.3. Purification of Astrocytes and Preparation of Feeding Layer

1. 10% FCS medium.
2. Vortex with a flask holder accessory (LabScientific, Livingston, NJ).
3. Aluminum foil.

4. 0.05% trypsin–EDTA solution (Invitrogen, Carlsbad, CA, store at −20°C).
5. Trypan Blue Stain (Invitrogen, Carlsbad, CA).
6. Hemocytometer (Fisher Scientific, Pittsburgh, PA).

2.2. Activity-Dependent Labeling of Synaptic Vesicles

1. CNQX-saline: 136 mM NaCl, 2.5 mM KCl, 10 mM glucose, 10 mM HEPES-NaOH pH 7.2, 1 mM $CaCl_2$, 3 mM $MgCl_2$, and 10 μM CNQX (Tocris, Ellisville, MO).
2. Stimulation saline: 96 mM NaCl, 40 mM KCl, 10 mM HEPES-NaOH pH 7.2, 2 mM $CaCl_2$, 1.3 mM $MgCl_2$, and 20 μM FM1-43 (Invitrogen, Carlsbad, CA) (*see* **Note 3**).
3. "Stimulation" chamber (*see* **Note 4**).
4. Fixative: 4% paraformaldehyde, 0.1–5% glutaraldehyde (*see* **Note 5**), 0.2 mM $CaCl_2$, 90 mM cacodylate–HCl pH 7.4.
5. Shaker.

2.3. Photoconversion

1. Tris-buffered saline (TBS): 25 mM Tris–HCl pH 7.5, 150 mM NaCl.
2. 0.15 mg/mL diaminobenzidine (DAB; *see* **Note 6**) in TBS.
3. Fluorescence microscope with camera (Olympus, Center Valley, PA).
4. Fluorescent microscope filter (super-wide blue excitation 420–480 nm, dichroic mirror 500 nm, emission 515 nm long pass (Olympus, Center Valley, PA) (*see* **Note 7**).
5. Neutral density filter (ND25) (Olympus, Center Valley, PA).
6. 24-well culture plate (BD Bioscience, San Jose, CA).
7. Karnovsky fixative (9): 4% paraformaldehyde, 5% glutaraldehyde, 0.2 mM $CaCl_2$, 80 mM cacodylate–HCl pH 7.4.

2.4. Histological Processing of Neuronal Cell Culture for Electron Microscopy

1. 4% Osmium tetraoxide (OsO_4, Electron Microscopic Sciences (EMS), Hatfield, PA, *see* **Note 8**). 80 mM cacodylate–HCl pH 7.4 with and without 1% potassium ferrocyanide.
2. 90 mM cacodylate–HCl pH 7.4, 0.2 mM $CaCl_2$.
3. 2% uranyl acetate in water.
4. Ethanol, 100%.
5. Ascending ethanol series 50, 70, and 90% prepared from 100% ethanol.
6. Propylene oxide (EMS, Hatfield, PA).
7. 8 mL snap-cap glass vials (EMS, Hatfield, PA).
8. Resin mixture for embedding. Prepare separately (a) Epon resin: 20 g EMbed-812, 11 g DDSA, 9 g NMA, 0.8 mL

DMP-30 and (b) Spurr resin: 10 g ERL, 6 g DER, 26 g NSA, 0.4 mL DMAE (all from EMS, Hatfield, PA). When both resins are completely mixed, combine Epon and Spurr resin batches and mix thoroughly.

9. Aclar film (SPI supplies, West Chester, PA).
10. BEEM capsules, size 00 (EMS, Hatfield, PA).
11. Concentrated hydrofluoric acid.

2.5. Sectioning

1. Diamond knife (EMS, Hatfield, PA).
2. Ultramicrotome (Leica, Bannockburn, IL).
3. Double edge stainless steel razor blades (EMS, Hatfield, PA).
4. Copper slot grids (2×1) Formvar coated (EMS, Hatfield, PA).
5. Forceps for grid handling.
6. Grid storage box (EMS, Hatfield, PA).
7. Hanaichi lead solution (10) (*see* **Note 9**): Calcined lead citrate 0.20 g, lead nitrate 0.15 g, lead acetate 0.15 g, sodium citrate 1.00 g, distilled water 41.00 mL.

   Prepare calcined lead citrate by heating crystal lead citrate for several hours in a melting pot (200–300°C) until the color changes to a light brownish yellow. Overheated lead citrate with a dark brownish or black color cannot be used.

   Place the cooled calcined lead citrate and other reagents in a 50 mL volumetric flask and mix well to produce a yellowish milky solution. Add 9.0 mL of 1 N NaOH to the solution and mix well until the solution becomes clear with a light yellowish color. Transfer the solution to an amber glass with a screw cap bottle for storage.
8. Parafilm.

3. Methods

3.1. Hippocampal Cell Culture

When preparing and working with cell cultures general sterile working practices have to be employed to avoid contamination of the cultures with fungi or bacteria. Sterile working practices include the use of a biosafety cabinet, sterile tools (e.g., the sterilization of the dissecting tools with 70% ethanol), sterile supplies (like culture wells and pipettes), and sterilized solutions. Many solutions can be bought sterile but it is a good working habit to sterile filter solutions before use.

Glia cells especially astrocytes promote neuronal survival and growth. We therefore prepare an astrocyte feeding layer first, on top of which neurons are seeded. The preparation of the feeding layer takes about 2 weeks and includes proliferation of glia cells and the purification of astrocytes. When astrocytes have grown to a confluent monolayer on coverslips neurons are seeded and the medium is changed to Neurobasal which promotes neuronal growth. We use neuronal cells after 14–21 days in vitro and it thus takes 4–5 weeks before experiments can be done.

In our lab we prepare cell cultures every week. The hippocampus is dissected and the cells are dissociated. From this batch of hippocampal cells some cells are placed into a flask with a medium that promotes astrocyte growth and some cells are seeded on top of the feeding layer that was prepared during the previous 2 weeks. When starting the cell culture routine, however, only coverslips and astrocytes are prepared during the first 2 weeks. Starting with the third week neurons can be plated. A wealth of additional information on neuronal cell culture can be found in reference (11).

3.1.1. Preparation of Coverslips

1. Several days before preparing a 24-well plate, clean coverslips in concentrated nitric acid on a shaker or rotator for 1 h. Replace nitric acid with concentrated hydrochloric acid and clean for at least 1 h. Concentrated acids are strong corrosives and therefore should be handled with caution and discarded properly.
2. Wash the coverslips thoroughly with demineralized water until the pH does not become acidic when the coverslips are swirled.
3. Wash the coverslips once in 95% ethanol and store in 95% ethanol.
4. The day before splitting astrocytes (**Section 3.1.3**, Step 4), prepare a 24-well plate by flaming coverslips and placing one coverslip into each well (*see* **Note 10**). This and all following steps are carried out inside the biosafety cabinet.
5. Place ~2 μL of coating solution in the center of each coverslip, then distribute evenly with a sterile blunt-ended plastic or wooden stick.
6. Dry the 24-well plate in a biosafety cabinet for 20 min.
7. Add ~500 μL of sterile demineralized water to each well to wash coverslips. Remove water.
8. Sterilize the 24-well plate with UV light in a biosafety cabinet for 20 min (lid removed).

9. Add a few drops of 10% FCS medium so that the coverslips are just covered. Place the 24-well plate into the incubator for use the next day.

3.1.2. Cell Culture Preparation

1. Set up a dissecting microscope in a biosafety cabinet and sterilize dissecting instruments with 70% ethanol.
2. Place three 60-mm petri dishes into three 100-mm petri dishes and surround the 60-mm dishes with ice. Add dissecting solution to the 60-mm dishes.
3. Decapitate the pup with the medium-sized pair of scissors and cut open the scalp with the scalpel. Using the fine scissors, cut the skull along the fissure between the hemispheres. Using the small spatula, remove the brain directly into the ice-cold dissection solution.
4. With the sterilized scalpel, cut off the brain stem and divide the cortices along the mid-sagittal plane. Transfer the cortices to the second ice-cold 60 mm dish.
5. Under the dissecting microscope, place a hemicortex on its mid-sagittal plane and remove the meninges with the two fine forceps starting at the anterior cortex. When the posterior cortex is reached, flip the cortex and clean the hippocampus, which is curled beneath the cortex. Cut both sides of the hippocampus and roll it out. Using the scalpel (or forceps), separate the hippocampus from the cortex.
6. Transfer the hippocampus into the third dish with ice-cold dissection solution and chop it into 0.5 mm pieces. Repeat for the second hemisphere.
7. Collect the pieces with a pipette and transfer them into the prepared enzyme solution.
8. Incubate in a water bath or in the incubator at 37°C with intermittent agitation for 60 min.
9. To remove the enzyme, wash the hippocampus pieces as follows: Pellet the pieces using gentle centrifugation ($\sim 10\times g$), remove the supernatant by aspiration, and replace it with Neurobasal medium. Centrifuge again and remove the supernatant.
10. With a plastic transfer pipette, add 2 mL of triturating solution and triturate the tissue carefully, avoiding air bubbles. Allow the bigger pieces to settle and transfer the upper portion of the supernatant (about 1 mL) to a fresh 15 mL centrifuge tube. The gentler this step is done, the higher the survival rate of the neurons. Bring the volume back to 2 mL by adding triturating solution.
11. Repeat Step 10 until no more tissue pieces are left.

12. Mix 50 μL of cell suspension with 50 μL of trypan blue solution and count the live cell density, cells not stained with trypan blue, using the hemocytometer (*see* **Note 11**). Count four squares: $N/4*2*10^4$ is the cell number per mL.
13. To grow neurons, plate cells in the prepared 24-well plate with an astrocyte feeding layer at a density of 15,000 per well in 500 μL of Neurobasal medium. For the preparation of the astrocyte feeding layer see steps under **Section 3.1.3**.
14. Seed 1 million cells into a 75 cm^2 plug-seal cap flask with 10 mL of 10% FCS medium to grow astrocytes.
15. Place the 24-well plate and the flask into a CO_2 incubator (5% CO_2, 37°C). The cap of the flask must be slightly unscrewed to allow for CO_2 exchange.
16. The next day, change the Neurobasal medium in the 24-well culture plate.
17. After 3–4 days in vitro add 12.5 μL FUDR to each culture well. FUDR blocks DNA synthesis which prevents cell proliferation.

3.1.3. Purification of Astrocytes and Preparation of Astrocyte Feeding Layer

1. Change the 10% FCS medium in the astrocyte flask (Step 14, **Section 3.1.2**) the day after seeding cells and allow the cells to grow for 1 week.
2. Tightly screw the cap and wrap the flask in aluminum foil. Place the flask on a vortexer or rotary shaker and vortex (~300–500 rpm) overnight to detach all cells except astrocytes from the bottom of the flask. The vortex speed can be increased if not all non-astrocyte cells detach from the bottom (*see* **Note 12**).
3. The next morning, remove the medium containing all detached cells and add 10 mL of 10% FCS medium. Allow the astrocytes to grow for another day in the incubator.
4. Split astrocytes: remove the medium by aspiration and wash once with 2 mL of warmed 0.05% trypsin/EDTA solution. Aspirate this solution and add an additional 2.5 mL of 0.05% trypsin/EDTA solution. Put the flask back into the incubator until most cells have detached from the bottom (a few min). Using a 5-mL sterile pipette, add 2.5 mL of 10% FCS medium and triturate a few times.
5. Transfer the suspension to a 15 mL conical centrifuge tube and gently pellet cells in a centrifuge at $\sim 10\times g$ for ~1 min.
6. Discard the supernatant and resuspend cells in 2–5 mL of 10% FCS medium.

7. Mix 50 μL of the cell suspension with 50 μL trypan blue. Count astrocytes with a hemocytometer and plate 15,000 cells in 500 μL 10% FCS medium into each well that contain the prepared coverslips of **Section 3.1.1**.
8. Allow astrocytes to grow for a few more days until they form a confluent feeding layer.

3.2. Activity-Dependent Labeling of Synaptic Vesicles

1. After 14–20 days in vitro place an individual cell culture coverslip into a "stimulation" chamber filled with CNQX-saline inside the biosafety cabinet (*see* **Note 13**). The following steps can be carried out at the regular lab bench.
2. Replace CNQX-saline with stimulation saline and incubate for 2 min. This will label all synaptic vesicles (SV) that are retrieved from the plasma membrane after neurotransmitter release with FM1-43.
3. Wash off the stimulation saline by carefully rinsing with CNQX-saline for 10–15 min before fixation. Use one thorough quick wash first to end stimulation followed by three additional changes of saline over the 15-min period.
4. Aspirate the CNQX-saline and add fixative to the stimulation chamber. Fix for 20 min with gentle agitation. The specimens may be protected from light with aluminum foil.

3.3. Photoconversion

1. Thoroughly wash coverslips three times in TBS. Residual fixative may cause DAB to precipitate!
2. Place coverslip back into the stimulation chamber that will fit under the microscope for photoconversion. Add 0.15 mg/mL of DAB in TBS and place on a shaker for 5 min.
3. Place the chamber with the coverslip under an epifluorescent microscope and identify a region of interest using a ×40 objective. To photoconvert, irradiate with blue excitation light for 6–12 min. (*see* **Note 14**).
4. Monitor the progress of photoconversion by occasionally switching to bright-field illumination. Although the small number of SVs that accumulate DAB polymer cannot be seen in the bright-field image, over time neurons and astrocytes take on a brownish color. When neurons show a light brownish tint, photoconversion is complete. Longer conversion times will cast a brown stain over the cells, which will also be visible as background in electron micrographs. A small amount of such background does not interfere with the SV labeling or with the fine structure. Extended conversion times, however, result in deeply stained cells under the light microscope. Such dark specimens have a strong

background in the EM and ambiguous SV labeling (*see* **Note 15**).

5. Photoconversion is stopped by turning off the excitation light.
6. With the photoconverted region still in place, take a reference image at 40x followed by a 10x image. These images will be used later to identify the photoconverted region after embedding.
7. Wash thoroughly three times with TBS and place coverslip in 24-well cell culture plate in Karnovsky fixative until cells are embedded for EM.

3.4. Histological Processing of Neuronal Cell Culture for Electron Microscopy

1. Wash coverslips in the culture wells three times in cacodylate buffer. During the following steps the coverslips remain in the original culture wells until they are transferred to a glass vial as described in **Section 3.4**, Step 10 (*see* **Note 16**).
2. Postfix in 500 μL of 1% OsO_4 plus 1.5% potassium ferrocyanide in 90 mM cacodylate–HCl pH 7.4. Place on a shaker and incubate at room temperature for 20–60 min. Longer fixation times may result in higher contrast of the final EM graphs.
3. Wash three times in cacodylate buffer.
4. Postfix in 1% OsO_4 in 80 mM cacodylate–HCl pH 7.4 on a shaker for 20–60 min.
5. Wash once in cacodylate buffer, and then wash thoroughly with demineralized water.
6. For block contrasting stain neurons in 400 μL of 2% uranyl acetate for 1 h.
7. Wash thoroughly with water.
8. Dehydrate in an ascending series of 50%, 70%, 90%, and 100% ethanol by successively changing solutions in the culture wells. Incubate each ethanol step for 5–10 min. Repeat the 100% ethanol step three times to ensure complete dehydration.
9. In the meantime, prepare 8-mL snap-cap glass vials with ~2 mL of propylene oxide, and place a pencil written paper label identifying the sample into each vial.
10. Transfer the coverslips from the 24-well plate to the glass vials containing propylene oxide. Ensure that each coverslip is on top of the paper label with the cells facing up and fully covered by the propylene oxide. Place on a rotator for 10 min.

11. Replace the propylene oxide with a mixture of propylene oxide and resin (1+1). Place on a rotator for 10–20 min.
12. For resin infiltration, replace the propylene oxide/resin mixture with pure resin. Place on a rotator for 15 min. Exchange pure resin two more times to ensure complete removal of propylene oxide. Extend the last infiltration step to 60 min.
13. Place a piece of Aclar film on a microscope slide lightly coated with resin so that the film is held in place by a small amount of resin between the slide and the film. Place a coverslip next to its label on top of the Aclar film, cells facing up. Under a stereo microscope confirm that the cells are facing up!
14. Add some more resin to the coverslip and cover the cells with a second piece of Aclar film. Complete the sandwich with a second glass slide and place it inside an oven. The used glass vials can be used as weights for even, flat embedding.
15. Polymerize at 60°C for 16–48 h.
16. Remove the top slide and the top piece of Aclar film and discard. Under a microscope, identify the photoconverted area using the reference images (*see* **Section 3.3**, Step 6). The arborization pattern of neurons can be used to identify the photoconverted neurons even when the area is not recognizable by the brownish DAB polymer. Surround the area with pencil marks and place a resin-filled beam capsule upside-down on top of the marked region. Polymerize at 60°C for another 16–24 h.
17. Break the BEEM capsule off of the glass slide. This leaves the coverglass and the cells attached to the BEEM capsule.
18. Clean the glass with a razor blade to remove any resin and place the block face into a small amount of concentrated hydrofluoric acid for 20 min or until the glass is completely dissolved. *Wash thoroughly!* (*see* **Note 17**). The monolayer of cells is now exposed and can be trimmed and sectioned.

3.5. Sectioning

1. Remove the block from the BEEM capsule and mount with the cells facing up into a specimen holder.
2. Under a microscope identify photoconverted neurons within the pencil marks and mark that region with a razor blade. The marked area should be a trapezoid with a height of about 0.4 mm and a width of about 0.7 mm. A trapezoid of this size allows short bands of sections to fit onto the slot grids.

3. Trim the block until only the trapezoid remains. For the final perpendicular cut use a new blade and cut perpendicular to the surface of the base and top of the trapezoid.
4. Fasten the block into the ultramicrotome and align the block face *exactly* parallel to the diamond knife edge (*see* **Note 18**).
5. Cut thin sections (50–60 nm) right from the start. Although they may not be complete the very first sections already contain cells.
6. Collect sections in series by placing short ribbons on Formvar-coated slot grids. The longitudinal axis of the ribbon should be parallel to the longitudinal axis of the 2 × 1 mm slot grid. None of the sections should be placed on the grid's rim. Also, ensure that no sections are lost between grids.
7. Let the sections dry before continuing with on-section staining.
8. Place a 40 μL droplet of the Hanaichi lead solution on a flat piece of parafilm. Place a slot grid with the sections facing down on top of the droplet and stain for 3 min. Longer staining time will increase contrast.
9. Wash thoroughly with distilled water and let dry.
10. Examine serial sections in the electron microscope: Align slot grids in the specimen holder such that the series of sections is aligned with the direction of movement of the specimen holder. Choose a section in the middle of a continuous series of sections. Photograph an area with several synapses that contain photoconverted SVs. Write down the perpendicular distance of that area toward the margins (top, bottom, left, and right) of the section. Note the branching pattern of dendrites and any landmark structures that may be helpful to identify the photographed area in the following sections. Forward the specimen holder to the next section and find the same area on that section by using the distance toward the margins of the section, structural landmarks, and the EM micrograph. Photograph the same synapses. Advance to the next section and repeat procedure until a sufficient number of synapses are fully contained in the series of micrographs. The series can be expanded in both directions of the first EM micrograph taken. Change to additional slot grids as needed to complete the series. **Figure 26.1** shows an example of synapses that contain synaptic vesicles containing the electron-dense DAB polymer that formed during photoconversion of FM1-43.

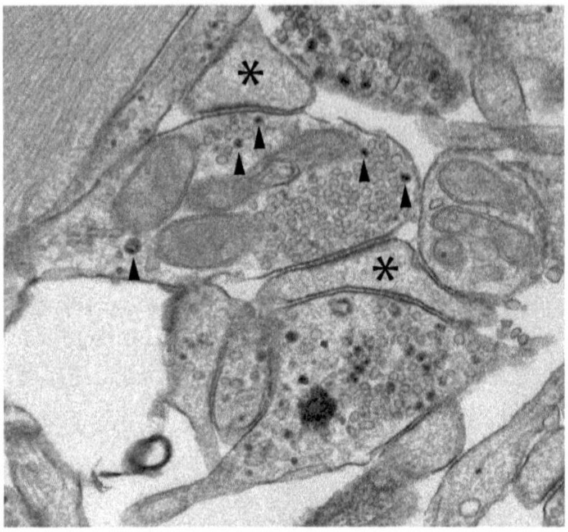

Fig. 26.1. Synapses of cultured hippocampal neurons are shown after photoconversion of FM1-43. Photoconverted synaptic vesicles that contain the DAB polymer are electron-dense (*arrowheads*), whereas unlabeled vesicles have the typical electron-lucent appearance. For clarity, the electron-dense synaptic vesicles are indicated by arrowheads in only one synapse. The asterisks indicate postsynaptic spines.

4. Notes

1. Both 10% FCS medium and Neurobasal medium are stable at 37°C for ~2–3 weeks. For extended storage, media should be refrigerated at 4°C.
2. Tissue-culture treated plastic provides excellent conditions for neuronal cell culture. However, glass provides superior optical characteristics for high-resolution light microscopic work and is resistant to organic solvents during EM processing. To ensure optimal cell growth on glass cover slips, selection of glass sold specifically for cell culture is recommended.
3. The saline can be stored at 4°C, but $CaCl_2$ and FM1-43 should be added just before use.
4. Any small plastic (e.g., petri dish) or glass dish can be used as a "stimulation chamber." The smaller the volume of the dish, the more cost-effective is the staining procedure. We have been using a small self-made plastic dish that tightly fits our cell-culture coverslips and fits into the slide holder of a microscopic stage. With these chambers we use less than 200 μL of stimulation saline per coverslip.
5. EM grade 16% paraformaldehyde solution (Electron Microscopic Sciences (EMS), Hatfield, PA) and 25% glutaraldehyde solution (EMS) can be purchased in 10 mL

ampoules. This allows for safe handling and highest quality of the reagent.

The concentration of glutaraldehyde in the fixative can be varied from 0.1 to 5%. Higher glutaraldehyde concentrations increase nonspecific fluorescent background. If FM1-43 fluorescence must be observed or measured before photoconversion, use 4% paraformaldehyde and 0.1% glutaraldehyde in cacodylate buffer to abolish fluorescent background. High glutaraldehyde concentrations, on the other hand, facilitate the preservation of the ultrastructure and do not interfere with photoconversion. The background fluorescence photoconverts more slowly than the FM dye, thus the background fluorescence will not be visible in the EM. However, very high levels of background fluorescence limit the visibility of the FM1-43 stain in the light microscope and may result in nonspecific DAB polymers that could occlude the fine structure.

6. DAB is very stable when dissolved in TBS and can be frozen (at −20°C) for several weeks. The slightly higher pH increases the stability of the DAB solution and reduces background staining. Successful photoconversion relies on high-quality DAB. Many groups have failed when using solid DAB purchased from Sigma-Aldrich. Our trusted vendor, unfortunately, recently discontinued the production of our preferred DAB tablets and no recommendation regarding vendors can be made at this point.
7. We have successfully used the recommended excitation filter (super-wide blue) for many years. The wide spectrum of the filter includes an intensity peak of the mercury-arc light source at 420 nm. Although this peak is placed somewhat away from the absorption maximum of FM1-43 (479 nm), at 420 nm absorption is still 40% of the maximum. Therefore the filter provides a higher light intensity for photoconversion than that of a narrower filter. However, narrower excitation filters have been used successfully by other groups. The duration of the photoconversion for each filter must be determined experimentally.
8. Osmium can conveniently be bought in 2-mL ampoules that ensure optimal quality and safety in the laboratory.
9. A stable lead solution (9) is based on calcined lead citrate. This lead solution is stable and free from precipitates during storage and staining when kept at room temperature or in the refrigerator for 1 year. It also provides high contrast on-section staining.
10. Flaming the glass removes residual alcohol and leaves a small film of carbon on the glass surface which facilitates the coating of the coverslips.

11. Trypan blue is a stain for live cells. Dead cells take up dye very heavily and thus should not be included when counting live cell density.
12. Astrocytes are amazingly well attached to the flask surface, and vigorous shaking does not harm them. However, they need to be protected from light, hence the aluminum foil. Also, the pH must not change or the cells will die. Therefore, a plug-seal cap flask should be used and tightly sealed to prevent leakage of CO_2 when vortexing cells.
13. Neurotransmitter release and SV recycling is an ongoing process triggered by action potentials in living neurons. To prevent nonspecific or unintended labeling of SV, neuronal activity must be suppressed. The suppression of neuronal activity is also necessary to abolish destaining of SV during the wash period after stimulation. Neuronal activity is suppressed by making three changes to the standard extracellular saline. First, the glutamate receptor antagonist CNQX is added which will suppress postsynaptic activity that may lead to neuronal activity. Second, the membrane potential is slightly hyperpolarized to prevent spontaneous action potentials by reducing the potassium concentration from 5 to 2.5 mM. Third, the calcium concentration is lowered to 1 mM and the magnesium concentration is raised to 3 mM. This change lowers the release probability which impedes vesicle fusion during activity.

    Once coverslips are immersed in CNQX-saline activity is suppressed almost instantaneously; thus experiments can be continued right away after transferring cells from the biosafety cabinet to the lab bench.
14. FM dyes bleach rapidly, it is therefore not advisable to examine the specimen for an extended period of time before photoconversion. During photoconversion the dye will bleach within the first few minutes, however, at that time photoconversion is not yet complete. The final conversion time depends on several factors. The duration of the photoconversion depends on the light intensity. When using a mercury-arc light source, converting for 8–10 min with a neutral density (ND25) filter in place gives reliable results. Omitting the neutral density filter results in much higher light intensity and shorter conversion times. Very high light intensities may produce additional nonspecific background and may make it more difficult to control staining intensity. Any changes that alter the excitation light intensity alter photoconversion times. For example, the mercury bulb produces less light with age, and photoconversion times should be adjusted accordingly. Light intensity also depends on the magnification of the objective. When using a 60x objective, the light intensity that

hits the specimen is higher (more focused on a smaller area) and photoconversion times will be shorter. When using a 20x lens, photoconversion will take ~30–40 min. A smaller objective than 20x is not recommended for FM1-43 photoconversion.

15. If an upright microscope with water immersion lenses is used, DAB polymer will also precipitate on the lens's surface. Therefore, the water immersion lens must be cleaned after every photoconversion session. Failure to clean the lens will reduce the intensity of the excitation light, which will eventually lead to unsuccessful photoconversion.
16. Handling the coverslips in a 24-well culture dish is easy and convenient. Lifting the coverslips off the bottom will ensure that trapped solution beneath the coverslip will mix properly when changing solutions. 400 μL of solution is needed to fully immerse a coverslip within a well of a 24-well plate.
17. Even residual amounts of hydrofluoric acid can blind the surface of optical lenses over time. It is therefore recommended to store treated blocks away from optical instruments.
18. It is very important that the block face is properly aligned with the diamond knife. This can be achieved by observing the backlight as it passes between the knife and the block face. This light band must have the exact same width from left to right. Also, the distance of the knife must be equal at the top and at the bottom of the block. This can be achieved by advancing the block at its bottom position toward the knife until the light band just disappears. Retract the block in 10 equal steps and move the top of the block to the knife edge. Advance the block toward the knife while counting the steps. When less than 10 steps are needed before the light band just disappears, the top part of the block face needs to be tilted away from the knife. When the light band is still visible after 10 steps, the top of the block face needs to be tilted toward the knife. Repeat these steps until the distance to the knife is equal at the bottom and at the top of the block face. During this procedure great caution must be taken not to damage the diamond knife by moving the block into the knife edge!

Acknowledgments

This work was supported by the NIH grants U54 NS039408 and R210263208.

References

1. Sandell, J. H. and Masland, R. H. (1988) Photoconversion of some fluorescent markers to a diaminobenzidine product. *J. Histochem. Cytochem.* **36**, 555–559.
2. Cochilla, A. J., Angleson, J. K., and Betz, W. J. (1999) Monitoring secretory membrane with FM1-43 fluorescence. *Annu. Rev. Neurosci.* **22**, 1–10.
3. Betz, W. J. and Bewick, G. S. (1992) Optical analysis of synaptic vesicle recycling at the frog neuromuscular junction. *Science* **255**, 200–203.
4. Harata, N. C., Aravanis, A. M., and Tsien, R. W. (2006) Kiss-and-run and full-collapse fusion as modes of exo-endocytosis in neurosecretion. *J. Neurochem.* **97**, 1546–1570.
5. Zakharenko, S. S., Zablow, L., and Siegelbaum, S. A. (2001) Visualization of changes in presynaptic function during long-term synaptic plasticity. *Nat. Neurosci.* **4**, 711–717.
6. Henkel, A. W., Lubke, J., and Betz, W. J. (1996) FM1-43 dye ultrastructural localization in and release from frog motor nerve terminals. *Proc. Natl. Acad. Sci. USA* **93**, 1919–1923.
7. Schikorski, T. and Stevens, C. F. (2001) Morphological correlates of functionally defined synaptic vesicle population. *Nat. Neurosci.* **4**, 391–395.
8. Harata, N. C., Ryan, T. A., Smith, S. J., Buchanan, J., and Tsien, R. W. (2001) Visualizing recycling synaptic vesicles in hippocampal neurons by FM 1-43 photoconversion. *Proc. Natl. Acad. Sci. USA* **98**, 748–753.
9. Karnovsky, M. J. (1965) A formaldehyde-glutaraldehyde fixative of high osmolarity for use in electron microscopy. *J. Cell Biol.* **27**, 137A.
10. Hanaichi, T., Sato, T., Iwamato, T, Malavasi-Yamashiro, J., Hoshino, M., and Mizuno, M. (1986) A stable lead by modification of Sato's method. *J. Electron Microsc.* **35**, 304–306.
11. Banker, G. and Goslin, K., eds. (1998) Culturing nerve cells. *Cellular and Molecular Neuroscience Series.* MIT Press, Boston, MA.

Subject Index

A

Adjuvant
Freund's 99
Titermax Gold 73, 90
Affinity chromatography
antipeptide antibody purification 103
ascites fluid 89
glutathione resin 6
GST fusion protein isolation 6
His-tagged tobacco etch virus (TEV) protease isolation 5
IgG subtype isolation 89
Ni-Sepharose 9
p-aminobenzamidine–agarose 9, 16–17
peptide affinity purification 100–101, 103–104, 106
peptide sepharose column 100, 103
polyhistidine (His_6) fusion protein isolation 30
protein binding 6, 30, 81
protein elution 19, 89
protein A resin 81, 89
thrombin isolation 16
Agrobacterium tumefaciens
culture 27
transformation 30
Alexa fluor conjugate 58–59, 142–143
Aluminum planchette 182, 188
AMIRA DEV 3D rendering software, *see* Image analysis software
Ammonium sulfate precipitation 66, 70, 72
Anesthetic 74, 176, 239, 242
Animal
fixation 138, 167–169, 172–173, 177–178
cryo 157
immersion 242
perfusion 242, 246
Antibiotics
8-azaguanine 77, 90
carbenicillin 7, 24
chloramphenicol 7
hygromycin 23
penicillin 47, 77–80, 317, 331
streptomycin 44, 77–80, 317, 331
Antibody
antibiotin 170, 288
antipeptide 93–107
peptide affinity purification 100–101, 103–104, 106
peptide selection 94–95
detection 60, 145, 177, 249
chemiluminescent 60
fluorescent microscopy 333
immunoelectron microscopy 54
dilution 73, 91, 102, 132, 161, 164, 208, 213, 273
monoclonal 63, 81, 88–89, 186, 245, 288
anti-biotin 175
ascites fluid
cell fusion 82–83, 86, 90–91
hybridoma, *see* Hybridoma
IgG isolation 89
immunization
nanogold conjugation 135, 146, 283
production 80, 88
splenocyte isolation 85–91
storage 106
phosphospecific 94
preimmune serum 132, 273
primary polyclonal 22, 43–44, 63–75, 93–94, 99, 104, 183, 185, 189, 231, 244–245
purification 43, 63–75
purification, ammonium sulfate precipitation ... 66, 70, 72
purification, IgG fractionation
secondary 54, 56–58, 65, 67, 69, 72, 95–96, 100, 102, 110, 136, 158, 161–164, 183, 185, 188, 195, 198, 200–201, 203, 208, 213, 229, 231–232, 252, 254–257, 271, 278, 281, 283, 293, 299, 302
Alexa fluor 142–143
chemiluminescent detection 49
dilution 73, 91, 102, 132, 161, 164, 208, 213, 273
Fab' nanogold 136, 147, 152, 227, 229, 233
fluorescent 44, 142, 329–330
fluoronanogold 136, 139
horseradish peroxidase (HRP) labeled 218, 316, 322
nanogold conjugation
storage 106
titer determination 100, 102–103, 257
Antigen colocalization 55, 57, 249–250, 256
Antigenicity prediction software 94
Antipeptide antibody, *see* Antibody
Arabidopsis thaliana
Agrobacterium transformation 22–26
calcium–PEG transfection 35, 37, 40
cell wall digestion 35, 37, 39
cultivation 23, 25
protoplasts 33–42
seed collection 24, 28
transfection 35
transformant selection 23

S.D. Schwartzbach, T. Osafune (eds.), *Immunoelectron Microscopy*, Methods in Molecular Biology 657,
DOI 10.1007/978-1-60761-783-9, © Springer Science+Business Media, LLC 2010

Ascites fluid
affinity chromatography 89
IgG isolation 89
isolation 89
production 80, 88

B

Beem capsule 131, 295, 319, 323, 334, 340
Benzil 169, 174, 176
Biotinylated DNA 288, 295
Biotinylated protein 170
Blocking agents 309
Block trimming 131
Blood collection
cardiac puncture 68
ear vein 67–68
mouse 91
rabbit 65, 67–69, 74
Brain 133–215, 246, 330, 336

C

Calcined lead citrate 138, 143, 319, 325, 334, 343
Carbohydrate localization 4, 217–224
Carbon coating 209, 343
Cardiac
perfusion 135, 242
puncture 65, 68–70
Castor bean
endosperm 124, 129
pericarp 124, 129
Cell counting 84, 326, 344
Cell fusion 82–83, 86, 90–91
Cell lines
astrocytes 335, 337–338
glia 335
HEK 57, 318, 320
HeLa 46, 51, 54–57
hippocampal cell 330–338
hybridoma 75, 79–80, 87–88
murine myeloma 77, 89
primary neuronal cells 330
rat kidney fibroblasts 276
splenocytes 85–86, 90–91
Cell lysis
mammalian cells 47–48, 52
dounce homogenizer 59
RIPA buffer 52
sonication 18, 59
E. coli 11, 14, 37
CELLocate 228, 232, 234
Cell wall digestion
Arabidopsis thaliana 33
cellulase 37
macerozyme 37
Centrifugation 9–14, 16, 27, 35, 38–41, 52, 86, 116, 118, 239, 241–242, 253, 263–264, 291, 300, 336
CsCl–ethidium bromide 35
Chemiluminescence 54
Chlamydomonas reinhardtii 267
Chromium coating 300, 302, 305–306, 311
Cloning 30, 44, 79, 87, 91
CNBr-activated Sepharose 4B 100, 103
CNQX-saline 333, 338, 344
Colloidal gold
antibody conjugation 110
preparation 110–113, 115, 117–118
protein A conjugation 109–119, 208, 213
size determination 44
streptavidin conjugation 109–119, 218
Colocalization 55, 57, 249–257
Computer image reconstruction 260, 263, 270
Contrast enhancement 186
Coomassie stain 6, 9
Correlated-fluorescent electron microscopy 135
Correlated-light electron microscopy 315–326
Coverslip 47, 51, 55–57, 59–61, 84, 226, 228, 230–232, 234, 276–277, 279–282, 285, 304, 310, 317–318, 320–326, 330–331, 335–336, 338–340, 342–345
Critical point drying 300, 302, 305
Cryobiopsy 168–170, 174, 178
Cryofixation 156–160, 177–178, 181, 186–187, 191, 198–199
Cryoforceps 169–170, 174, 176, 178
Cryogen preparation 168, 171–172
Cryoknife 169, 173–174, 178
Cryosections 191, 250, 256
Cryostat 219–220
Cytoplasmic membrane leaflet 210
Cytoskeleton 275–276, 282, 285, 300–301, 304

D

DAB, *see* Diaminobenzidine
Deep etching 285
Dehydration 123, 135, 137, 140, 143, 153, 167–169, 176, 191–192, 232, 238–239, 242–243, 246–247, 295, 299, 323, 339
Detergents 11, 14–15, 59, 134–135, 142, 215, 298–299
Dialysis 8, 15, 66
Diaminobenzidine
horseradish peroxidase substrate 218, 318
photoconversion 329, 333
synaptic vesicle label 333, 341–342
Diamond knife 125, 127, 137, 142, 151, 183, 185, 228, 231, 240, 251, 254, 262, 265, 319, 324, 326, 334, 341, 345
3D image reconstruction 271
Dimethylformamide (DMF) 98, 100–102, 239, 242, 246
DNA
biotinylation 288, 295
concentration 39, 47, 58
hybridiaztion 288–289, 291–294
isolation 35, 39
CsCl–ethidium bromide gradient 35
Dry fracture 305

E

Electron microscopy *in situ* hybridization 290, 292–294
Electrophoresis 6, 9, 25, 48, 53–55, 66–67, 70–73, 107
SDS-polyacrylamide gel 48–49
Electroporation 23, 26, 30, 58–59
ELISA, indirect 64–65, 68–69, 76–77, 79, 82–83, 87, 91
Embedding
algae 261
brain slices 133–144
cells 123–124, 134, 146, 148–152
Epon 137, 183, 319

epoxy 131, 148, 150–153, 157, 183, 185–187, 189, 226, 228, 230–234, 246–247, 260, 272
freeze substituted tissue 174
gelatin 127, 131, 158, 160–161, 194, 221, 239, 242–243, 253, 256, 264
hydrogen peroxide etching 131, 273
Lowicryl HM20 .... 157, 182, 186, 202
Lowicryl K4M 182, 189
LR-gold 169
LR White 160, 163, 238, 246–247, 289, 291, 295
mammalian 133–144
plant 123–125, 127
polymerization 131, 157, 189, 246, 272, 295
resin 151, 155, 186–187, 225, 238, 292
Saccharomyces cerevisiae 192, 196, 198–199
spurr 137, 319, 334
tissue 238
tissue culture 57, 145–154, 181–190, 317
attached 148
cover slip 228, 230–232, 234, 276, 279
culture dish 47, 51–52, 77, 182, 207, 210, 250, 253, 276, 279
Endocytosis 58, 275, 329–345
Epitope 5, 13, 21–31, 33–61, 63, 75, 93–96, 101–103, 107, 156–157, 191, 238, 298
Epitope-tag
c-Myc 22, 30
C-terminal 30
FLAG 22, 30, 44
green fluorescent protein (GFP) 22, 34, 44
hemagglutinin (HA) 35, 44–46
N-terminal 5, 45
polyhistidine (His_6) 22, 30
Epitope-tagged protein expression 21–31, 33–61
Epon 125, 137, 157, 183, 219, 221–222, 260, 262, 264–265, 267, 319, 333–334
Epoxy resin 131, 148, 150–153, 157, 183, 185–187, 189, 226, 228, 230–234, 246–247, 260, 272
ER marker 316
Escherichia coli
cultivation 10–11, 13
protein induction
large scale 5–6, 12, 37
small scale 6
Rosetta™ (DE3) pLysS 5, 7
sonication 10–11
transformation 10
Ethidium bromide 35–36, 38–39
Euglena 72–73, 261, 268–271
Exoplasmic membrane leaflet 208, 210

F

Fab' nanogold 147, 152, 227, 229, 233
$F(ab')_2$ nanogold 111, 116–117
Fixation
cells
algae 261
attached 148
cover slip 226, 228
tissue culture 145–153, 250
yeast 192
chemical 123–132, 152, 156, 167, 191–192, 206, 208, 210, 270
physical 168, 206
post 153, 219, 221–222, 226, 233, 241
pre
tissue 134–135, 146, 148, 159, 219–220, 247
brain slices 140
plant 123–132
Fluorescence 44, 60, 137, 143, 214, 333, 343
Fluorescent conjugate 135
Fluorescent dyes 329–330
Fluoro-nanogold conjugate 135
FM dyes 330, 343–344
Formaldehyde 208, 238–239, 256
Freeze etching 278, 281
Freeze-fracture replica immunolabeling 205–215
Freeze substitution 156–157, 159–160, 164, 169, 172, 175, 182, 184–188, 192–194, 197–198, 200–202
Freund's adjuvant 99
FRIL, *see* Freeze-fracture replica immunolabeling
FT-capsules 193–194, 197–198
FT-chamber 193–194, 197, 202

G

G-capsules 193–194, 197
G-chamber 193–194, 197
Gelatin capsule 125, 127, 131, 158, 160–161, 194, 219, 221–222, 239, 243, 246, 262, 264–265, 289, 291, 295
GFP, *see* Green fluorescent protein
Glass knife 198, 243, 251, 253, 289, 292
Glutaraldehyde 124, 126, 130, 135, 137, 142, 147, 156–164, 169, 176, 182, 184, 188, 192, 194–195, 208, 219–221, 223, 227, 229, 231, 233, 238, 241, 247, 250, 256, 261, 263, 270–271, 277, 299, 302, 318, 325, 333, 342–343
Glutathione sepharose 7, 9, 11–12, 14–16
Glutathione-*S*-transferase fusion protein 5
Glycan localization 217–224
Gold intensification 225–234
Green fluorescent protein (GFP) 22, 34–35, 42, 44, 54–57, 60, 199–200, 231, 315–316
fusion 22
Grids
etching 289, 295
formvar/carbon coated 125
formvar coated 113, 137, 142, 151, 158, 161, 170, 175, 183, 185, 195, 198, 208–209, 213, 282, 319, 324, 334, 341
gold 190, 270
mesh 125, 151, 183, 185, 195, 208, 228, 231, 240, 251, 254, 278, 282, 289
nickel 113, 125, 158, 161, 164, 170, 175, 177, 183, 185, 195, 198, 240, 243, 251, 254, 262, 270, 289, 292, 295
slot 137, 142, 261–262, 266, 272, 319, 324, 334, 340–341

H

Hanaichi lead solution 138, 319, 324, 334
Heat induced antigen retrieval 237–247
Hemocytometer 37, 40, 77, 79–81, 84, 318, 321, 332–333, 337–338
HIAR, *see* Heat induced antigen retrieval
High-pressure cryofixation 157–159
High-pressure freezing 156–160, 168, 182, 184, 186–188, 192–193, 196, 200–201, 210

Horseradish peroxidase (HRP)
  conjugate ... 49, 218–220, 223
  conjugated lectin ... 217–224
  reporter gene ... 315–326
HRP, *see* Horseradish peroxidase (HRP)
Hyaluronic acid binding protein (B-HABP) ... 170, 175
Hybridoma
  cloning ... 79, 91
  culture ... 88
  formation ... 75, 87–88, 90–91
  selection ... 91
  splenocyte–myeloma fusion ... 90
  storage ... 87
Hydrogen peroxide ... 65, 77, 131, 219, 263, 266, 273, 318

I

Image analysis software ... 271
Immersion fixation ... 138, 242, 246
Immunization
  antigen ... 4, 6, 19, 99
    injection ... 68
    preparation ... 64, 67–68, 76, 82, 99
  mouse ... 76, 82, 99
  rabbit ... 64, 67–68
Immunocytochemistry ... 4, 134–136, 143, 205, 315–316
Immunoelectron microscopy
  scanning ... 297–313
  transmission ... 156, 162, 213, 260, 297
Immunofluorescence microscopy
  cell fixation ... 52, 55
  cell permeabilization ... 55, 60
  coverslip mounting ... 57
  nuclear stain ... 57
  protein localization ... 43
Immunoglobulin purification ... 89
Immunogold labeling
  blocking ... 227, 229, 292
  multiple label ... 238
  post-embedding ... 135
  pre-embedding ... 134
  scanning electron microscopy ... 297–313
  transmission electron microscopy ... 297
Immunoreplica technique
  rip-off ... 276–277, 279–280, 283, 285
  unroof ... 276–277, 279–280, 282–283, 285
IMOD, *see* Image analysis software
In situ hybridization ... 288, 290, 292–293
In vivo cryoapparatus ... 169, 173–174, 178
In vivo cryotechnique ... 167–178
Isopentane–propane (IP) cryogen preparation ... 168, 171
IVCT, *see In vivo* cryotechnique

K

Karnovsky's fixative ... 137, 139–140
Keyhole limpet hemocyanin (KLH) ... 94, 99–105
$KMnO_4$ contrast enhancement ... 186

L

Lead solution ... 138, 142–143, 183, 241, 319, 324–325, 334, 341, 343
Leaf mesophyll protoplasts ... 33–42
Lectin ... 217–224
Liquid nitrogen ... 13–14, 16–17, 51, 80, 84, 88, 157, 159, 163, 168–174, 176, 178, 182, 184–185, 193, 207–208, 212, 233, 251, 253, 278, 281, 285, 303, 306
Lowicryl HM20 ... 157, 160–161, 182, 186, 188, 192, 202
Lowicryl K4M ... 131–132, 182, 186, 188–189, 272–273
LR-gold ... 169, 174, 176
LR White ... 131–132, 156–157, 160, 162–164, 238–239, 242–247, 260, 272, 273, 289, 291, 295

M

Mammalian cells
  culture ... 184, 187
    astrocytes ... 332–333
    glia ... 335
    HEK ... 318, 320
    HeLa ... 46
    hippocampal ... 330–338
    hybridoma ... 87–88
    murine myeloma ... 78–79, 83–84
    primary neuronal ... 330
    rat kidney fibroblasts ... 276
    renal ... 152
    splenocytes ... 85–86, 90–91
  transfection ... 46–47
  viability assay ... 84–88, 91
Methyl cellulose ... 191, 251–252, 255, 257
Microtome ... 127, 151, 161, 259–260, 265, 271, 282
Moisture chamber ... 208, 211, 213, 215, 219–220, 224
Monoclonal antibody, *see* Antibody

N

Nanogold
  conjugation ... 135
    antibody, *see* Antibody
    $F(ab')_2$ ... 109–119
    IgG ... 116
    protein A ... 115
    streptavidin ... 117, 218
  intensification ... 225–234
  size determination ... 146
  synthesis ... 105
Nembutal ... 135, 138, 239, 242
Neuron specific label ... 316–317
Nick translation ... 288
Nuclear envelope ... 198–199, 299–300, 303, 307, 311, 322
Nuclear pore complex ... 198–200, 306–307

O

OCT-compound ... 219–220, 251, 253
Organ preparation ... 246
Osmium black ... 218, 223
Osmium tetroxide ... 137, 148, 156, 164, 185, 187–189, 192, 194, 201, 218, 299, 305, 310

P

Paraformaldehyde, preparation ... 58, 176, 233, 294, 309
PBS (phosphate buffered saline) ... 9, 47, 50, 64–67, 76, 100–101, 111, 126, 146, 158, 169, 207–208, 218, 226, 228, 234, 250, 263, 278, 290, 301

PCR, *see* Polymerase chain reaction
Peptide
affinity purification................100–101, 103, 106
antigenicity estimation.........................95–96
antigen sequence selection.....94–96, 98–99, 104, 107
carrier protein conjugation...............94, 100–101
immunization...............................99–100
purification......................................94
solubility prediction..........................95–97
solubilization...................................98
synthesis.....................................94–96
Perfusion fixation................135, 138, 239, 242, 246
Permeabilization........60, 134, 136, 139, 142, 145, 226, 228–229, 233
Phosphospecific antibody, *see* Antibody
Photoshop...............................303, 307–308
Physical fixation...............................168, 206
Plant
Allium cepa....................................157
Arabidopsis thaliana.............................21
castor bean................................124, 129
embryo.....................................123–132
mesophyll cells..............................33–34
onion...........................157–160, 162–163
poplar...............................159, 162–163
protoplast.......................................33
tissue cryofixation..........................158–159
Plasma membrane.......58, 159–160, 163, 199, 206, 209, 211–212, 222–223, 275–285, 298, 330, 338
Plasmid
Arabidopsis thaliana.............................21
binary vector....................................29
isolation..35
CsCl–ethidium bromide gradient.............35
large scale..................................37
mammalian cell expression..................5, 17, 26
pCMV-HA..46
pCMV-myc......................................46
pGEX..5, 7
Tet On..46
Tet Off...46
Ti..21
Platinum carbon replica........................282, 285
Platinum coating..................................300
Plunging quick freezing...................168, 169–172
PMSF.......................................8, 14, 17
Polyclonal antibody, *see* Antibody
Polymerase chain reaction..........................30
Post-embedding immunolabeling....................155
Pre-embedding
carbohydrate labeling...........................217
immunolabeling.....134, 139, 146, 149, 225, 232–233
lectin labeling..............................217–224
Primary antibody, *see* Antibody
Pristane..................................80, 88–91
Protease
inhibitors...............................16, 47–48, 52
proteinase K............288, 290, 292–293, 295–296
thrombin................................7, 9, 16–17
tobacco etch virus (TEV)..........................5
trypsin.....................................47, 51
Protein
antigen...................................3–20, 191
assay...................................8, 15, 17, 48
BCA...................................48, 52, 59
Bradford.....................................59
expression..............3, 5, 7, 10–11, 17–18, 35, 46
Arabidopsis thaliana..........................21
induction......................6, 10–13, 16, 18
insoluble protein.........................4, 6, 12
integral membrane protein.......................4
large scale........................6, 8, 13, 35, 37
mammalian cells............................43–61
peripheral membrane protein....................4
small scale....................................6
soluble protein...............4–6, 11–12, 14, 16
transient.....................21–23, 43–61
misfolding.......................................45
mislocalization...............................34, 45
purification..................................13–14
affinity, GST/His tag...................4–5, 17
inclusion body...................4, 6, 12, 14–16
insoluble........................4, 6, 12, 14–15
integral membrane...............................4
oligomeric.....................................4
peripheral membrane............................4
soluble.....................................14–16
structure analysis............................94–95
tag...45, 60
glutathione-*S*-transferase.........................5
green fluorescent protein (GFP)........22, 34, 44, 231, 315
removal.....................................5–6, 19
Protein A
gold conjugation...........................109–119
gold labeling..............................218, 271
post-embedding...................135, 155, 218
pre-embedding....................263, 266, 288
Protoplast
Arabidopsis thaliana..............................21
mesophyll...................................33–42
transfection........................35, 37, 40, 41–42

Q

Quick-freezing method............................168

R

Rapid freezing.................192, 206–207, 210–211, 278, 281
rDNA immunolocalization......................287–296
Replica immunolabeling, *see* Freeze-fracture replica immunolabeling
Resin, *see* Embedding, resin
Reynold's lead citrate solution.........195, 198, 202, 203
Ringer's solution.........................276–277, 279
RIPA buffer................................48, 52, 59
RNase A..................................8, 290, 295
Root tip........................160, 290–291, 293–294

S

Saccharomyces cerevisiae........................191–203
Sample preparation
cells....................124, 210, 238, 261–264, 300
dehydration.....................123, 126, 264, 290
tissue..................123, 126–127, 130, 210, 238
Saponin............................227, 229, 232–233
Scanning electron microscopy..................297–313
SDS-digested freeze-fracture replica labeling.........206
SDS-FRL, *see* SDS-digested freeze-fracture replica labeling

Sections
cryo 219
etching 132
serial 218, 260–263, 265, 268, 271–272, 324, 341
staining
lead citrate 152, 219, 222, 238
osmium tetroxide (OsO_4) 241, 244
thick 175
ultrathin 244
uranyl acetate 186
SEM, *see* Scanning electron microscopy
Serial image reconstruction 263
Serum preparation 65, 69
Shadowing 212, 282
Silver intensification 226–227, 229–234
Sodium meta-periodate 183, 185–187, 189
Software
image analysis 271
photoshop 303, 307–308
protein analysis
antigenicity prediction 94
secondary structure 95–96
solvent accesibility 95–96
sequence analysis 94–95
Specimen carrier 157, 182, 184–185, 188, 193, 196–197, 200, 251, 253, 256
Splenocyte preparation 85–86
Sørensen's phosphate buffer 288, 290
Staining 56–57, 130, 135, 137, 139, 141, 143–144, 148, 150, 153, 164, 169, 175, 177, 183, 186, 222, 223, 228, 231, 233, 240, 243, 246–247, 249, 257, 270, 309, 325–326, 341–344
Streptavidin 108–119, 218, 224
Subcutaneous injection 68
Synaptic vesicles 141–142, 330, 333, 338, 341–342

T

Tannic acid 188, 238, 241, 247, 299, 302, 305, 310
TEM, *see* Transmission electron microscopy
Tissue culture
astrocytes 335
cell 57, 145–153, 181–190, 317
count 321, 337
lysis 18, 47–48, 52, 78, 85–86
viability 84
coverslip 51
dish 47, 51, 77, 182, 207, 210, 250, 253, 276, 279
glia cells 335
gold foil 206–207
HEK cells 57, 318, 320–325
HeLa cells 46
hippocampal cells 330–338
hybridoma, *see* Hybridoma
media
DMEM 51
HAM-F12 .... 276, 279
HAT 78
HT 78–79
MEM 47, 87
neurobasal 342
murine myeloma cells 78–79, 83–84
neuronal cells 330, 333, 335, 339, 342
rat kidney fibroblasts 276
renal cells 152
sectioning 146, 148, 150–152
splenocytes 85–86, 90–91
transfection 46–47, 51, 321
trypsin release 318, 320
Titer determination 100, 102
Tokuyasu method 191–192, 250
Toluidine blue 239, 243, 247, 254
Tomography 225, 275–285, 297
Transfection
efficiency 35, 41, 45–46, 56, 59
mammalian cells 45–47, 51–52, 56–57
mesophyll protoplasts 35, 37, 41–42
reagents 57
calcium-PEG 35, 37, 40
Fugene 6 56
Lipofectamine 2000 57
TransIT-LT1 47
Transformation
Arabidopsis thaliana 21
mammalian cells 47
Transient expression 34, 43–61
Transmission electron microscopy 156, 218, 260, 297
Triton X-100 .... 50, 101, 111, 117, 136, 142, 147, 149, 152, 301
Trypan blue 77, 78–79, 81, 84, 86–88, 318, 321, 326, 332–333, 337, 344

U

Ultramicrotome 137, 142, 175, 183, 195, 219, 222, 228, 231, 251, 253–254, 292, 319, 324, 334, 341
Uranyl acetate 126, 128

V

Vacuum
evaporation 209, 214
infiltration 31, 124
Vibratome 135–136, 139, 143

W

Western blotting 29, 43, 46, 48–49, 53–54, 59, 66, 70, 73, 91, 107
Wet chamber 198, 302, 304

X

Xenopus laevis oocytes 299

Y

Yeast culture 192, 196

MIX
Papier aus verantwortungsvollen Quellen
Paper from responsible sources
FSC® C105338

If you have any concerns about our products,
you can contact us on
ProductSafety@springernature.com

In case Publisher is established outside the EU,
the EU authorized representative is:
Springer Nature Customer Service Center GmbH
Europaplatz 3, 69115 Heidelberg, Germany

Printed by Libri Plureos GmbH
in Hamburg, Germany